Phytochrome and
Photoregulation in Plants

Phytochrome and Photoregulation in Plants

Edited by

Masaki Furuya

University of Tokyo
and
National Institute of Basic Biology
Okazaki, Japan

1987

ACADEMIC PRESS, INC.
Harcourt Brace Jovanovich, Publishers
Tokyo Orlando San Diego New York
Austin Boston London Sydney Toronto

ACADEMIC PRESS JAPAN, INC.
Ichibancho Central Bldg., 22-1 Ichibancho, Chiyoda-ku, Tokyo 102

United States Edition published by
ACADEMIC PRESS, INC.
Orlando, Florida 32887

United Kingdom Edition published by
ACADEMIC PRESS, INC. (LONDON) LTD.
24/28 Oval Road, London NW1 7 DX

Library of Congress Cataloging-in-Publication Data

Phytochrome and photoregulation in plants.

   Proceedings of the Sixteenth Yamada Conference
held October 13-17, 1986 at Okazaki National Institute in Japan.

   Includes index.
   1. Phytochrome—Congresses.   2. Plants—Photo-
morphogenesis—Congresses.   I. Furuya, M. (Masaki),
DATE
QK898.P67P47   1987   581.19′297   87-72460
ISBN 0-12-269970-X (alk. paper)

Printed in Japan
87 88 89 90 9 8 7 6 5 4 3 2 1

Contents

III. PROBLEMS AND PROSPECTS IN SPECTROPHOTO-METRICAL AND BIOPHYSICAL APPROACHES

IV. PROBLEMS AND PROSPECTS OF PHYSIOLOGICAL APPROACHES

V. PROBLEMS AND PROSPECTS OF PHOTOMORPHOGENETIC STUDIES

Preface

The life of green plants depends absolutely upon the utilization of sunlight not only as the source of energy for photosynthesis but also for environmental information required during growth and development. Phytochrome, a phototransducer chromoprotein involved in red/far-red reversible photomorphogenetic reactions in plants, was discovered in 1959 by members of the U.S.Department of Agriculture, who were working in Beltsville, Maryland. Since then, phytochrome has been extensively investigated by plant physiologists, biochemists and biophysicists. It is now clear that phytochrome plays a central and crucial role in many aspects of photoregulation at the molecular, cellular and organ levels in plants.

The recent and rapid development of molecular approaches to the study of phytochrome has resulted in an enormous increase in our knowledge of the molecular properties of phytochrome. Hence, not only students in biology but also workers in the field have had difficulty in keeping abreast of the progress. Taking this state of affairs into account, the Yamada Science Foundation asked me to organize the Sixteenth Yamada Conference, designed to investigate the consensus and reveal any disagreements over the latest advances in this rapidly moving field, and to stimulate discussion of the prospects for future studies of phytochrome and photomorphogenesis in plants. The meeting was held on October 13-17, 1986, at the Okazaki National Research Institute in Japan, and about 130 specialists, including most of the scientists who are actively working on these subjects, from 17 countries, partcipated. Approximately half of the conference was devoted to round-table discussions and the exchange of ideas and criticisms, and the essence of their conclusions is clearly apparent throughout this book.

This book is not a compilation of the abstracts submitted by the participants in this conference but, rather, it consists of chapters on selected topics by plenary and review lecturers. In addition, such chapters as II-3, III-6, IV-3, IV-7, V-1 and V-4 are summaries of the round-table discussion sessions which were written by each chairperson, so that those who did not attend the sessions will find a vivid account of what was discussed and concluded there. Special thanks should go to the chairpersons of these sessions. As can be seen from this book, this meeting was highly successful, and I wish to thank all the participants who created a superb atmosphere for the communication of ideas.

On behalf of the organizing committee of this conference, I wish to express my gratitude to the Yamada Science Foundation for their very generous financial support, and to the Okazake National Research Institute for providing a very efficient and comfortable location for this meeting. I also wish to record my appreciation for the creative and thoughtful planning and management of the conference by my colleagues Drs. Y. Inoue, A. Kadota, K. Manabe, Y. Miyoshi, N. Sato, S. Tokutomi, M. Wada, M. Watanabe and K. T. Yamamoto. The success of this conference was also the result of the excellent assistance and patient work at my office by our secretary, Mrs. Kimie Mori, to whom I am most grateful. Finally it is a pleasure to thank the Academic Press in Tokyo and its staff for their helpful expertise in the preparation of this book.

MASAKI FURUYA

April 1987, Tokyo

Acknowledgments

Yamada Science Foundation was established in February 1977 in Osaka through the generosity of Mr. Kiro Yamada. Mr. Yamada was president of Rohto Pharmaceutical Company Limited, a well-known manufacturer of medicines in Japan. He recognized that creative, unconstrained, basic research is indispensable for the future welfare and prosperity of mankind and he has been deeply concerned with its promotion. Therefore, funds for this Foundation were donated from his private holdings.

The principal activity of the Yamada Science Foundation is to offer financial assistance to creative research in the basic natural sciences, particularly in interdisciplinary domains that bridge established fields. Projects which promote international cooperation are also favored. By assisting in the exchange of visiting scientists and encouraging international meetings, this Foundation intends to greatly further the progress of science in the global environment.

In this context, Yamada Science Foundation sponsors international Yamada Conferences once or twice a year in Japan. Subjects to be selected by the Foundation should be most timely and stimulating. These conferences are expected to be of the highest international standard so as to significantly foster advances in their respective fields.

The editor wishes to acknowledge the executive members of Yamada Science Foundation, including Leo Esaki, Kenichi Fukui, Osamu Hayaishi, Noburo Kamiya, Jiro Kondo, Shigekiyo Mukai, Takeo Nagamiya, Shuntaro Ogawa, Shûzô Seki, Tomoji Suzuki, Jin-ichi Takamura and Yasusada Yamada.

I. INTRODUCTION

THE HISTORY OF PHYTOCHROME

Masaki Furuya[1]

Biology Department
Faculty of Science
University of Tokyo
Hongo, Tokyo, and

Division of Biological Regulation
National Institute for Basic Biology
Okazaki, Japan

I. GENESIS (THE BELTSVILLE ERA: 1920–1963)

In prehistoric times, man may have realized that light greatly influences growth and development in green plants. Priestly discovered in 1772, that light is utilized by green plants as the source of energy for the production of complex organic substances through photosynthesis, and a century later it was found that light plays another important role in plant growth by providing environmental information. Darwin (1881) separated the photoreceptive site from photo-tropically responsible region in a monocot seedling. His was excellent pioneering work on "photoregulation" in plants. Klebs (1916–17) published enormous amounts of data on the effects of visible light on cell growth and development of fern protonemata. The elucidation of the molecular basis of the effect of light signals on plant life, however, was not made until quite recently.

The physiological capacity to adjust its life cycle to the seasonal change of light conditions, which results from the motion of the Earth, is crucial for a plant's survival. In this connection, the appropriate timing of seed germina-

---

[1] Present address: Frontier Research Programs, The RIKEN Institute, Wako City, Japan 351-01

3

tion and flowering is essential. The control of flowering by measurement of day or night length (Garner and Allard,1920) is a typical example of how light acts as an environmental signal to control the developmental step; and the red and far-red reversible effect on seed germination (Borthwick et al.,1952) is another example.

The inside-story of the discovery of phytochrome has been described by the original investigators (Borthwick, 1972; Butler, 1980). In brief, after the discovery of the red and far-red photoreversible effect, the group at Beltsville had the unique idea that changes in optical density of an appropriate sample of plants should be the result of irradiating it alternately with actinic red and far-red light. Using a 2-wavelength difference spectro-photometer built by K. Norris, they succeeded in measuring photoreversible changes in optical density at 660 nm and 730 nm in etiolated maize seedlings and crude extracts (Butler et al., 1959). The term "phytochrome" was half jokingly used by Warren Butler in their laboratory shortly after the discovery (Borthwick, 1972), and then published first in the journal SCIENCE (Borthwick and Hendricks, 1960).

In the following few years, the Beltsville group dis-covered fundamental properties of phytochrome in terms of spectrophotometry and biochemistry (Siegelman and Butler, 1965) and they developed a procedure for the isolation and purification of phytochrome. They found its average molec-ular weight to be 90,000–150,000 daltons, and demonstrated the action spectra for the photochemical conversion of Pr and Pfr in vitro and the non-photochemical transformation of phytochrome such as reversion and decay of Pfr in the dark.

We should applaud that the key phenomena such as photo-periodism, the red and far-red reversible effects and the photoreceptor phytochrome were discovered by members of the same institution at Beltsville. Theirs was the greatest-contribution in this field and opened a new research area not only in plant physiology but in photobiology, develop-mental biology and molecular biology.

II. THE ERA OF DISAPPOINTMENT (1963–1970)

The year 1963 was an important year in the history of phytochrome. Since 1963 phytochrome both in vivo and in vitro began to be measured in several laboratories in the United States with the commercially available, 2-wavelength difference spectrophotometer, Ratiospect R2 (Hillman, 1967), in addition to custom-made machines (DeLint and Spruit,

1963). Naturally, the number of articles on spectrophotometric measurements of phytochrome increased exponentially,but by 1966 they reached a plateau of ca.20 papers/year and stayed there for the following decade.

One of the major reasons for stagnation was that most experiments failed to correlate photometrically measured phytochrome content, initial Pfr state and non-photochemical transformation of Pfr in vivo, with photoreversible responses of plants to red and far-red light (Hillman, 1967). At the first symposium on phytochrome at the ASPP annual meeting in 1964, all the speakers, such as Edwards and Klein, Furuya and Hillman, and Hopkins and Hillman, demonstrated no, or little, such correlation. Typical examples were the Pisum and Zea paradoxes; In the former, growth response to far-red light is obtained in the absence of optically detectable Pfr (Hillman, 1965) and, in the latter, response to red light is saturated by an detectably small percentage of Pfr, in spite of the fact that it is reversible by far-red filters which generate a detectable amount of Pfr (Chon and Briggs, 1966). These paradoxes remain unresolved. Only a few papers reported a correlation, like the Pfr-dependent inhibition of growth in oat mesocotyl (Loercher, 1966) and in the rice coleoptile (Pjon and Furuya,1968). Although our knowledge of phytochrome accumulated steadily during those days (Furuya, 1968), we were still a long way from unravelling precisely the chain of molecular events from the light signal to each response in plants (Mitrakos and Shropshire, 1972).

III. THE SEVENTIES: THE ERA OF GROPING

Since the early 1970's, the phytochrome family has met regularly in Europe, and we can take a look at the progress of phytochrome studies in this decade in the proceedings of these meetings (Mitrakos and Shropshire, Jr., 1972; Smith, 1976; DeGreef, 1980) and also in textbooks (Mohr, 1972; Smith, 1975). In this period, useful new physical and chemical techniques were introduced in this field, so that molecular properties of phytochrome, such as the chemical structure of its chromophore (Rudiger and Correll, 1969; Chapters III 2 and 3) and apoprotein (Briggs and Rice, 1972; Pratt, 1982) and the spectrophotometric and immunochemical properties of large and small phytochrome (Pratt, 1979; Furuya, 1983) were revealed step by step, although our understanding of native phytochrome was still far from complete.

Since its discovery (Butler et al., 1959), phytochrome was long believed to be easily extractable from plant tissues by a simple buffered solution. However, a small amount of phytochrome was recovered in a pelletable fraction when extracted in the light (Rubinstein, 1969). This paper was ignored until Quail et al (1973) found that the pelletability of phytochrome from crude extract was enhanced by a brief irradiation of the etiolated oat tissue with red light. Unfortunately, no biological implication of this photoinduced phytochrome binding to particulate fractions has been reported so far (Chapter IV 2).

In contrast, microbeam irradiation proved to be a useful technique for finding the intracellular localization of physiologically active photoreceptors. Good examples are the dichroic orientation of chloroplasts in <u>Mougeotia</u> (Haupt, 1971; Chapter IV 4) and the spacial separation of the actions of phytochrome from blue-UV photoreceptors in fern protonema (Furuya et al., 1980; Chapter IV 5).

Quick responses, such as nyctinasty (Fondeville et al., 1966), root tip adhesion and electric potentials (Tanada, 1968), and enhancement of enzymatic activities in isolated organelles (Manabe and Furuya, 1975; Jose and Smith, 1976) were thought to be nice experimental systems for the study of the primary action of phytochrome but no crucial evidence has yet been obtained in any experimental system.

IV. THE EIGHTIES : THE ERA OF MOLECULAR BIOLOGY

There is no need to describe here the history of phytochrome studies in the 1980's, as the experts in this field have summarized the recent progress in this book.

The primary structure of phytochrome apoproteins was revealed by Quail and his colleagues (Chapter II 2) from a consideration of cDNA sequences and gene structure. The knowledge of the primary structure has greatly accelerated the study of functional sites in chromopeptide domains (Chapters II 4 and 5). Molecular biological approaches to the photocontrol of gene expression have also opened a promising new field (Chapters II 1 and 3), and will result in further great advances in our knowledge of photomorphogenesis in plants (Chapter V 5).

Spectrophotometrical approaches continue everlastingly in the study of phytochrome because of its very nature (Chapters III 1-4), and the application of modern techniques seems useful (Chapters III 5 and 6). Considering that the primary action of the phytochrome molecule remains a mystery (Chapter V 1), one can see why physiological approaches designed to separate the elementary processes of the photo transduction chain in membranes are crucial (Chapter IV).

V. PROBLEMS FOR THE FUTURE

Many diverse effects of light on the molecular, cellular and developmental processes in plants have been ascribed to phytochrome in the literature (Shropshire and Mohr, 1983). Are there several functional sites in a phytochrome molecule which induce different primary actions? Five possible functional sites have been reported :N-terminus(#1-51) for absorption spectrum of Pfr (Vierstra and Quail, 1982), #115-414 domain for Pr spectra (Chapter II 5), #402-414 for protein kinase activity (Chapter II 4), #415-623 for Pfr spectra (Chapter II 5), and #624-800 for dimer formation (Jones and Quail, 1986). Alternatively, are there several different molecular species of phytochrome with different effects? We know of at least two different apoproteins of phytochrome (chapters II 6 and 7) and a few different cDNAs (chapter II 2). Perhaps both possibilities hold true?

Irrespective of the number of the functional sites on the phytochrome molecule, the partner molecule(s) for the transduction of the phytochrome signal is crucial, although, unfortunately, little is known about such partner(s). For example, where is the partner in plant cells (chapter IV 1). Are the partners located in the cytosolic fraction or in organelles? Do different organelles each have different transduction systems?

Future work to solve these questions is warranted, because the answers to these questions are not only important for photomorphogenesis in plants but for the demonstration of an essential molecular model of biological regulation.

REFERENCES

Briggs, W.R., and Rice, H.V. (1972). Ann.Rev.Plant Physiol.23:293.

Borthwick, H. (1972). In "Phytochrome" (Mitrakos, K., and Shropshire, W.Jr., eds), p.3, Academic Press, London.

Borthwick, H.A., Hendricks, S.B., Parker, M.W., Toole, E.H. and Toole, V.K. (1952). Proc.Nat.Acad.Sci. U.S. 38:662.

Butler, W.L. (1980). In "Photoreceptors and plant development" (De Greef, J., ed), p.3. Antwerp Univ.Press, Antwerp.

Butler, W.L., Norris, K.H., Siegelman, H.W., and Hendricks, S.B. (1959). Proc.Nat.Acad.Sci.U.S. 45:1703.

Darwin, C., and Darwin, F. (1881). The power of movement in plants. The Appleton Co., London.

DeGreef, J. (1980). "Photoreceptors and plant development". Antwerp University Press, Antwerp.

DeLint, P.J.A.L., and Spruit, C.J.P. (1963). Meded. Landbouwhogesch.Wageningen 63:1.

Fondeville, J.C., Borthwick, H.A., and Hendricks, S.B. (1966). Planta 69:357.

Furuya, M. (1968). Progr.Phytochem. 1:347

Furuya, M. (1983). Phil.Trans.Roy.Soc.London, B303:361.

Furuya, M., Wada, M., and Kadota, A. (1980). In "The Blue light Syndrome" (H.Senger, ed), p.119. Springer, Berlin.

Hendricks, S.B., and Borthwick, H.A. (1967). Proc.Nat. Acad.Sci.U.S.A. 58:2125.

Hillman, W.S. (1967). Annu.Rev.Plant Physiol. 18:301.

Klebs, G. (1916, 1917). Sitz-Ber.Heidelberg.Akad.Wiss., Math.-Naturw.Kl.Abhandle. 4 (1916), 3 and 7 (1917).

Loercher, L. (1966). Plant Physiol. 41: 932.

Marme, D. (1977). Ann.Rev.Plant Physiol. 28:173.

Mohr, H. (1972). "Lectures in photomorphogenesis", Springer, Berlin.

Mitrakos, K., and Shropshire,Jr., W. (ed.) (1972) "Phyto-chrome", Academic Press, London.

Pjon, C.J., and Furuya, M. (1968). Planta 81: 303.

Pratt, L.H. (1979). Photochem.Photobiol.Rev, 4:59.

Pratt, L.H. (1982). Ann.Rev.Plant Physiol. 33:557.

Priestley, J. (1772). Phil.Trans.Royal Soc.London, 147.

Quail, P.H., Marme, D., and Schafer, D. (1973) Nature, New Biol., 245:189.

Rudiger, W., and Correll, D.L. (1969). Liebigs Ann.Chem., 723:208.

Senger,H.(ed.)(1980). "The Blue Light Syndrome", Springer, Berlin.

Shropshire, Jr., W., and Mohr, H. (ed.) (1983) Encycl.Plant Physiol. 16A and 16B. Springer, Berlin.

Siegelman, H.W., and Butler, W.L. (1965). Ann.Rev.Plant Physiol. 16: 383.

Smith, H. (1975). "Phytochrome and photomorphogenesis", McGraw Hill Co., U.K.

Smith, H. (ed.) (1976). "Light and plant development", Butterworths, London.

Tanada, T. (1968). Proc.Nat.Acad.Sci.U.S. 59:376.

II. PROBLEMS AND PROSPECTS OF
MOLECULAR APPROACHES

HOW USEFUL ARE MOLECULAR TECHNIQUES IN ADDRESSING PHYSIOLOGICAL PROBLEMS IN PHOTOMORPHOGENESIS?

Department of Plant Biology
Carnegie Institution of Washington
Stanford, California

Winslow R. Briggs

INTRODUCTION

During the past decade there have been significant advances in our understanding of light regulation of the developmental processes occurring when etiolated plants are moved from darkness into light. More and more studies have probed this light regulation at the molecular level, with a strong emphasis on nuclear encoded proteins destined for the chloroplast. This choice of material is a logical one since the mRNAs for many such proteins - e. g. the small subunit of the enzyme ribulose bisphosphate carboxylase/oxygenase (SS) and the light-harvesting chlorophyll a/b-binding protein (LHCP) - are present in low abundance in the dark, and show dramatic increases in abundance in the light. At least one mRNA, that for the NADPH-dependent protochlorophyllide oxidoreductase (reductase), shows a sharp decrease in abundance in the light, at least in monocots. As will become clear later, the magnitude of these changes provides sufficient resolution to address at least some questions about regulation. The earlier studies looked first at mRNA abundance as determined by in vitro translation, then moved to measurement of mRNA hybridizable to the appropriate cDNA clones, and finally made direct measurements of RNA produced by run-on transcription by nuclei isolated from plants given a variety of light treatments. The extensive literature developing out of these efforts has been recently reviewed several times (e. g. Ellis, 1986; Schäfer and Briggs, 1986; Sharma, 1985;

11

Thompson et al., 1985; Tobin and Silverthorne, 1985; see also Kuhlemeier et al., 1987).

In attempting to understand light regulation of a developmental process such as greening of etiolated plants, a common approach is to investigate at which point or points light might act in a known sequence of events involved in that process. If light treatment affected a single step in this sequence, there would be predictable consequences downstream from that step (and undoubtedly upstream as well). Current studies of phytochrome action on greening have focused on the chain of events starting with transcription, progressing through pre-mRNA processing and transport, mRNA abundance (obviously a function both of rates of transcription and mRNA stability), appearance of protein, and where appropriate, appearance of final product. If, for example, light regulation is just at the transcriptional level, one to predict a specific temporal sequence of tightly coupled events: transcriptional changes should be followed by appropriate changes in mRNA, protein, and product, and all changes should respond with similar photobiological properties: fluence requirements, far red reversibility (if phytochrome is the photoreceptor), obedience to the reciprocity law, etc.

One detailed study of such a transduction chain is that of Chapell and Hahlbrock (1984) on the effect of ultraviolet light on flavonoid biosynthesis in cultured parsley cells. A simple kinetic model, based on measured rate constants, was sufficient to describe the system: regulation by ultraviolet light was clearly transcriptional. In this case the kinetic relationships were sufficiently clear and straightforward that an elaborate photobiological analysis was not essential for this conclusion.

The vast majority of studies to date have focused on regulation by phytochrome, and phytochrome will therefore be the focus of the remainder of this article. Here the picture is neither as clear nor as accessible as for the parsley cells. The kind of pulse - chase experiments available to Chapell and Hahlbrock (1984) in their cell culture system, essential for determining the in vivo stability of mRNAs and proteins, have thus far not been possible with intact seedlings. Hence rate constants for mRNA degradation have remained elusive. Likewise, reliable in vivo transcription measurements, possible in cell culture, have to date been replaced by the cumbersome in vitro "run-on transcription" assay. Both Thompson et al. (1985) and Schäfer and Briggs

(1986) address the limitations of this technique. Finally, as will become clear below, regulation solely at the level of transcription is not adequate to account for all of the complex changes involved in greening: the Chapell - Hahlbrock model is simply not sufficient.

It is against this background that we address the question: how sharp are the molecular tools for dissecting the complex transduction process involved in phytochrome regulation of greening? Is the physiological approach useful? In the classical approach, the physiologist alters the light stimulus and asks how a given response changes. This process will normally include studying fluence dependence, following induction kinetics, testing for far-red reversibility and escape from reversibility, varying fluence rate and exposure time for a given fluence to test for reciprocity, etc. One then uses the results to try to deduce the nature of the transduction pathway. To study phytochrome phenomena at the molecular level, one begins by treating plants with red light to see whether a particular gene product increases or decreases, and many workers have simply sought a yes or no answer. A number of studies have gone considerably beyond this primitive stage, however, and the question at hand is whether they are at present sufficient to help us learn at least some details of the sequence of events from phytochrome phototransformation to the final response. The reader is referred to Schäfer and Briggs (1986) for a recent summary of such studies.

The following paragraphs attempt to establish the present limits of the molecular methodologies available in studying the transduction chain(s) involved in phytochrome potentiation of greening. It has been known for many years that a brief irradiation with red light, followed by an appropriate dark period, will reduce or eliminate the lag period for greening in subsequent white light, and that the effect only showed limited far red reversibility (see Virgin, 1972, for a review of the earlier literature). More recently it has become clear that both for pea (Raven & Shropshire, 1975; Horwitz et al., submitted) and barley (Briggs et al., in preparation) the effect consisted of a very low fluence (VLF, see Briggs et al., 1985) component that was not far red-reversible and a low fluence (LF) component that was. Time course data and data on escape from photoreversibility for the LF component were available for both systems. Thus it was feasible to ask photobiological questions at the levels of transcription, pre-mRNA processing and transport, mRNA abundance, etc. to determine where the phytochrome transduction chain might intercept the sequence of events from gene transcription to chlorophyll accumulation, and indeed whether there was more than one such interception.

Studies with Barley

Recently Mösinger et al. (1985, 1987) have presented time courses for red light-induced changes in transcriptional activity of isolated barley nuclei both for LHCP reductase mRNAs, so clearly Pfr may have an effect at the transcriptional level. If transcriptional regulation alone determined mRNA abundance in these cases, there should be a simple mathematical relationship between rate of transcription and mRNA abundance: abundance should be a simple integral of transcription rate. In the case of the LHCP, the time course published by Batschauer and Apel (1984) for accumulation of the mRNA could indeed be a simple integral of the transcription time course presented by Mösinger et al. (1985, 1987). Since there is no information on LHCP mRNA stability, however, and since there is inevitable scatter in the data, it is not possible to exclude Pfr control at the level of mRNA stability as well.

Accepting that Pfr does regulate the LHCP mRNA levels at least partially at the transcriptional level, however, one can then ask whether these mRNA levels limit chlorophyll accumulation. This hypothesis is reasonable in that chlorophyll is thought to be a fairly precise indicator of the concentration of LHCP polypeptides, and is thought to be required for stabilization of the LHCP apoprotein (Apel and Kloppstech, 1980; Bennett, 1981; Harpster et al., 1984). A first step is therefore to compare the Pfr dependence of mRNA and chlorophyll accumulation, respectively. The fluence-response curve for chlorophyll accumulation (Briggs et al., in preparation) is sufficiently similar to that for LHCP mRNA accumulation (Mösinger et al., in preparation) that the hypothesis is reasonable. Both curves show VLF and LF components and the fluences required for threshold or saturation responses are similar.

Time course studies, however, present a somewhat different picture. Mösinger et al. (1987) have found that the initial rate of increase in the apparent rate of transcription of LHCP mRNA is the same for the first hour and a half following either a saturating LF or VLF light treatment. By contrast, Briggs et al. (in preparation) find that the initial rates of change for LF and VLF potentiation of greening are clearly different. Hence the conclusion is inescapable that chlorophyll accumulation per se cannot be strictly limited by the abundance of LHCP mRNA. Indeed, as suggested elsewhere (Briggs et al., in preparation), Pfr regulation of synthesis of the chlorophyll precursor δ-aminolevulinic acid may be what imposes the overall limitation on greening. At present, the mechanism by which phytochrome controls the synthesis of δ-aminolevulinic acid is unknown.

Studies of the down-regulated reductase present a somewhat different picture. Though there is good evidence that transcription from the reductase gene is decreased by Pfr (Mösinger et al., 1985, 1987), neither kinetic nor fluence-response studies support the hypothesis that the abundance of reductase mRNA is limited by transcription. Although there is scatter in the data, it is clear that red light treatments in the VLF range have little effect on the abundance of reductase mRNA 3 or 4.5 hours after light treatment, although they reduce measured in vitro transcription by isolated nuclei significantly. On the other hand, treatments in the LF range dramatically decrease reductase mRNA abundance with only a small effect on transcription (Mösinger et al., 1987). Some effect of Pfr in reducing reductase mRNA stability seems inevitable.

Studies with Pea

At present, there is insufficient information at the transcriptional level to address the question of exclusive transcriptional control of greening processes in pea directly, although Gallagher et al. (1985) have clearly shown that in vitro transcriptional activity of isolated pea nuclei for both SS and LHCP mRNAs is strongly promoted by prior irradiation of the seedling. However, it has been possible to inspect other parts of the sequence of events following gene activation to look for possible phytochrome control. These include mRNA partitioning between cytoplasm and nucleus (Sagar et al., in preparation), presumably a reflection of mRNA processing and transport, overall abundance of specific mRNAs (Kaufman et al., 1984, 1985, 1986; Horwitz et al., submitted), and chlorophyll accumulation per se (Horwitz et al., submitted).

Sagar et al. (in preparation) separately assayed cytoplasmic and nuclear fractions for light effects on the

abundance of a number of mRNAs known to be increased in pea
by red light treatment. Although in every light-sensitive
case examined the light-induced increase was quantitatively
confirmed, light had no effect whatsoever on the ratio of
the cytoplasmic increment to the nuclear. For these mRNAs,
including those for SS, LHCP, and ferredoxin, at least,
phytochrome regulation could not be occurring at the mRNA
processing or transport step. Although the data show the
usual scatter, the negative case is a very strong one -
indeed a solid negative case is easier to make than a solid
positive one - cf. the case for exclusive regulation of LHCP
mRNA abundance at the transcriptional level.

It should be mentioned parenthetically that the Sagar
et al. results point to another sort of regulation: plots of
nuclear versus cytoplasmic increase in mRNA abundance
following the various experimental treatments yielded
straight lines of widely differing slopes, ranging from
about 3 for the SS and LHCP mRNAs to over 17 for ferredoxin.
Thus each of the pre-mRNAs must have special structural
properties regulating their steady-state partitioning
between nucleus and cytoplasm.

Although data are not yet available to evaluate
transcriptional regulation by Pfr as the limiting factor in
greening in pea, one can examine whether mRNA abundance
limits chlorophyll accumulation. As with barley, both
kinetic and fluence-response data for pea are at hand
(Horwitz et al., submitted). First, Horwitz et al. compared
the time courses for changes in mRNA abundance following
either VLF or LF treatments with those for potentiation of
chlorophyll accumulation. In both cases, the initial slope
was higher following LF treatment, and the overall kinetics
for chlorophyll accumulation and change in mRNA abundance
were similar in the VLF and LF cases. Likewise, escape from
far red photoreversibility showed a similar time course for
both chlorophyll and LHCP mRNA effects.

Hence, at first glance, the hypothesis that chlorophyll
accumulation was limited by LHCP mRNA abundance at the onset
of greening appears tenable. However, and unlike the
situation in barley, a comparison of the fluence-response
relationships for chlorophyll accumulation with those for
increase in LHCP mRNA abundance reveals a significant
discrepancy. The two fluence-response curves are similar in
shape, both showing a VLF component, a plateau, and an LF
component (the latter photoreversible in both cases).
However, the curve for chlorophyll lies at fluences roughly

one full log unit higher than that for LHCP mRNA abundance. Thus LHCP mRNA abundance in pea cannot be what limits chlorophyll accumulation, and again one must look elsewhere - as was the case with barley, Pfr regulation of δ-aminolevulinic acid synthesis presents a viable alternative (Horwitz et al., submitted).

Horwitz et al. (submitted) present another kind of evidence that LHCP mRNA abundance need not limit chlorophyll accumulation in pea. They grew plants under a wide range of fluences of continuous red light and examined both chlorophyll levels and mRNA abundance. Lowering the fluence rate by about two orders of magnitude dramatically lowered chlorophyll levels but had no statistically significant effect on LHCP mRNA abundance. In this case also, something else must be limiting chlorophyll accumulation.

Another interesting case in pea concerns the behavior of an mRNA responsive to Pfr, and recently shown to code for ferredoxin (Dobres et al., 1987). This mRNA increases sharply to a maximum within about 2 hours and remains at that level for the ensuing 22 hours (Kaufman et al., 1986). If at any time within the first seven hours the Pfr is removed, the mRNA descends to reach the level again (Kaufman et al., 1986). The kinetics of this descent are presently unknown. By contrast, the mRNA for the LHCP climbs steadily over the entire twenty four hours following a saturating red light pulse (Kaufman et al., 1986); far red at any time simply stops that component of the increase representing the LF portion of the response. While one is not formally able to make any statements as to the relative role of mRNA transcription and stability changes in these two cases, one is justified in stating that there are major quantitative differences in the ways in which these two messages are regulated by Pfr.

Conclusions

The molecular tools available at present are sufficient to draw at least a few conclusions concerning phytochrome regulation of greening. The level of Pfr in both barley and pea almost certainly has an effect on transcription of LHCP mRNA. In barley, the data are consistent with the hypothesis that transcriptional regulation is a major factor in determining mRNA abundance, though the data are insufficient to state that Pfr regulation of mRNA stability is not also involved. The barley data also indicate that chlorophyll accumulation in white light is not limited by the rate of transcription of LHCP mRNA - the initial rate of change for transcription is saturated by Pfr levels far lower than those saturating the potentiation of chlorophyll accumulation.

In barley, the case for the reductase also seems clear cut. The fluence-response relationships for transcription by isolated nuclei differ significantly from those for mRNA abundance, and Pfr regulation at the level of mRNA stability in addition to transcriptional regulation, seems highly probable.

In pea, data are not available to determine whether LHCP mRNA abundance is limited at the transcriptional level, nor can anything be said concerning mRNA stability. However, it is clear that mRNA abundance does not limit chlorophyll accumulation under the experimental conditions used to date. It is also clear that, at least for the mRNAs examined, regulation by Pfr does not occur at the level of pre-mRNA processing and transport. Finally, at minimum there must be large quantitative differences in the relative roles of transcription and mRNA degradation in determining mRNA abundance in different cases.

From the relatively limited data discussed above, it is clear that Pfr regulation of the processes involved in chloroplast development and greening in the light must occur coordinately at several levels: transcription, mRNA stability, and chlorophyll synthesis per se. For chlorophyll biosynthesis, the mechanism is unknown, but of course could be at the level of transcription or mRNA stability for a limiting enzyme in the synthesis pathway). Other levels are in no way excluded. It is also clear that the negative case is far easier to support than the positive. For the positive case, for example, unequivocal statements that transcription is the exclusive step at which Pfr acts to regulate mRNA abundance for the the LHCP await data on mRNA stability and the influence of Pfr thereon.

We return now to the question posed at the start of this article: how good are the molecular tools for the sort of analysis just described? While the answer is obviously going to be different in different cases, some generalities can be stated. In our experience, the kind of slot blot analysis of mRNA abundance used for most of the pea studies described above has sufficient variability that a genuine 30% change might well be lost in noise. If discrimination at that level is required to answer a question, then the technique is inadequate. Run-on transcription experiments are even more variable. Here even doublings are dubious in some cases, and many repetitions are required. Only when changes are several fold can one comfortably draw conclusions. The difficulties are compounded when one is comparing transcriptional changes, mRNA abundance changes, and chlorophyll changes. One requires many repetitions, for example, to be certain that the time course for one change is the same as (or different from) that for the other. Likewise, when the difference between two fluence-response curves is an order of magnitude or more, certain conclusions are possible. It would be risky, however, to draw conclusions with a difference of even as much as half an order of magnitude. The molecular data, at least, still show too much scatter.

In the future, advances in at least three areas are essential to enable one to sort out Pfr effects in regulating not just chloroplast development, but a whole host of other processes in photomorphogenesis. First, methods are needed to measure mRNA stability directly and quantitatively. Second, methods are needed for reliable measurement of in vivo transcription rates. Third, systems must be developed for plant nuclear DNA to investigate regulation of transcriptional initiation in soluble systems in vitro. In addition to these three urgent needs is a broad requirement for improved reproducibility in all of the assays. All of these areas are currently under vigorous attack, giving hope that in the near future, the greater resolution available will enable us to make not just qualitative but quantitative statements about the ways in which Pfr regulates development.

ACKNOWLEDGEMENTS

A portion of the work described in this article was carried out while the author was on sabbatical leave at the University of Freiburg as U. S. Senior Scientist Awardee of the Alexander von Humboldt-Foundation. The author is very grateful for this award. This is Carnegie Institution of Washington Department of Plant Biology Publication Number 974.

REFERENCES

Apel, K., and Kloppstech, K. (1980). Planta 150:426.
Batschauer, A., and Apel, K. (1984). Eur. J. Biochem. 143:593.
Bennett, J. (1981). Eur. J. Biochem. 118:61.
Briggs., W. R., Mandoli, D. F., Shinkle, J. R., Kaufman, L. S., Watson, J. C., and Thompson, W. F. (1985). In "Sensory Perception and Transduction in Aneural Organisms" G. Columbetti and P.-S. Song, eds.), p. 265, Plenum Publishing Corp, New York.
Briggs, W. R., Mösinger, E., Batschauer, A., Apel, K., and Schäfer, E. In 1987). "Molecular Biology of Plant Growth Control" (J. E. Fox and M. Jacobs, eds.), in press, Alan R. Liss, Inc., New York.
Briggs, W. R., Mösinger, E., and Schäfer, E (1987). Submitted to Plant Physiology.
Chappell, J., and Hahlbrock, K. (1984). Nature 311:76.
Dobres, M. S., Elliott, R. C., Watson, J. C., and Thompson, W. F. (1987). Plant Mol. Biol., in press.
Gallagher, T. F., and Ellis, R. J. (1982). EMBO J. 1:1493.
Harpster, M. H., Mayfield, S. P., and Taylor, W. C. (1984). Plant Mol. Biol. 3:59.
Horwitz, B. A., Thompson, W. F., and Briggs, W. R (1987). Submitted to Plant Physiol.
Kaufman, L. S., Thompson, W. F., and Briggs, W. R. (1984). Science 226: 1447.
Kaufman, L. S., Briggs, W. R., and Thompson, W. F. (1985). Plant Physiol. 78:388.
Kaufman, L. S., Roberts, L. I., Briggs, W. R., and Thompson, W. F. 1986). Plant Physiol. 81:1033.
Kuhlemeier, C., Green, P. J., and Chua, N.-H. (1987). Annu. Rev. Plant Physiol. 38:221.
Mösinger, E., Batschauer, A., Apel, K., Schäfer, E., and Briggs, W. R. (1987). Submitted to Plant Physiol.
Mösinger, E., Batschauer, A., Schäfer, E., and Apel, K. (1985). Eur. J. Biochem. 133:309.
Mösinger, E.., Batschauer, A., Vierstra, R., Apel, K., and Schäfer, E. (1987). Planta, in press.
Raven, C. W., and Shropshire, W. (1975). Photochem. Photobiol. 21:423.
Sagar. A., Briggs, W. R., and Thompson, W. F. (1987). In preparation.
Schäfer, E. and Briggs, W. R. (1986). Photobiochem. Photobiophys. 12:305.
Sharma, R. (1985). Photchem. Photobiol. 41:747.

Thompson, W. F., Kaufman, L. K., and Watson, J. C. (1985). BioEssays 3:153.
Tobin, E. M., and Silverthorne, J. (1985). Annu. Rev. Plant Physiol. 36:569.
Virgin, H. I. (1972). In "Phytochrome" (K. Mitrakos and W. Shropshire, Jr., eds.), p. 371. Academic Press, New York.

MOLECULAR BIOLOGY OF PHYTOCHROME

Peter H. Quail[1]
Christiane Gatz[2]
Howard P. Hershey[3]
Alan M. Jones[4]
James L. Lissemore[5]
Brian M. Parks
Robert A. Sharrock[6]

Departments of Botany and Genetics
University of Wisconsin
Madison, WI

Richard F. Barker[7]
Kenneth Idler
Michael G. Murray

Agrigenetics Advanced Research Laboratory
Madison, WI

Maarten Koornneef
Richard E. Kendrick

Departments of Genetics and
Plant Physiological Research
Agricultural University
Wageningen, The Netherlands

[1]Supported by NSF grant DMB8302206 and USDA grant 85-CRCR-1578
[2]Supported by a Feodor-Lynen Forschungsstipendium
[3]Present address: E.I. du Pont de Nemours & Co., Wilmington, Delaware 19898
[4]Present address: Department of Biology, University of North Carolina, Chapel Hill, NC 27514
[5]Supported by National Science Foundation Graduate Fellowship
[6]Supported by NSF Postdoctoral Fellowship DMB 8508836
[7]Present address: Plant Breeding Institute, Trumpington, Cambridge, CB2 2LQ England

I. INTRODUCTION

Studies using the tools of molecular biology and
immunochemistry have yielded significant new information on a
number of aspects of the phytochrome system in recent times
(Lagarias, 1985; Quail, et al., 1986a,b). In particular our
understanding of the structure of the phytochrome molecule has
greatly improved and questions regarding the structure,
organization and expression of the genes for the phytochrome
polypeptide are beginning to be answered.

II. STRUCTURE - FROM GENE TO PROTEIN

Restriction-map polymorphism among existing cDNA clones
indicates that at least four phytochrome genes are expressed
in etiolated Avena seedlings (Hershey et al., 1985). Sequence
analysis of two complete and one partial coding region shows
98% homology between these sequences at both the nucleotide
and amino acid levels. Figure 1 depicts the structural
relationship between the Type 3 Avena phytochrome gene, its
mature mRNA and the native phytochrome monomer (Daniels and
Quail, 1984; Hershey et al., 1985; 1987). The gene is 5.94
kbp in length and consists of 6 exons and 5 introns, one each
of the latter in the 5' and 3' untranslated regions of the
sequence. The transcription start site, located by S_1
nuclease mapping, is 35 bp downstream of tandem TATA boxes.
The 5' flanking DNA contains sequences related to CAAT and GC
boxes (Dynan and Tijan, 1985), elements that have regulatory
functions in other systems, but their significance to
phytochrome-gene expression is unknown. The mature mRNA is
3.78 kb long, before poly(A)$^+$ addition, and consists of 142 b,
3387 b and 252 b of 5'-untranslated, coding and
3'-untranslated sequence, respectively. In addition to the
ATG translation-initiation codon, the sequence contains tandem
TGA stop codons and an AATAAA poly(A)-addition signal
starting 35 b upstream of the poly(A)-addition site.
The encoded, mature phytochrome polypeptide is 1128 amino
acids long, with a single chromophore attachment site at
Cys-321 (Hershey et al., 1985). Biochemical studies indicate
that, in the native chromoprotein, the polypeptide is folded
into two principal domains: a globular, chromophore-bearing,
NH_2-terminal domain of ~ 74 kDa and a more open, COOH-terminal
domain of ~ 55 kDa (Jones and Quail, 1986; Lagarias and
Mercurio, 1985; Vierstra et al., 1984). These two domains are
linked by a proteolytically vulnerable segment of the chain,
with no evidence of other stable interactions between the two

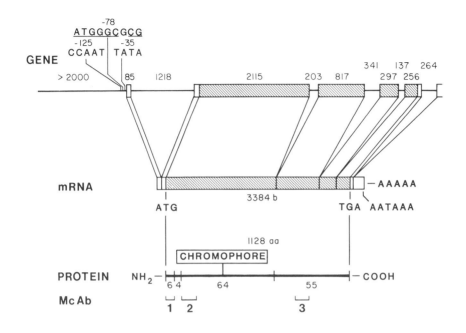

FIGURE 1.   Schematic representation of the Type 3 Avena
phytochrome gene (top), its mature transcript (middle) and
polypeptide product (bottom) (Hershey et al., 1985; 1987).
Gene: Exons are indicated as boxes (open = noncoding; shaded
= coding); introns and 5' flanking DNA are indicated as lines.
The length of each segment is indicated above in bp.   The
position of CAAT (-125), GC (-78) and TATA (-35) boxes are
indicated relative to the transcription start site.   mRNA:
The overall length of the mature transcript is indicated
below, with the locations of the original introns marked by
vertical lines.   Initiation (ATG) and stop (TGA) codons are
indicated together with the putative poly(A)-addition signal
(AATAAA) and poly(A)+ tail.   Protein: The mature phytochrome
chromoprotein (1128 aa long = 124 kDa) with chromophore
covalently linked at Cys-321 is aligned with the mRNA coding
region.   Vertical lines indicate the locations of three
proteolytically vulnerable sites in the polypeptide chain
leading to peptides of the size (in kDa) indicated below.
McAb: Square brackets indicate the locations of epitopes for
monoclonal antibodies designated Types 1, 2 and 3 (Daniels and
Quail, 1984; Jones and Quail, 1987).

regions of the polypeptide. The 74-kDa, NH_2-terminal domain
released by proteolysis corresponds almost exactly to the
major, 5' exon in the gene (Fig. 1). This observation is
consistent with the pattern commonly found with other
multidomain proteins (Gilbert, 1985). This pattern has led to
the hypothesis that multidomain proteins frequently may have
arisen during evolution by the bringing together of existing
sequences (exons) to form a contiguous stretch of DNA encoding
a single transcript (Gilbert, 1985). Whether the 3' exons
encode three structurally distinct subdomains within the
COOH-terminal region of the polypeptide is presently unclear.
Quaternary structure analysis indicates that the phytochrome
polypeptides are associated as non-globular dimers under
physiological conditions (Jones and Quail, 1986; Lagarias and
Mercurio, 1985), with evidence that the dimerization site is
in the COOH-terminal domain, within 40 kDa of the terminus
(Jones and Quail, 1986; Vierstra et al., 1984).

Examination of sequence homologies between proteins from
evolutionarily divergent species can be expected to provide
insight into conserved regions potentially important for
function. Recent sequence analysis of the complete coding
region of a cDNA clone from <u>Cucurbita</u> (Sharrock et al., 1986)
has enabled a comparison to be made between monocot and dicot
phytochromes (Fig. 2), long known to have poor immunochemical
cross-reactivity (Cordonnier and Pratt, 1982; Pratt, 1973;
Vierstra and Quail, 1985). Together with biochemical studies
and comparative hydropathy and secondary structure analysis
(Fig. 3), these data have provided information on some
conserved structural features of the phytochrome molecule.
Overall the higher level of sequence homology in the
NH_2-terminal two-thirds of the polypeptide than in the
COOH-terminal one-third (Fig. 2B), suggests that the
NH_2-terminal domain is more likely to be involved in critical
structural and functional properties of the photoreceptor.

Sequences involved in at least two properties of the
phytochrome molecule are expected to be conserved: (a) Those
involved in interactions with the chromophore, since the
spectral properties of the photoreceptor from diverse sources
are highly conserved; and (b) Those involved in the
biologically "active site" of the molecule, since it is
reasonable to expect that the molecular mechanism of
phytochrome action will be conserved between species.

All the sequence information necessary for correct
protein-chromophore interaction, in both the Pr and Pfr forms,
is located in the 74-kDa, NH_2-terminal domain of <u>Avena</u>
phytochrome, since the spectral properties of the
photoreceptor are unaltered by the proteolytic cleavage of
this domain from the molecule (Jones et al., 1985).
Indications that the polypeptide segment between residues 150

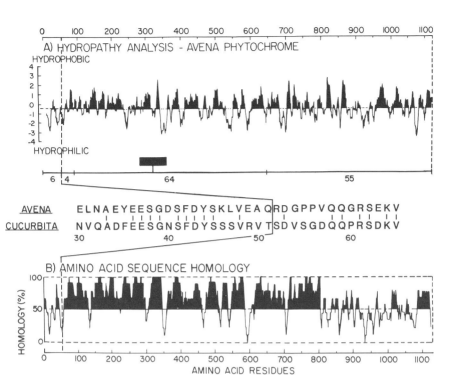

FIGURE 2. Phytochrome structure and sequence comparison.
(A) Hydropathy profile of Avena phytochrome (Hershey et al.,
1985). (B) Amino acid sequence homology between Avena and
Cucurbita phytochromes (Sharrock et al., 1987). (Center)
Amino acid sequence comparison between Avena and Cucurbita
phytochromes, residues 30 to 65. The location is indicated of
the site at Arg-52 that is proteolytically cleaved only in the
Pr form.

and 345 might interact with the chromophore are provided (a)
by the relatively high level of sequence (Fig. 2B) and
predicted secondary structure (Fig. 3) homology, between Avena
and Cucurbita phytochromes (Sharrock et al., 1986), and (b) by
the extensive, hydrophobic nature of this region (Fig. 2A).
This latter observation is consistent with the notion, derived
from chemical probe studies (Song, 1985; Hahn et al., 1984),
that the chromophore resides in a hydrophobic cavity within
the protein. More direct evidence of protein-chromophore
interactions in this region is provided by recent proteolytic
mapping studies. A 16-kDa peptide segment extending
approximately from residue 190 to 325 appears to carry a

CHOU - FASMAN SECONDARY STRUCTURE COMPARISON
AVENA AND CUCURBITA PHYTOCHROME

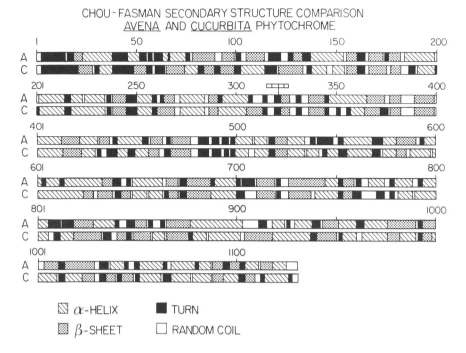

☒ α-HELIX   ■ TURN
▦ β-SHEET   ☐ RANDOM COIL

FIGURE 3. Comparison of the predicted secondary
structures of Avena and Cucurbita phytochromes obtained by
Chou-Fasman analysis (Chou and Fasman, 1978). A = Avena; C =
Cucurbita. Numbers indicate amino acid residue positions.
Locations of predicted α-helix, β-sheet, β-turn and random
coil are indicated as defined in the legend. Location of the
chromophore is indicated at Cys-321 above the Avena sequence.

substantial portion of the sequence involved in stabilizing
the chromophore in the Pr form, as indicated by spectral
analysis (Jones and Quail, 1987). Moreover, a highly
conserved, hydrophobic stretch of 20 amino acids, between
residues 190 and 210 (Fig. 2B) appears crucial to this
interaction, since its removal abolishes the Pr spectrum
(Grimm et al., 1986; Quail et al., 1987). In contrast,
sequences required to stabilize the chromophore in the Pfr
form appear to lie outside the region from residue 190 to 426,
since chromopeptides generated from this region lack Pfr
absorbance (Grimm et al., 1986; Jones and Quail, 1987). It
has been known for some time that the 6-kDa, hydrophilic
segment at the NH$_2$-terminus of the polypeptide is involved in
establishing the Pfr (but not the Pr) form of the molecule
(Eilfeld and Rüdiger, 1984; 1985; Jones et al., 1985; Kelly

and Lagarias, 1985; Litts et al., 1983; Vierstra and Quail, 1982b; 1983a,b), but additional, critical sequences are clearly required, since cleavage of this segment causes only a partial loss of absorbance and a shift in the λmax for Pfr. These additional sequences remain to be identified but would appear to lie between residues 52 and 190 and/or residues 426 and ~ 650 (Fig. 2). The relatively low level of sequence homology and secondary structure conservation in the 6-kDa, NH_2-terminal segment, with the exception of that for residues 36 to 45 (Fig. 2, 3), might indicate that it is this 10-amino acid stretch that is involved in the protein-chromophore interaction in this region of the polypeptide.

Sequences potentially involved in the "active site" of phytochrome have been sought by identifying regions of the polypeptide chain involved in photoconversion-induced conformational changes. One such region is localized in a 4- to 10-kDa segment at the NH_2-terminus as indicated by (a) its differential sensitivity to proteolysis (Daniels and Quail, 1984; Jones et al., 1985; Lagarias and Mercurio, 1985; Vierstra and Quail, 1982a,b), (b) its differential affinity for a spatially mapped monoclonal antibody (Cordonnier et al., 1985; Quail et al., 1986b), and (c) its differential susceptibility to phosphorylation by exogenous protein kinases (Wong et al., 1986) in the Pr and Pfr forms. We have now shown by NH_2-terminal sequence analysis of a proteolytic fragment, that one of the sites that is cleaved in the Pr, but not in the Pfr form, is located at Arg-52 (Jones and Quail, 1987). This residue is in a strongly hydrophilic segment of the polypeptide (Fig. 2A) and is therefore likely to be exposed on the surface of the molecule, available for intermolecular contact. On the other hand, as indicated above, the amino acid sequence of the 6-kDa, NH_2-terminal segment is poorly conserved overall between Avena and Cucurbita phytochromes, especially that immediately surrounding Arg-52 (Fig. 2). This observation might mean that this segment is unlikely to be directly involved in an "active site", or that the short stretch of amino acids adjacent to Arg-52, between positions 36 and 45, showing the highest conservation in the segment, are critical to this function. This latter possibility might receive some support from the observation that, although sequence homology is low in the 6-kDa segment, a relatively high proportion of the amino acid substitutions are conservative. As a result the overall hydropathic (Sharrock et al., 1986) and secondary structural properties (Fig. 3) are reasonably well conserved, perhaps providing the correct structural context for functionally critical residues. Other regions of the molecule that differ in accessibility to proteases in the Pr and Pfr forms occur at 74, 83-84 and 100 kDa from the NH_2-terminus, indicating

additional sites of photoconversion-induced conformational
change (Jones et al., 1985; Lagarias and Mercurio, 1985;
Vierstra et al., 1984). The fact that these changes are
conserved in phytochrome from a number of plant species
(Kerscher and Nowitzki, 1982; Vierstra et al., 1984) is
indicative of functional importance.

III. EPITOPE LOCATION

    Monoclonal antibodies have played a significant role in
studies of phytochrome structure. In particular they have
aided identification of regions of the polypeptide involved in
photoinduced conformational changes (Cordonnier et al., 1985;
Quail et al., 1986b; Shimazaki et al., 1986) and in
dimerization (Jones and Quail, 1986). Integral to these
studies has been the mapping of the epitopes for these
antibodies within the polypeptide chain. Three principal
antigenic regions have been defined in various studies: one
within the 6- to 10-kDa, NH_2-terminal subdomain, one within
the 64-kDa, chromophore-bearing subdomain, and one within the
55-kDa, COOH-terminal domain. Monoclonal antibodies having
epitopes in these three regions were initially defined for
operational convenience as Type 1, 2 and 3 monoclonals
respectively (Daniels and Quail, 1984; Jones and Quail, 1987).
The mapped epitope locations of the panel of antibodies
obtained in this laboratory are indicated in Fig. 1.
    Recently, Grimm et al. (1986) obtained data which led
them to question whether all the monoclonal antibodies that we
had classified as Type 1 in our initial studies do indeed map
to the NH_2-terminus, as opposed to the COOH-terminus. While
published data showing recognition by some of these antibodies
of the proteolytically-released 74-kDa, NH_2-terminal domain
(Daniels and Quail, 1984; Jones et al. 1985) verifies the
original localization, it remained possible that some of our
Type 1 monoclonals had been incorrectly mapped. To test this
possibility directly we have examined the ability of a
spectrum of our antibodies to bind to a synthetic peptide with
a sequence identical to the first 65 amino acids at the
NH_2-terminus of <u>Avena</u> phytochrome (Hershey et al., 1985). It
is clear that all the Type 1 monoclonals tested recognize the
synthetic sequence, whereas Type 2 and 3 antibodies do not,
thereby serving as a convenient internal control (Fig. 4).
These data unequivocally demonstrate that the complete
sequence information for the epitope(s) for all our Type 1
antibodies is located within 7 kDa of the NH_2-terminus of the
phytochrome polypeptide.

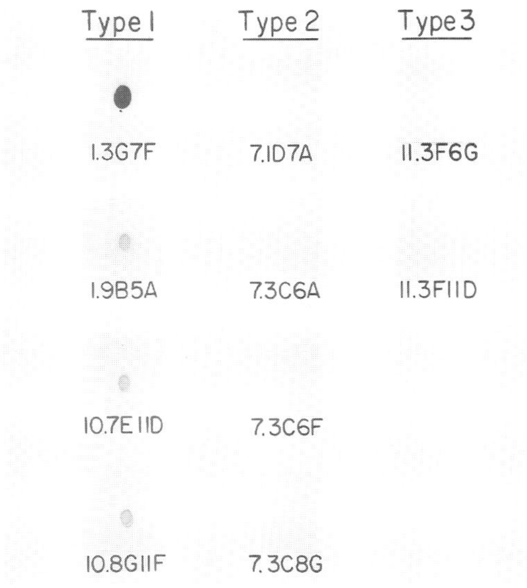

Type 1	Type 2	Type 3
●		
I.3G7F	7.ID7A	II.3F6G
●		
I.9B5A	7.3C6A	II.3FIID
●		
IO.7E IID	7.3C6F	
●		
IO.8GIIF	7.3C8G	

FIGURE 4. Immunodetection of synthetic phytochrome
peptide by three types of monoclonal antibodies specific for
124-kDa phytochrome from etiolated Avena. A peptide,
identical in sequence to the first 65 residues of the amino
terminal end of the etiolated Avena phytochrome polypeptide,
was synthesized by Dr. D. Davis, Applied Biosystems, Foster
City, CA. Pure, lyophilized synthetic peptide was rehydrated
in Tris-buffered saline (20 mM Tris pH 7.0, 170 mM NaCl, 0.02%
NaN_3) and made 10% glycerol (final peptide concentration 1
mg/ml). Replicate spots of 2 μl of this sample were applied
to dry nitrocellulose (Millipore Corp., Bedford, MA) and
allowed to dry at room temperature. The nitrocellulose sheet
was then hydrated in Western transfer buffer, transferred to
0.2% milk buffer to block remaining binding sites, and then
probed with 2 ml of individual Type 1, 2 and 3 monoclonal
antibodies (5 μg/ml) as described previously (Parks et al.,
1987).

IV. REGULATION OF PHYTOCHROME GENE EXPRESSION

There are at least two major reasons for investigating
the expression of the genes for phytochrome. First, since the
plant responds quantitatively to the level of the active form
of the photoreceptor, an overall understanding of how

phytochrome regulates development requires definition of how
absolute levels of the molecule are controlled in the cell.
Second, since in Avena phytochrome has been found to
down-regulate the expression of its own genes very rapidly
upon Pfr formation, phytochrome genes themselves provide a
model system for approaching the general question of the
molecular mechanism(s) by which phytochrome regulates gene
expression.

A number of recent publications have dealt with various
aspects of autoregulated phytochrome gene expression and so
will not be elaborated in detail here (Colbert et al., 1983;
1985; Lissemore et al., 1987; Quail et al., 1986a,b; 1987;
Sharrock et al., 1987). The principal findings from these
studies can be summarized as follows: (a) Down-regulation of
Avena phytochrome gene transcription can be detected within 5
min of Pfr formation using nuclear run-off transcription
assays, suggesting direct action of the phytochrome
signal-transduction chain without the intervention of
expression of other genes (Quail et al., 1986a); (b) The
magnitude of the decrease in transcription (3-fold) is
insufficient to account for the 10- to 50-fold reduction
observed in steady-state phytochrome mRNA levels, thereby
indicating the involvement of post-transcriptional as well as
transcriptional regulation of expression (Quail et al.,
1986a); (c) Attempts to establish a transient gene-expression
system using electroporation of Avena protoplasts (Fromm et
al., 1985; 1986) for analyzing the cis-acting nucleotide
sequences involved in autoregulation of expression, have thus
far been unsuccessful. Osmotic stress appears to abolish
phytochrome-regulation of gene expression, despite the
presence of apparently normal levels of the photoreceptor
molecule in the protoplasts (R.A. Sharrock and A.M. Jones,
unpublished); (d) Attempts to identify a plant system suitable
for analysis of phytochrome-gene expression using stable
transformations have led to the somewhat unexpected finding
that the magnitude of the light-induced decrease in endogenous
phytochrome mRNA levels is less in a number of species than
in Avena (Lissemore et al., 1987; Quail et al., 1987; Sharrock
et al., 1987; A.H. Christensen, unpublished). This
observation indicates that different species have adopted
different strategies for regulating phytochrome levels and
that care is needed in selection of a system suited to
analysis of this problem.

In parallel with the above approaches to identifying
cis-acting regulatory elements, we have begun attempts to
detect and identify trans-acting cellular factors involved in
phytochrome gene expression. The most immediately obvious and
attractive candidate for a role in autoregulation of the
phytochrome gene is the photoreceptor molecule itself. We

have tested for the interaction of purified 124-kDa <u>Avena</u> phytochrome with DNA fragments derived from the Type 3 phytochrome gene (Fig. 1) using two different assays: (a) Immunoprecipitation of phytochrome from mixtures of the purified photoreceptor and labeled restriction fragments to test for phytochrome-bound DNA (Scheller et al., 1982); (b) A gel-retardation assay in which labeled DNA fragments that bind phytochrome are expected to exhibit reduced migration relative to the free DNA upon electrophoresis through nondenaturing polyacrylamide gels (Fried and Crothers, 1981; Singh et al., 1986). No detectable binding of the photoreceptor was observed in either the Pr or Pfr forms, to any phytochrome gene sequence tested, from 1 kbp upstream to ~ 130 bp downstream of the transcription start site (C. Gatz, unpublished). We must conclude therefore that current evidence is strongly against the possibility that phytochrome interacts directly with cis-acting nucleotide sequences in the 5'-flanking DNA of its own gene to bring about decreased transcription. We are currently testing for the presence of other proteins in nuclear extracts capable of specific binding to phytochrome-gene sequences.

V. PHYTOCHROME MUTANTS

Koornneef et al. (1985) reported that the <u>aurea</u> (au^W) mutant of tomato lacks spectrally detectable phytochrome as determined by <u>in vivo</u> spectrophotometry. This mutant exhibits phenotypic characteristics consistent with a phytochrome deficiency, namely, reduced inhibition of hypocotyl and stem elongation and reduced anthocyanin and chlorophyll synthesis in response to light. We have now shown by immunoblot analysis that the $\underline{au}^W$ mutant lacks detectable levels of the phytochrome polypeptide (Parks et al., 1987). Given the detection limits of the spectrophotometric and immunoblot assays it is concluded that the mutant can contain no more than 5% of the wild type level of phytochrome.

To determine the molecular basis for this lesion we have examined the levels of phytochrome mRNA in the $\underline{au}^W$ mutant. Northern analysis indicates that the mutant contains the same amount of phytochrome mRNA as the wild type, and <u>in vitro</u> translation assays establish that this mRNA is equally translatable from the two tissues (Sharrock et al., 1987). As these data indicate that the mutant has the capacity to synthesize normal levels of the phytochrome polypeptide, the absence of detectable levels of this polypeptide in the cell would appear to result from a high rate of turnover. There are a number of possible mechanisms that would produce an unstable molecule. These include: (a) An amino acid addition, substitution or deletion in the phytochrome polypeptide leading to intrinsic instability of the protein; (b) A lesion in a component of the chromophore biosynthetic pathway which would block synthesis or lead to an aberrant chromophore, either of which might lead secondarily to instability of the polypeptide; (c) A lesion in the phytochrome polypeptide or a putative "chromophore-attachment enzyme" blocking normal covalent attachment of the tetrapyrrole and therefore inducing instability; and (d) A mutation in the cellular mechanism responsible for the 50- to 100-fold difference in turnover rates of Pr and Pfr such that the Pr form is now subject to abnormally high turnover. Experiments designed to test some of these possibilities are in progress.

What are the consequences of severe phytochrome deficiency to the expression of genes normally under phytochrome control? To approach this question, we have examined the levels of the mRNAs for phytochrome (phy), small subunit of ribulose bisphosphate carboxylase (rbcS) and chlorophyll a/b binding protein (cab) in mutant and wild type tomato tissue (Sharrock et al., 1987). Unexpectedly, given the situation in other species, the levels of neither phy nor rbcS mRNAs changed in wild-type etiolated tissue in response to saturating pulses of red light. Consistent with this apparent lack of phytochrome control of these genes in tomato, however, we find that phy and rbcS mRNA levels are the same in wild type and au^W mutant tissue. The deficiency of phytochrome does not affect the expression of these genes. cab mRNA levels, in contrast, are strongly increased in a red-far red reversible fashion in the wild type, but are virtually unaffected by the light pulses in the au^W mutant. These data show clearly that the expression of a gene that is normally phytochrome-regulated is greatly reduced when the photoreceptor is deficient.

Interestingly, these observations bear on two long-standing issues in phytochrome research. First, the data provide the first direct evidence that the chromoprotein that is detected spectrally and immunochemically in plant extracts is responsible for the biological function of the photoreceptor in the cell. This conclusion is based on the observation that the molecule detected in vitro is necessary for normal, phytochrome-regulated gene expression. Second, the data provide compelling evidence that Pfr is the active form of phytochrome. The question is often raised as to whether the photoconversion driven increase in Pfr or the decrease in Pr is responsible for phytochrome action. Although indirect arguments based on physiological experiments have been developed, none are definitive. Table 1 illustrates

TABLE 1. Relative levels[a] of Pr and Pfr forms[b] of phytochrome estimated to be present in dark-grown tomato seedlings before and immediately after a saturating pulse of red light.

Tomato line	Dark		Red-irradiated	
	Pr	Pfr	Pr	Pfr
Wild type	100	0	14	86
$\underline{au^W}$	<5	0	<1	<4

[a]Dark control set to 100%; from Parks et al. (1987).
[b]From Vierstra and Quail (1983).

that the relative level of Pr in etiolated $\underline{au^W}$ mutant seedlings is less than that in wild-type seedlings following saturating red irradiation. If the absence of "inhibitory" effects of Pr were responsible for phytochrome-induced cab gene expression in the wild type following red irradiation, full expression of this gene would be expected in the dark in the mutant. No such level of expression is observed. Conversely, the absence of a red-light induced response in the mutant, indicates a requirement for formation of sufficient Pfr to alter gene expression.

REFERENCES

Chou, P.Y. and Fasman, G.D. (1978). Annu. Rev. Biochem. 47: 251.
Colbert, J.T., Hershey, H.P. and Quail, P.H. (1983). Proc. Natl. Acad. Sci. USA 80:2248.
Colbert, J.T., Hershey, H.P., and Quail, P.H. (1985). Plant Mol. Biol. 5:91.
Cordonnier, M.M. and Pratt, L.H. (1982). Plant Physiol. 70:912.
Cordonnier, M.M., Greppin, H. and Pratt, L.H. (1985). Biochemistry 24:3246.
Daniels, S.M. and Quail, P.H. (1984). Plant Physiol. 76:622.
Dynan, W.S. and Tijan, R. (1985). Nature 316:774.
Eilfeld, P. and Rüdiger, W. (1984). Z. Naturforsch. 39c:742.

Eilfeld, P. and Rüdiger, W. (1985). Z. Naturforsch. 40c:109.

Fried, M. and Crothers, D.M. (1981). Nuc Acids Res. 9:6505.

Fromm, M., Taylor, L.P. and Walbot, V. (1985). Proc. Natl.
    Acad. Sci. USA 82:5824.

Fromm, M., Taylor, L.P. and Walbot, V. (1986). Nature 319:791.

Gilbert, W. (1985). Science 228: 823.

Grimm, R., Lottspeich, F., Schneider, H.A.W. and Rüdiger, W.
    (1986). Z. Naturforsch. 41c:83.

Hahn, T.-R., Song, P.S., Quail, P.H. and Vierstra, R.D.
    (1984). Plant Physiol. 74:755.

Hershey, H.P., Barker, R.F., Idler, K.B., Lissemore, J.L. and
    Quail, P.H. (1985). Nuc. Acids Res. 13:8543.

Hershey, H.P., Barker, R.F., Idler, K.B., Murray, M.G. and
    Quail, P.H. (1987). Nuc. Acids Res., submitted.

Jones, A.M. and Quail, P.H. (1986). Biochemistry 25:2987.

Jones, A.M. and Quail, P.H. (1987). Eur. J. Biochem., submitted.

Jones, A.M., Vierstra, R.D., Daniels, S.M. and Quail, P.H.
    (1985). Planta 164:501.

Kelly, J.M. and Lagarias, J.C. (1985). Biochemistry 24:6003.

Kerscher. L. and Nowitzki, S. (1982). FEBS Lett. 146:173.

Koornneef, M., Cone, J.W., Dekens, R.G., O'Herne-Robers, E.G.,
    Spruit, C.J.P. and Kendrick, R.E. (1985). Plant Physiol.
    120:153.

Lagarias, J.C. (1985). Photochem. Photobiol. 42:811.

Lagarias, J.C. and Mercurio, F.M. (1985). J. Biol. Chem.
    260:2415.

Lissemore, J.L., Colbert, J.T. and Quail, P.H. (1986). Plant
    Mol. Biol., in press.

Litts, J.C., Kelly, J.M. and Lagarias, J.C. (1983). J. Biol.
    Chem. 258:11025.

Parks, B.M., Jones, A.M., Adamse, P., Koornneef, M., Kendrick,
    R.E. and Quail, P.H. (1987). Plant Mol. Biol., submitted.

Pratt, L.H. (1973). Plant Physiol. 51:203.

Quail, P.H., Colbert, J.T., Peters, N.K., Christensen, A.H.,
    Sharrock, R.A. and Lissemore, J.L. (1986a). Phil. Trans.
    R. Soc. London. B314:469.

Quail, P.H., Barker, R.F., Colbert, J.T., Daniels, S.M.,
    Hershey, H.P., Idler, K.B., Jones, A.M. and Lissemore,
    J.L. (1986b). In "Molecular Biology of Plant Growth
    Control, UCLA Symposia on Molecular and Cellular Biology"
    (J.E. Fox and M. Jacobs, eds.), New Series, Vol. 44, Alan
    R. Liss, Inc., New York, in press.

Quail, P.H., Christensen, A.H., Jones, A.M., Lissemore, J.L.,
    Parks, B.M. and Sharrock, R.A. (1987). In "Integration
    and Control of Metabolic Processes: Pure and Applied
    Aspects", 4th FOAB Congress (W.J. Whelan ed.), ICSU Press,
    Miami, in press.

Sharrock, R.A., Lissemore, J.L. and Quail, P.H. (1986). Gene
    47:287.

Sharrock, R.A., Parks, B.M., Koornneef, M. and Quail, P.H.
     (1987). Molec. Gen. Genetics, submitted.
Scheller, A., Covey, L., Barnet, B. and Prives, C. (1982).
     Cell 29:375.
Shimazaki, Y., Cordonnier, M.M. and Pratt, L.H. (1986). Plant
     Physiol. 82:109.
Singh, H., Sen, R., Baltimore, D. and Sharp, P.A. (1986).
     Nature 319:154.
Song, P.S. (1985). In "Optical Properties and Structure of
     Tetrapyrroles" (G. Blauer and H. Sund eds.), p. 331. W.
     de Gruyter, Berlin.
Vierstra, R.D., Cordonnier, M.-M., Pratt, L.H. and Quail, P.H.
     (1984). Planta 160:521.
Vierstra, R.D. and Quail, P.H. (1982a). Proc. Natl. Acad. Sci.
     USA 79:5272.
Vierstra, R.D. and Quail, P.H. (1982b). Planta 156:158.
Vierstra, R.D. and Quail, P.H. (1983a). Biochemistry 22:2498.
Vierstra, R.D. and Quail, P.H. (1983b). Plant Physiol. 72:264.
Vierstra, R.D. and Quail, P.H. (1985). Plant Physiol. 77:990.
Wong, Y.S., Cheng, H.C., Walsh, D.A. and Lagarias, C.J.
     (1986). J. Biol. Chem. 261:12089.

PHOTOCONTROL OF GENE EXPRESSION

Elaine M. Tobin

Biology Department
University of California
Los Angeles, California

I. INTRODUCTION

In this summary of the discussion section held on the photocontrol of gene expression, I hope to include many of the significant points and some of the data that were presented at the time. I've given less space to details of topics that are covered in other chapters of this volume. To some extent I have reorganized the order and grouped topics that were discussed at several separate times. Some references and a little background that wasn't explicitly discussed, but which I hope will aid in giving a more complete picture of the area, are also included. I have tried to represent others' views accurately, but I think that it would be unfair to strictly attribute my version to them. Therefore, I warn the reader that this is a necessarily biased and incomplete account of a very lively and stimulating discussion led by Hans Mohr, Peter Quail, and Eberhard Schäfer in addition to myself.

Support for work from my laboratory has come from grants from N.I.H. (GM 23167-11), N.S.F. (DCB-84-17606), and U.S.D.A. (85-CRCR-1-1622).

39

II. PHYTOCHROME REGULATION OF RNA LEVELS

Phytochrome action has now been shown to affect the RNA levels for specific nuclear gene products in a number of species (reviewed by Tobin and Silverthorne, 1985). In the cases of the small subunit (SSU) of ribulose 1,5-bisphosphate carboxylase (Rubisco)(encoded by rbcS genes), the major light-harvesting chlorophyll a/b-proteins (LHCP) (encoded by cab genes), ferredoxin, and a number of thus far unidentified proteins, the RNA levels increase in response to brief red illumination. In other instances, including protochlorophyllide oxidoreductase and phytochrome itself, the RNA levels are negatively regulated, i.e. they decrease in response to the action of phytochrome. One such instance has recently been found in my laboratory by S. Flores, who isolated a Lemna cDNA clone (p106) which is much more abundant in light-grown than in dark-treated plants. The regulation of this RNA by phytochrome, unlike the other two negatively regulated RNAs studied in barley and oat, does not involve the rapid disappearance of the RNA in response to phytochrome action.

In cases where transcription rates for cloned gene sequences have been assayed by in vitro nuclear run-off experiments, phytochrome action has been shown to increase or decrease transcription of the specific genes in the plants. These instances include Lemna rbcS, cab, and rRNA genes (Silverthorne and Tobin, 1984), soybean rbcS genes (Berry-Lowe and Meagher, 1985), barley cab, protochlorophyllide reductase and rRNA genes (Mösinger et al., 1985), and oat phytochome genes (Quail et. al., 1986).

A.   Use of the in Vitro Transcription Assay

The question of what exactly is being measured by an in vitro run-off experiment and what the limits of interpretation of such data are were discussed. In such experiments, nuclei are isolated and incubated in a reaction mix that includes a radioactive nucleotide. The labeled RNA is then isolated and hybridized to a cloned sequence. Comparisons between samples are made by hybridizing equal amounts of radioactivity to the cloned sequence of interest; thus, values relative to other transcribed sequences are obtained. The fact that RNA polymerase II does not reinitiate transcripts in the isolated nuclei in animal systems and has not been shown to do so in plant systems means that all radioactivity is incorporated into transcripts that had already been initiated and are in the process of elongating when the nuclei were isolated. This limitation is perhaps useful in that one gets a look at a single time point when the plants are harvested and rapidly chilled. However, the products that will hybridize to a single cloned sequence are of varying lengths, suggesting that truncated transcripts may be present. These could arise by polymerases terminating prematurely or as a result of ribonuclease activity. Such problems make it impossible to measure directly the number of polymerases engaged in transcription. There is evidence that transcription is not random in the system; for example in some cases that have been examined (e.g. J. Silverthorne) transcripts will hybridize only to the coding strand of a known genomic sequence. The lack of reinitiation also raises a problem in interpreting experiments in which factors such as phytochrome are added to isolated nuclei and have an effect on the amount of radioactivity incorporated, perhaps even into specific sequences (Ernst and Osterheld, 1984; Mösinger et al., cited in Schäfer and Briggs, 1986). It is known that pH and salt concentrations can greatly influence the results of such in vitro experiments and can even affect the activity of the different polymerases differently (e.g. Chappell and Hahlbrock, 1986). It may prove difficult to relate the in vivo situation to effects of factors added to the in vitro system as it currently exists. It is to be hoped that an initiating in vitro transcription system will be developed for RNA polymerase II in plant systems.

B. Light Regulation in Transgenic Plants

A number of laboratories have also looked at the differences in transcription of specific genes in light-grown compared to dark-grown plant tissue either by the in vitro run-off experiments (e.g. Gallagher and Ellis, 1982) or in plants transformed with either cab or rbcS promoters fused to reporter genes (reviewed by Silverthorne and Tobin, 1987). In some instances the inserted promoter region was shown to be sufficient for phytochrome regulation of the RNA level, and in other circumstances a blue light-receptor seemed to be important in the regulation of the RNA level. Additionally, the state of development of chloroplasts seems to be an important factor in the expression of the inserted sequence (as has been shown to be the case in untransformed plants)(see Silverthorne and Tobin, 1987, and further discussion below). Thus far, the in vitro run-off nuclear transcription experiments have not been attempted with transgenic plants, and the question of whether some of the differences which have been ascribed to transcriptional differences might reflect different RNA stabilities was raised. It was pointed out that in some cases, the expression level of a test construct (promoter plus reporter gene) was compared to a control construct having a constitutive promoter attached to the same reporter gene. The identical transcripts from both constructs should have identical stabilities. However, in many studies to date absolutely identical transcripts haven't been used, and for studies of possible RNA stability differences under different illumination conditions, this would be essential. It was also pointed out (A. Cashmore) that experiments comparing the expression of the reporter gene in transgenic plants to its expression in in vitro run-off experiments with nuclei isolated from the same plants might be useful in seeing whether the two approaches give consistent results.

A problem exists in comparing different transgenic plants because the levels of expression of an inserted sequence can vary widely among individual transformants. It has been suggested that the site of insertion may be important in the expression of an inserted gene, but even the expression of constructs containing two genes in tandem has been found to vary relative to each other. Since the site of insertion is apparently random, there is no way to control it; therefore, it is important to pool many individual plants transformed with a particular construct to minimize such effects. Clearly it is not yet possible to use comparisons between different transgenic plants for quantitative experiments.

On a positive note, however, several laboratories have been able to confer phytochrome regulation on a reporter gene using 5'upstream sequences from phytochrome regulated genes. These experiments have unequivocally demonstrated the importance of cis-regulatory elements in the phytochrome regulation of transcription. Furthermore, experiments in which regulatory elements of one species function in another species (including a monocot element in a dicot plant, Nagy et al, 1986) make it reasonable to think that there is remarkably good conservation of the mechanism of the transcriptional response to phytochrome action. On the other hand, it was noted that studies to date have been limited to a few genes, and the complexities of the phytochrome responses in the transgenic plants have not yet been fully explored.

Another factor that may need to be considered in assaying the levels of expression of inserted genes in transgenic plants is that the time of day that plant tissue is harvested may be important, since there seems to be in some instances a circadian rhythm influence on the expression level of some light-regulated genes (Kloppstech, 1985; S. Kay and N.-H. Chua, unpublished results).

C. Can Phytochrome Action Affect the Stabilities of Specific RNAs?

Some experimental observations have suggested that phytochrome action might affect mRNA stabilities. In particular, the decline of phytochrome mRNA in oats (Quail et al, 1986) and of protochlorophyllide reductase mRNA in barley (Mösinger et al, 1985) after a brief red illumination is so rapid that a phytochrome-mediated destabilization of these RNAs (in addition to the transcription decline which has been demonstrated in both cases) seems a reasonable possible interpretation of the results. Another intriguing observation that is consistent with such a possibility is the work of Kaufman et al. (1986) on the regulation of RNA for ferredoxin in pea seedlings. In this case, the RNA level reaches a plateau within about 2 hours after a brief red illumination, but the response does not escape from phytochrome control during the first seven hours (longer times were not assayed) when the RNA levels are measured 24 hr after the red light.

It has not yet been experimentally determined whether mRNA stability changes do play a role in these responses; since the RNA level is a balance between synthesis and degradation, they could be explained solely by transcriptional changes and a short half-life for the particular RNAs. However, there are many examples in the animal literature (e.g. Palmiter and Carey, 1974; Guyette, et al, 1979; Brock and Shapiro, 1983) in which both transcription and RNA stabilities are regulated. Recent experiments have suggested that 5' or 3' untranslated regions of mRNAs are involved in determining stability (e.g. Morris et al, 1986; Shaw and Kamen, 1986; Brawerman, 1987). J. Silverthorne reported that different members of the Lemna gibba rbcS gene family are expressed differently in the roots of the plant, with 2 of the 6 sequences for which a specific probe is available being present at very low levels. Yet when root nuclei are assayed in vitro for transcriptional activity, these two genes are being actively transcribed, suggesting that there is a difference in processing or stability of these RNAs in root tissue. It is, however, difficult to measure RNA lifetimes in vivo by pulse labeling in higher plants. One possibility that was discussed to examine the possible role of such regulation was transformation experiments with transcribed sequences altered on different constructs and placed under the contol of identical constitutive promoters. In such experiments, differences in steady state RNA levels might reflect different stabilities rather than differential transcription. However, the limitations discussed already on being able to compare different transformation events could prove a problem.

D. How Can Phytochrome Action Lead to Altered Transcription?

Another topic that was briefly explored was possible mechanisms by which phytochrome action could lead to altered transcription. It seems clear that one can expect protein factors to interact with cis-regulatory DNA sequences (see, e.g. Fluhr et al, 1986a) (and perhaps similarly, if there is also post-transcriptional regulation, protein or RNA factors might interact with specific mRNAs). However, there is little/no evidence about how phytochrome action might be involved in such regulation. Questions were raised about whether such regulation would be positive or negative, whether one would expect a different factor for each different gene or gene family regulated, and whether regulation of positively and negatively regulated genes would be by similar mechanisms. However, these questions remain unanswered now. It was suggested that protein phosphorylation/dephosphorylation would be an attractive hypothesis for regulating the function of DNA binding proteins, but once again there is no experimental evidence available to support or disprove it. There were several posters (Otto and Schäfer; Datta and Roux) that raise the possibility that phytochrome action can affect protein phosphorylation/dephosphorylation, but there is no consensus amongthese studies as to which proteins are affected. In the case of chalcone synthetase induction in cultured parsley cells, preliminary results presented at a recent meeting (Kaulen et al, 1986) were reported to suggest that a DNA binding protein moves from the cytosol to the nucleus after induction by UV light.

Suggested approaches to understand further the possible transduction chain included development of an initiating in vitro transcription system, transformation/mutagenesis ex-

periments using upstream regions of phytochrome regulated
genes shown to bind protein factors, and the study of mutants
in the photoreceptor and in the response transduction pathway.

E. What Is the Pfr Requirement for the Transcriptional Response?

In general it was apparantly assumed by the group that
there would be an intermediate step between Pfr formation and
transcriptional responses, and we briefly considered what was
known about the need for the continued presence of Pfr.
Kaufman et al. (1986) have shown that the escape times for
the responses of different pea RNAs can vary widely, but they
measured RNA levels rather than transcription. It was also
pointed out that different members of gene families may
respond differently to phytochrome action (Tobin et al, 1985;
Fluhr et al, 1986b). In cases where transcription has been
assayed with the in vitro run-off system, a number of
interesting results have been obtained. Briggs and his
collaborators in Freiburg found that barley nuclear run-off
transcription increased in response to either low fluence
(LF) or very low fluence (VLF) illumination. However, the
response was more transient after a VLF pulse, and if assayed
three hours later, no increase in transcription could be
detected.   In another approach, several groups have looked
at transcription of specific coding sequences after light-
grown plants are transfered to the dark.   In pea,
transcription of rbcS seems to drop rapidly (Gallagher,
Jenkins and Ellis, 1985), while in soybean (Berry-Lowe and
Meagher, 1985) and Lemna (Silverthorne and Tobin,
unpublished) the decline is much slower.   However, in
soybean, if far-red light is given at the end of the light
period, the drop in transcription is much more rapid,
suggesting the need for the continuous presence of Pfr for
the transcriptional response. The reversal of this far-red
effect with red light was not tested.

III. MULTIPLE PHOTORECEPTORS ARE INVOLVED IN REGULATING GENE EXPRESSION

It is clear that the in vivo situation for regulation of
the expresion of the genes studied to date is more
complicated than a simple on/off switch involving
phytochrome. One important aspect that has been extensively

studied in H. Mohr's group is the developmental readiness of the tissue to respond (e.g. Oelmüller et al, 1986). Recent experiments have indicated that a blue light receptor interacts with the phytochrome pathway under certain circumstances. Fluhr and Chua (1986) have found that while the regulation of expression of two pea rbcS genes (3A and 3C) is mediated by phytochrome in etiolated plants, a blue light receptor is required in mature green tissue. What this might mean in terms of two phytochrome pools ("green" and "etiolated") is not known. Simpson et al. (1986) have also described an apparent loss of phytochrome regulation in mature tissue which would be consistent with these observations. Further evidence for the importance of a blue photoreceptor has been provided by Kaufmann et al (1985). In addition, there are responses which can be affected by phytochrome action, but which require blue or UV light first. Examples include chalcone synthatase in cultured parsley cells, hypocotyl extension in sesame seedlings, and anthocyanin synthesis in Sorghum. In these cases H. Mohr said that blue or UV light serve to potentiates the system to allow phytochrome regulation. With the exception of phototropism, which is a true blue light response, Mohr said that in higher plants all the light effects can be explained using phytochrome as the effector through which the other photoreceptors act. How this interaction might be achieved is not known.

The requirement for multiple photoreceptors was discussed extensively. H. Senger reminded us that green algae can assemble the complete photosynthetic apparatus in darkness, yet blue light can increase the synthesis of Rubisco, LHCP, and chlorophyll. Perhaps such responses may be a way of measuring shading in natural conditions. It was suggested (J. Shinkle) that blue light could be used as a way of counting photons, while the phytochrome system gives "color vision." It has also been proposed that the use of multiple photoreceptors allows a more effective coverage of the whole light spectrum.

The various light receptors seem to be interdependent, and can ultimately act via phytochrome in many cases. The blue/UV/phytochrome system is apparently elastic. A wheat mutant described by H. Mohr has been shown to have lost the blue response and to have replaced its function with a UV receptor. This replacement is insufficient to allow growth in the laboratory, but is functional in the field. G. Richter's work (Richter, 1984; Richter and Wessel, 1985) with cultured tobacco cells, in which blue light, but not red, can stimulate the accumulation of a number of RNAs for chloroplast proteins was brought up, and Mohr explained why he thinks that even in such a case one cannot rule out the involvement of phytochrome.

While phytochrome is now relatively well characterized, the blue and UV photoreceptors are not. They have been identified by their characteristic action spectra--a 370 nm peak for the blue receptor and a 290 nm peak for the UV receptor. However, they do not show photoreversibility and so biochemical characterization has proven almost impossible. Almost any flavoprotein can act as a photoreceptor, but it is difficult to pinpoint which ones are important for regulatory events in vivo.

IV. ROLE OF FUNCTIONAL PLASTIDS

Many of the genes whose expression is known to respond to phytochrome action encode chloroplast proteins. In general such imported proteins function in combination with chloroplast synthesized components, and it has been a long-standing question of how such synthesis is coordinated. Y. Sasaki presented data on the response of both the chloroplast encoded large and nuclear encoded small subunits of Rubisco in pea plants. She has found that phytochrome action increases the expression of both kinds of genes and, in addition, can affect the amounts of chloroplast DNA found in these seedlings. Experiments in which pea plants were treated with α-amanitin have suggested that the regulation of the rbcL gene may involve translational control (Sasaki, 1986). Several other reports of translational control of chloroplast RNAs have been published (see Klein and Mullet, 1986, and references therein).

There is increasing evidence that expression of some nuclear genes encoding chloroplast polypeptides requires the presence of functional chloroplasts. Several lines of evidence support this conclusion. In potato tubers, the expression of some sequences was not observed unless the tissue was green (Eckes et al, 1985). Similarly, transgenic callus transformed with a rbcS-CAT fusion only expressed CAT when the tissue was allowed to synthesize chloroplasts upon removal of sucrose and the addition of cytokinin (Herrera-Estrella et al, 1984). Examination of transgenic tobacco transformed with similar rbcS and cab constructs revealed

that these were mainly expressed in tissues which contained
mature plastids (Simpson et al, 1986; also see Fluhr et al,
1986a). Evidence for the role of plastids also comes from
work with maize and tomato mutants and with normal plants
bleached by the herbicide norflurazon (Mayfield and Taylor,
1984; Batschauer et al, 1986; Oelmüller et al, 1986;
Oelmüller and Mohr, 1986).

One mechanism for how this regulatory role of the plastid
might be effected was elaborated by H. Mohr. He proposed
that a factor is produced by functional plastids which is
required for expression of nuclear-encoded chloroplast genes.
In mustard it is proposed that this signal is continuously
produced and has a short half life because in experiments in
which the plastids are rendered non-functional because of
photooxidation, the amount of translatable mRNA for the
Rubisco small subunit decreases. This could be reflecting a
decrease in transcription of the RNA. In a tomato "ghost"
mutant, which has white tissue lacking functional plastids,
both cab and rbcS genes are apparently not transcribed in the
white cells (P. Skolnick, unpublished results). However, the
role of a putative plastidic factor in other processes such
as mRNA translatability has not been ruled out.

IV.   CONCLUSIONS

The understanding of the control of gene expression by
the action of phytochrome and other photoreceptors has made
great strides in recent years. Most of the needed techniques
are available for rapidly expanding this knowledge,
particularly in the area of transcriptional regulation. The
complexities of responses to light need to be fully
appreciated and studied, including such areas as the possible
roles of post-transcriptional regulation, influence of a
chloroplast factor, and developmental readiness to respond.
These studies should, in the near future, lead us closer to
the ultimate goal of understanding in detail the mechanism(s)
of light-receptor response transduction.

ACKNOWLEDGMENT

I thank Dr. Jane Silverthorne for helping me to formulate
this report and to remember some of the details of the
discussion.

REFERENCES

Batschauer, A., Mösinger, E., Kreuz, K., Dorr, I. and Apel, K.(1986) Eur. J. Biochem. 154: 625.

Berry-Lowe, S.L. and Meagher, R.B. (1985) Mol. Cell. Biol. 5: 1910.

Brawerman, G. (1987) Cell 48: 5.

Brock, M.L. and Shapiro, D.J. (1983) Cell 34: 207.

Chappell, J. and Hahlbrock, K. (1986) Plant Cell Rep. 5: 398.

Eckes, P., Schell, J. and Willmitzer, L. (1985) Mol. Gen. Genet. 199: 216.

Ernst, E. and Oesterheld, D. (1984) EMBO J. 13: 3075.

Fluhr, R. and Chua, N.-H. (1986) Proc. Natl. Acad. Sci. USA 83: 2358.

Fluhr, R., Kuhlemeier, C., Nagy, F. and Chua, N.-H. (1986a) Science 232: 1106.

Fluhr, R., Moses, P., Morelli, G., Coruzzi, G. and Chua, N.-H. (1986b) EMBO J. 5: 2063.

Gallagher, T.F. and Ellis, R.J. (1982) EMBO J. 1: 1493.

Gallagher, T.F., Jenkins, G.I. and Ellis, R.J. (1985) FEBS Lett. 186: 241.

Guyette, W.A., Matusik, R.J. and Rosen, J.M. (1979) Cell 17: 1013.

Herrera-Estrella, L., Van den Broeck, G., Maenhaut, R., Van Montagu, M., Schell, J., Timko, M. and Cashmore, A. (1984) Nature 318: 579.

Kaufman, L.S., Roberts, L.L., Briggs, W.R., and Thompson, W.F. (1986) Plant Physiol. 81: 1033.

Kaufman, L.S., Watson, J.C., Briggs, W.R., and Thompson, W.F. (1985) In: Molecular Biology of the Photosynthetic Apparatus, ed. Steinbeck, K.E., Bonitz, S., Arntzen, C.J., and Bogorad, L. (Cold Spring Harbor, N.Y.) p. 367.

Kaulen, H., Staiger, D. and Schell, J. (1986) In: Abstracts of the First International Symposium on Chromatin Structure of Plant Genes, Frankfurt am Main, p. 17.

Klein, R.R. and Mullet, J.E. (1986) 261:11138.

Kloppstech, K. (1985) Planta 165: 502.

Mayfield, S. and Taylor, W.C. (1984) Eur. J. Biochem. 144: 79.

Morris, T., Marashi, F., Weber,L., Hickey, E., Greenshaw, D., Bonner, J., Stein, J. and Stein, G. (1986) Proc. Natl. Acad. Sci. USA 83: 981.

Mösinger, E., Batschauer, A., Schäfer, E. and Apel, K. (1985) Eur. J. Biochem. 147: 137.

Nagy, F., Kay, S.A., Boutry, M., Hsu, M.-Y. and Chua, N.-H. (1986) EMBO J. 5: 1119.

Oelmüller, R., Dietrich, G., Link, G. and Mohr, H. (1986) Planta 169: 260.

Oelmüller, R. and H. Mohr (1986) Planta 167: 106.
Palmiter, R.D. and Carey, N.H. (1974) Proc. Natl. Acad. Sci. USA 71: 2357.
Quail, P.H., Colbert, J.T., Peters, N.K., Christensen, A.H., Sharrock, R.A. and Lissemore, J.L. (1986) Phil. Trans. R. Soc. Lond B 314: 469.
Richter, G. (1984) Plant Mol. Biol. 3: 271.
Richter, G. and Wessel, K. (1985) Plant Mol. Biol. 5: 175.
Sasaki, Y. (1986) FEBS Lett. 204:279.
Schäfer, E. and Briggs, W.R. (1986) Photochem. Photobiophys. 12: 305.
Shaw, G. and Kamen, R. (1986) Cell 46: 659.
Silverthorne, J. and Tobin, E.M. (1987) BioEssays. In press.
Simpson, J. Van Montagu, M. and Herrera-Estrella, L. (1986) Science 233: 34.
Tobin, E.M. and Silverthorne, J. (1985) Ann. Rev. Plant Physiol. 36: 569.
Tobin, E.M., Wimpee, C.F., Karlin-Neumann, G.A., Silverthorne,J. and Kohorn, B.D. (1985) In: Biosynthesis of the Photosynthetic Apparatus, ed. K.E. Steinback, S. Bonitz, C.J. Arntzen, L. Bogorad (Cold Spring Harbor,N.Y.) p. 373.

STRUCTURE-FUNCTION STUDIES ON AVENA PHYTOCHROME[1]

J. Clark Lagarias
Yum-Shing Wong
Tom R. Berkelman
Daniel G. Kidd
Robert W. McMichael, Jr.

Department of Biochemistry and Biophysics
University of California
Davis, California (USA)

I. INTRODUCTION

Transduction of a light stimulus by the photosensory chromoprotein phytochrome into changes of cellular physiology is accomplished in three steps - photoperception, primary signal transmission and signal amplification. Our ongoing investigations have focused on phytochrome photoperception for the purified photoreceptor from etiolated _Avena sativa_ _L._ seedlings. Our progress in three areas of research is outlined in the discussion which follows. First, we have devised a methodology for quantitation of the rates of the two photoconversion processes under constant red and far-red illumination. These investigations are a prelude to a second study whose goal is to identify structural features of the phytochrome molecule which govern the relative efficiency of Pr to Pfr photoconversion and _vice_ _versa_. Using biochemical probes of protein structure, we have localized conformational changes that occur upon photoactivation to discrete regions within the 124 kDa subunit of _Avena_ phytochrome. These

[1]This research was supported by Grant No. PCM84-09074 from the National Science Foundation (to J.C.L.).

results implicate a structural basis for the light-dependent
interaction of phytochrome with other components of the plant
cell. Third, a polycation-dependent protein kinase was found
to be associated with purified Avena phytochrome preparations.
This finding raises the possibility that phytochrome may
itself be a protein kinase and/or that the primary signal
transmission pathway involves protein phosphorylation.

II. RESULTS AND DISCUSSION

Efficiency of Phytochrome Phototransformations

It is generally accepted that Pfr is the active form of
phytochrome because of observations that Pfr levels correlate
well with the magnitude of a given biological response. Since
the relative efficiency of the two photoconversions of
phytochrome determines the amount of Pfr produced by a given
fluence of light, we devised a method for quantitating these
photochemical parameters for 124 kDa Avena phytochrome (Kelly
and Lagarias, 1985). Our approach involves 1) calibration of
a diode array spectrophotometer for use as a spectroradiometer
2) determination of the light fluence using solution
actinometry and 3) measurement of light-induced absorbance
changes as a function of absorbed fluence to obtain
photochemical cross sections (i.e. rate constants) for the two
phytochrome phototransformations. The quantum yields for both
phototransformations, 0.152 ± 0.008 and 0.069 ± 0.004 for Pr
to Pfr and Pfr to Pr photoconversions under constant red and
far red illuminations respectively, were determined with this
method. The photoequilibrium mole percent of Pfr under
constant red light was estimated to be $87.6 \pm 0.6\%$.

Little is known of the structural features that influence
the overall quantum yields of the two photoconversions. We
therefore compared the relative efficiency of the photoconversions for two highly purified phytochrome preparations
from both etiolated oat and annual rye seedlings (Lagarias et
al., submitted). The quantum yields for both Pr to Pfr and
Pfr to Pr photoconversions were determined to be 15 to 20%
larger for rye phytochrome than for oat preparations in two
different buffer systems. Quantum yields were calculated
using new estimates of molar absorption coefficients of
132 mM^{-1} cm^{-1} for Pr at the red absorption maximum for both
rye and oat phytochromes based on quantitative amino acid
analyses (Lagarias et al., submitted). Because these
measurements were performed under identical experimental
conditions, the observed differences in quantum efficiencies
for oat and rye phytochromes reflect structural differences
between the two protein moieties. In view of these

differences, the possibility that the efficiency of one or
both phototransformation processes can be regulated by
biochemical modification of protein structure or by regulation
of the expression of a structurally different species of
phytochrome remain interesting questions to be addressed.

Light-induced Conformational Changes

Recent studies have provided compelling evidence for
conformational changes in the phytochrome molecule upon Pr to
Pfr photoconversion (for review see Lagarias, 1985). In our
laboratory, endoproteases have proven to be particularly
useful probes of light-induced changes in the folding of the
phytochrome polypeptide (Lagarias and Mercurio, 1985). These
studies show that Pr to Pfr photoconversion alters the
susceptibility of several specific sites within the 124 kDa
subunit of Avena phytochrome towards proteolysis. Two major
sites involved in the light driven structural changes were
mapped to an N-terminal 10 kDa polypeptide region and to at
least one site near the central hinge region between the two
major structural domains of the holoprotein, the larger
chromophore domain and a smaller nonchromophore domain (see
Fig. 1 for peptide map of the Avena phytochrome subunit).
These structural assignments of regions of light-induced
conformational changes on phytochrome have been made possible
with two new experimental tools - site-directed peptide
antibodies and zinc-dependent fluorescence of bilin-linked
peptides.

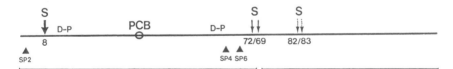

Figure 1.   Peptide map of Avena phytochrome. The major
sites of cleavage by subtilisin (S) are indicated with arrows
(Lagarias and Mercurio, 1985). The bold face arrow at 8 kDa
indicates a peptide bond more susceptible to subtilisin
cleavage for the Pr form, while the dotted arrows at 83/82 kDa
represent bonds more easily cleaved for Pfr (Lagarias and
Mercurio, 1985). Predicted asp-pro cleavages (D-P) are based
on the data of Hershey $\underline{et}$ $\underline{al}.$, 1985. PCB represents the
phytochromobilin prosthetic group. Filled triangles represent
the synthetic peptides, SP2 (R_{12}NRQSSQARVLAQTTLD$_{28}$), SP4
(H_{549}PRLSFKAFLEVVKM$_{563}$) and SP6 (K_{592}PKREASLDNQIGDLK$_{607}$).

An antiserum towards a synthetic peptide which encompasses the chromophore binding site on oat phytochrome was prepared and has proven invaluable for identification of chromophore-bearing peptide fragments obtained by limited proteolysis (Lagarias and Mercurio, 1985). Cross-reactivity studies with this antiserum also showed that the local environment of the bilin prosthetic group is structurally conserved between representative monocots and dicots (Mercurio et al., 1986). Based on the deduced amino acid sequence of one of three characterized cDNA clones (Hershey et al., 1985), new site-directed peptide antisera have been prepared. Antisera against two such synthetic peptides, SP2 and SP6, which correspond to regions near the N-terminus and near the central hinge region between chromophore and nonchromophore domains of Avena phytochrome respectively (see Fig. 1), cross-react with phytochrome and with selective polypeptides produced by partial proteolysis of phytochrome with subtilisin on Western blots (Fig. 2). SP2 antiserum detects 124, 83/82 and 72/69 kDa polypeptides but not 116, 75, 62/60, 52, 41 and 40 kDa polypeptides obtained by limited subtilisin digestion of Avena phytochrome and is therefore in good agreement with structural assignments made in our previous study (Lagarias and Mercurio, 1985). In addition, the epitope recognized by SP6 antiserum appears to lie within the chromophore binding domain, just N-terminal to the subtilisin cleavage site(s) between chromophore and nonchromophore domains (see Fig. 1). This conclusion is based on the strong cross-reactivity with the 62/60 kDa chromophore-bearing peptide and poor cross-reactivity towards the 52 kDa nonchromophore peptide (Fig. 2). These results illustrate the utility of site-directed peptide antibodies for structural analysis of Avena phytochrome.

A method to visualize bilin-linked polypeptides in SDS-polyacrylamide gels has greatly facilitated peptide mapping of phytochrome (Berkelman and Lagarias, 1986). This method relies on the formation of a highly fluorescent zinc complex with the linear tetrapyrrole prosthetic group under mildly oxidizing conditions at basic pH. This method is quite sensitive, permitting detection of as little as 50 ng (0.4 pmol) phytochrome in SDS-polyacrylamide gels. Chromophore bearing fragments obtained by limited proteolysis of phytochrome with trypsin and subtilisin have been identified directly by zinc-dependent fluorescence under UV light (Berkelman and Lagarias, 1986). These experiments corroborate the peptide map of Avena phytochrome shown in Fig. 1.

Phosphorylation by three mammalian protein kinases (i.e. cAMP-dependent, cGMP-dependent and Ca^{+2}-activated, phospholipid dependent protein kinases) also support the hypothesis of a significant protein conformational change in

Avena phytochrome upon Pr to Pfr photoconversion (Wong et al., 1986). In the Pr form, phytochrome is phosphorylated by all three protein kinases; only the cAMP-dependent protein kinase can phosphorylate the Pfr form. Using the technique of limited proteolysis developed in our earlier study, the regions on phytochrome phosphorylated by the cAMP-dependent protein kinase were shown to be different for Pr and Pfr (Wong et al., 1986). In addition to providing clear evidence for light-induced conformational changes, phosphorylation also gave us the means to specifically radiolabel these sites which will facilitate their molecular characterization.

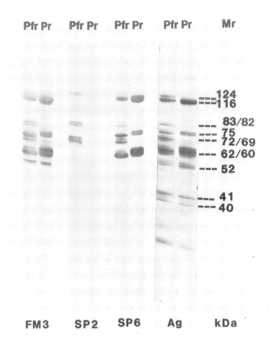

Figure 2. Peptide mapping of a subtilisin digest of Avena phytochrome with site-directed peptide antibodies: Pr vs Pfr analysis. Phytochrome samples were digested with subtilisin (1:500 subtilisin:phytochrome) for 1 h at 4°C and resolved on a 7.5-15% gradient SDS polyacrylamide gel (Lagarias and Mercurio, 1985). Immunoblots with antisera against two synthetic peptides, SP2 and SP6 (see Fig. 1), and against the denatured phytochrome holoprotein (FM3) are shown in the left six lanes. The two lanes on the right labeled Ag are silver-stained gels of the Pr and Pfr subtilisin digests.

Recent work to characterize the sites on phytochrome which
are differentially phosphorylated by the cAMP-dependent
protein kinase has involved two lines of investigation. Asp-
pro cleavage experiments of phytochrome phosphorylated as Pr
or Pfr by the cAMP-dependent protein kinase yields 4 major
products with molecular masses of 112, 60, 47 and 13 kDa (Fig.
3, lane 1). Except for the lack of a 65 kDa species that
appears to be unstable, these results correspond well with the
expected cleavage pattern based on the deduced amino acid
sequence of <u>Avena</u> phytochrome type 3 cDNA clones (Hershey <u>et
al.</u>, 1985). With radiolabeled Pr phytochrome samples,
extensive labeling of 60 and 13 kDa fragments is observed
(Fig. 3, lane 2). That these fragments retain the N-terminus
is revealed by immunoblot analysis with an antiserum raised to
the N-terminal synthetic peptide SP2 (Fig. 3, lane 3). A
similar analysis of a Pfr-phosphorylated sample has revealed
that the site(s) of phosphorylation is either more acid labile
than the N-terminal phosphorylation site on Pr or occurs on a
small peptide fragment liberated during acid treatment which
migrates off the SDS gel. The asp-pro cleavage experiments
support the N-terminal assignment (i.e. site 1) for the major
phosphorylation site on Pr while the major site phosphorylated
on Pfr by the cAMP-dependent protein kinase is different. In
addition, experiments are in progress to isolate
phosphopeptides by HPLC and to characterize the site(s) of
phosphorylation by amino acid sequence determination. This
work has been facilitated by the use of synthetic peptides as
substrates for the cAMP-dependent protein kinase. In this
regard, three good peptide substrates, SP2, SP6 and SP4,
correspond well to the phosphorylation sites 1, 3 and 2
identified previously (Wong <u>et al.</u>, 1986; see Fig. 1).

Polycation-dependent Protein Kinase Activity

In the absence of exogenous protein kinase,
phosphorylation of phytochrome on a serine residue(s) was
observed with ATP as the phosphoryl donor (Wong <u>et al.</u>, 1986).
This kinase activity is stimulated 10- to 100-fold when
polycationic molecules such as histone H1, protamine and
polylysine are present. Polylysine-stimulation of the
endogenous kinase activity in purified <u>Avena</u> phytochrome
preparations is shown in Fig. 4 (compare lanes 1 and 3).
Since polylysine, a synthetic polycation, serves to stimulate
this protein kinase activity, the kinase must be of plant
origin. Is phytochrome a protein kinase? Is it functionally
associated with a protein kinase? Or is this activity due to
a contaminant in the phytochrome preparations?

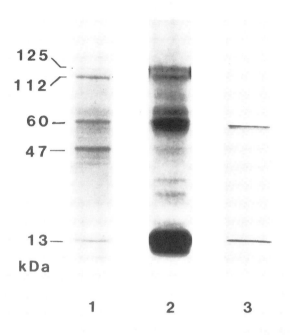

Figure 3. Asp-pro cleavage of <u>Avena</u> phytochrome.
Phytochrome, as Pr, was labeled by cAMP-dependent protein
kinase in the presence of [γ-^{32}P]ATP (Wong <u>et al.</u>, 1986).
Protein was precipitated with 80% ethanol at 0°C,
resolubilized in 70% formic acid, and incubated at 37°C for
24 h. Samples were then analyzed on a 5-20% gradient SDS
polyacrylamide gel. A Coomassie blue stained gel is shown in
lane 1. Lane 2 represents an autoradiograph of the same gel
shown in lane 1. Lane 3 is an immunoblot of an identical gel
probed with a polyclonal serum directed against synthetic
peptide SP2. Zinc-induced fluorescence analysis also shows
that the fragments with M_r's of 125, 112, 60, and 47 kDa
contain the site for chromophore binding (data not shown).

To test the hypothesis that phytochrome itself may be a
protein kinase, we have undertaken a number of experiments.
Preliminary evidence supporting this hypothesis is as follows:
1) all phytochrome preparations examined (including eight
independently purified oat phytochrome preparations and two
rye phytochrome preparations) exhibit this polycation-
stimulated protein kinase activity, 2) an ATP analog, 5'-p-
fluorosulfonylbenzoyl adenosine, inhibits this protein kinase
activity through covalent modification of phytochrome (Fig. 4,
compare lanes 2 and 4) - an effect reversed in part by
co-incubation with ATP (unpublished data) and 3) amino acid
sequence homology between phytochrome and known protein

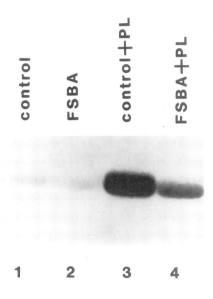

**Figure 4. Polylysine-stimulated protein kinase activity
of purified <u>Avena</u> phytochrome and inhibition with 5'-p-
fluorosulfonylbenzoyl adenosine (FSBA).** The autoradiograph
shows the (auto)phosphorylation of phytochrome after
incubating with $[\gamma-^{32}P]ATP$ at $30°C$ for 30 min in the absence
(lanes 1 and 2) or presence of 0.1 mg/ml polylysine (lanes 3
and 4). Phytochromes in lanes 2 and 4 were pretreated with
0.5 mM FSBA at $0°C$ for 30 min. Mole percent incorporation of
^{32}P into untreated and FSBA-treated phytochrome in the
presence of polylysine was 1.8% (lane 3) and 0.7% (lane 4).

kinases (including representatives of the oncogenic tyrosine protein kinases) lies within their ATP binding domains (Fig. 5). These results strongly implicate the presence of an ATP binding site on <u>Avena</u> phytochrome.

By contrast with the evidence supporting the intrinsic protein kinase hypothesis, observations that the specific activity is low and that gel exclusion chromatography separates pools of phytochrome with different specific activities could argue that the protein kinase is a co-purifying contaminant. We cannot yet dismiss the explanation that these results are due to pre-attached phosphate on phosphorylable serine residue(s) since our polycation-dependent protein kinase assay involves phytochrome itself as a substrate. It is therefore necessary to characterize the

Protein	Sequence	Ref.
phytochrome (403–415)	E L E K Q L R E K N I L K	a.
kinase A (84–96)	Q I E H T L N E K R I L Q	b.
kinase G (20–32)	E L E K R L S E K E E E I	c.
kinase G (401–413)	Q Q E H I R S E K Q I M Q	c.
phos b kinase (66–78)	L R E A T L K E V D I L R	d.
kinase C-I (392–404)	D V D C T L V E K R V L A	e.
kinase C-II (383–395)	D V E C T M V E K R V L A	e.
v-src (303–315)	S P E A F L Q E A Q V M K	f.
p56 tck (281–293)	S P D A F L A E A N L M K	g.
v-abl (400–412)	E V E E F L K E A A V M K	h.
SNF1 (96–106)	M Q G R I E R E I S Y L R	i.
mdr (610–622)	N H D E L M R E K G I Y F	j.

Figure 5. Amino ʌcid sequence homology between <u>Avena</u> phytochrome, known protein kinases and putative protein kinases. With the exception of the cGMP-dependent protein kinase region encompassing residues 20 to 32, all regions shown are found in the ATP-binding domain of the known protein kinases. Abbreviations used: kinase A, cAMP-dependent protein kinase; kinase G, cGMP-dependent protein kinase; kinase C, calcium-activated, phospholipid-dependent protein kinase; phos b kinase, ɣ subunit of phosphorylase b kinase. The boxed residues indicate identity with phytochrome while the circled residues represent conservative amino acid changes. References: a. Hershey <u>et al.</u>, 1985; b. Shoji <u>et al.</u>, 1984; c. Takio <u>et al.</u>, 1984; d. Reimann <u>et al.</u>, 1984; e. Knopf <u>et al.</u>, 1986; f. Schwartz <u>et al.</u>, 1983; g. Voronova and Sefton, 1986; h. Reddy <u>et al.</u>, 1983; i. Celenza and Carlson, 1986; j. Gros <u>et al.</u>, 1986.

in vivo phosphorylation sites on phytochrome and to determine
the stoichiometry of phosphate esters on the purified
photoreceptor.

Preliminary experiments have been undertaken to determine
the properties of this phytochrome-associated protein kinase.
Interestingly, while basic polypeptides including histone H1,
histone type IIA, polylysine, polyarginine and protamine
stimulate the protein kinase activity, the polyamines
spermine, spermidine and putrescine do not (Fig. 6). In
addition, recent experiments show that addition of phytochrome
to crude oat extracts enhances phosphorylation of several low
molecular weight proteins in a polylysine-dependent manner
(unpublished data). These studies raise a number of important
questions. What role does the basic polypeptide serve in the
activation of this protein kinase? What is the physiological
activator(s) and/or substrate(s) for this protein kinase? Are
phytochrome responses in plants mediated by protein
phosphorylation/dephosphorylation events? These questions are
the focus of current investigations in our laboratory.

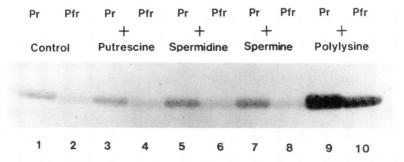

**Figure 6. Effect of polyamines on the endogenous
phytochrome-associated protein kinase.** Phytochrome (both Pr
and Pfr) was incubated with various polyamines in the presence
of [γ-^{32}P]ATP and subjected to SDS-PAGE and autoradiography
as described previously (Wong et al., 1986). The
concentration of polyamines was 0.1 mg/ml. For comparative
purposes, a sample incubated in the presence of polylysine
(0.1 mg/ml) is shown.

ACKNOWLEDGEMENTS

We thank Dr. Richard A. Houghten, Scripps Clinic and Research Foundation, La Jolla CA, for the synthetic peptides; Dr. William O. Smith, Jr., Smithsonian Environmental Research Center, Rockville MD, for the rye phytochrome preparations; Dr. Donal A. Walsh, Department of Biological Chemistry, University of California, Davis CA, for the cAMP-dependent protein kinase preparations; and Dr. David Glass, Department of Pharmacology, Emory University School of Medicine, Atlanta GA, for the cGMP-dependent protein kinase preparations.

REFERENCES

Berkelman, T.R., and Lagarias, J.C. (1986). Anal. Biochem. 156:194.
Celenza, J.L. and Carlson, M. (1986). Science 233:1175.
Gros, P., Croop, J., and Housman, D. (1986). Cell 47:371.
Hershey, H.P., Barker, R.F., Idler, K.B., Lissemore, J.L., and Quail, P.H. (1985). Nucleic Acids Res. 13:8543.
Kelly, J.M., and Lagarias, J.C. (1985). Biochemistry 24:6003.
Knopf, J.L., Lee, M-H., Sultzman, L.A., Kriz, R.W., Loomis, C.R., Hewick, R.M., and Bell, R.M. (1986). Cell 46:491.
Lagarias, J.C. (1985). Photochem. Photobiol. 42:811.
Lagarias, J.C., and Mercurio, F.M. (1985). J. Biol. Chem. 260:2415.
Lagarias, J.C., Kelly, J.M., Cyr, K.L., and Smith, W.O. Jr., submitted to Photochem. Photobiol.
Mercurio, F.M., Houghten, R.A., and Lagarias, J.C. (1986). Arch. Biochem. Biophys. 248:35.
Reddy, E.P., Smith, M.J., and Srinivasan, A. (1983). Proc. Natl. Acad. Sci. USA 80:3623.
Reimann, E.M., Titani, K., Ericsson, L.H., Wade, R.D., Fischer, E.H., and Walsh, K.A. (1984). Biochemistry 23:4185.
Schwartz, D.E., Tizard, R., and Gilbert, W. (1983). Cell 32:853.
Shoji, S., Parmelee, D.C., Wade, R.D., Kumar, S., Ericsson, L.H., Walsh, K.A., Neurath, H., Long, G.L., Demaille, J.G., Fischer, E.H., and Titani, K. (1981). Proc. Natl. Acad. Sci. USA 78:848.
Takio, K., Wade, R.D., Smith, S.B., Krebs, E.G., Walsh, K.A., and Titani, K. (1984). Biochemistry 23:4207.
Voronova, A.F., and Sefton, B.M. (1986). Nature 319:682.
Wong, Y-S., Cheng, H-C., Walsh, D.A., and Lagarias, J.C. (1986). J. Biol. Chem. 261:12089.

STRUCTURE-FUNCTION RELATIONSHIP OF PEA PHYTOCHROME AS DEDUCED FROM AMINO-TERMINAL SEQUENCE ANALYSIS OF PHYTOCHROME CHROMOPEPTIDES[1]

Kotaro T. Yamamoto

Division of Biological Regulation
National Institute for Basic Biology
Okazaki, Japan

I. INTRODUCTION

Rice and Briggs (1973) showed that the 62-kDa chromopeptide prepared by tryptic digestion of rye 120-kDa phytochrome exhibited photoreversible phototransformation between Pr and Pfr like its source 120-kDa phytochrome. This finding has made it evident that the native phytochrome molecule consists of functional domains in respect to its phototransformation. Stoker et al. (1978) observed that tryptic digestion of a 120-kDa oat phytochrome produced 60-, 40- and 22-kDa chromopeptides sequentially. We prepared several chromopeptides from the pea 114-kDa phytochrome. After limited proteolysis and separation by size-exclusion chromatography, we found that chromopeptides of 39[2] and 33 kDa also showed photoreversible

[1]Supported in part by Grants-in-Aid (6110111, 6123421) from the Ministry of Education, Science and Culture of Japan.

[2]The 39-kDa chromopeptide was designated as a 40-kDa fragment in the previous report (Yamamoto and Furuya, 1983). The difference in apparent molecular mass probably resulted from the different electrophoresis systems used to estimate the molecular mass. In the previous study, Na dodecylsulfate polyacrylamide gel electrophoresis with a continuous buffer system, as described by Weber et al. (1972), was utilized. In the present study a discontinuous buffer system, according to Laemmli (1970), was used.

phototransformation between new spectral forms, P660[3] and Pbl (Yamamoto and Furuya, 1983). On the other hand, Vierstra and Quail (1982) showed that a short segment at the amino terminus of native phytochrome was required for native spectral properties of Pfr. These observations indicate that the photoreversible phototransformation unique to phytochrome is achieved through interactions between its chromophore and multiple domains in the apoprotein.

Although the functional domains with respect to photoconversion have been localized using monoclonal antibodies (Daniels and Quail, 1984; Nagatani et al., 1984; Jones and Quail, 1986a) and studying kinetics of limited proteolysis (Lagarias and Mercurio, 1985), the precise location of these functional domains on the phytochrome polypeptide backbone remains unknown. Since the complete base sequence of oat phytochrome cDNA was determined recently (Hershey et al., 1985), any phytochrome chromopeptides can be localized exactly by comparing their amino-terminal amino acid sequence with the amino acid sequence predicted from the cDNA.

In the present study, I localized several phytochrome fragments by determining their amino-terminal sequences. The roles of these fragments in reversible phototransformation of phytochrome are discussed.

II. PHOTOCHROMIC CHROMOPEPTIDES OF PHYTOCHROME

Several chromopeptides of different size have been prepared from phytochrome by proteolysis with commercially available proteases, and their phototransformation has been characterized (Table 1).

The 62- and 56-kDa chromopeptides obtained from the pea 114-kDa phytochrome showed photoreversible transformation between Pr and Pfr, while the 39-, 33- and 25-kDa chromopeptides showed photoreversible transformation between P660 and Pbl (Yamamoto and Furuya, 1983). P660 is a spectral form similar to Pr, but has an absorption maximum at ca. 660 nm; Pbl is a spectral form having a broad and bleached absorption in the red and far-red region, and was first described as a spectral form of phytochrome denatured by urea (Butler et al.,

---

[3]Abbreviations: cDNA, complementary DNA; $P\lambda$, a spectral form of phytochrome having an absorption maximum at λ nm; PAGE, polyacrylamide gel electrophoresis; Pbl, a bleached form of phytochrome; Pfr(λ), a Pfr form of phytochrome having an absorption maximum at λ nm; SDS, Na dodecylsulfate; TFA, trifluoroacetic acid; Tris, tris(hydroxymethyl)aminomethane.

1964). An extinction of phytochrome in Pr at the absorption
maximum decreased by 14 % during the tryptic digestion from
the 114-kDa phytochrome to the 39-kDa chromopeptide (Yamamoto,
K. T., unpublished). Reiff et al. (1985) prepared a 39-kDa
chromopeptide by digesting oat native 124-kDa Pfr with trypsin, and confirmed its photoreversible transformation between
P660 and Pbl. The short chromopeptide of 2.5 kDa was photoactive when prepared from Pfr. Its absorption maximum was at
610 nm, and it showed irreversible transformation to P660 by
white light in acidic methanol. The chromopeptide was not
photoactive if prepared from Pr (Thummler et al., 1981; Thummler and Rudiger, 1984).

Absorption spectra were determined also in the lysate of
oat native 124-kDa phytochrome. When the phytochrome was
digested as Pfr with protease(s) existing in crude extracts,
the lysate containing a 74-kDa chromopeptide showed photoreversible transformations between Pr and Pfr(730) (Jones et

TABLE I. Phototransformation of photochromic chromopeptides of phytochrome in neutral buffer

Size of chromo-peptide (kDa)	Source phytochrome	Protease	Photo-transformation[a]	Reference[b]
121 (native)	Pea	–	Pr(667)<-->Pfr(730)	1
114	Pea	Endogenous protease(s)	Pr(667)<-->Pfr(722)	2
62	Rye 120-kDa Pr	Trypsin	Pr(666)<-->Pfr(725)	3
62	Pea 114-kDa Pr	Trypsin	Pr(667)<-->Pfr(722)	2
56	Pea 114-kDa Pr	Thermolysin	Pr(666)<-->Pfr(722)	2
39	Pea 114-kDa Pr	Trypsin	P(660)<-->Pbl	2
39	Oat 124-kDa Pfr	Trypsin	P(660)<-->Pbl	4
39	Pea 114-kDa Pr	Chymotrypsin	P(658)<-->Pbl	2
33	Pea 114-kDa Pr	Thermolysin	P(657)<-->Pbl	2
25	Pea 114-kDa Pr	Trypsin	P(657)<-->Pbl	5
2.5	Oat 60-kDa Pfr	Pepsin	P(660)<--P(610)[c]	6

[a]<-->, -->, reversible and irreversible phototransformation, respectively; number in parentheses shows wavelength of
an absorption maximum in nm in each spectral form.
[b]References: 1, Lumsden et al., 1985; 2, Yamamoto and
Furuya, 1983; 3, Rice and Briggs, 1973; 4, Reiff et al., 1985;
5, Yamamoto, K. T., unpublished; 6, Thummler et al., 1981;
Thummler and Rudiger, 1984.
[c]Measured in acidic methanol.

al., 1985). The lysate of Pr digested more extensively with subtilisin, which contained only one kind of chromopeptide of 12 kDa, showed photoreversible transformation between P658 and Pbl (Jones and Quail, 1986b).

III. LOCALIZATION OF PHYTOCHROME CHROMOPEPTIDES

Purification and amino-terminal sequence analysis of 6 tryptic or thermolytic fragments as well as the 114-kDa pea phytochrome are described in this chapter.

A. Purification of Chromopeptides by High-Performance Liquid Chromatography

Two tryptic preparations and a thermolytic preparation of pea phytochrome prepared by conventional size-exclusion chromatography (Yamamoto and Furuya, 1983) were further purified with high-performance liquid chromatography (Tri Rotor-VI, Japan Spectroscopic Co., Hachi-oji, Japan) prior to amino acid sequence analysis. An absorption spectrum between 240 and 700 nm and an absorbance at 220 nm of the effluent were monitored continuously by a photodiode array detector (MCPD-350PC, Otsuka Electronics, Osaka, Japan) and UVIDEC-100-VI detector (Japan Spectroscopic Co.), respectively.

1. 114-kDa PHYTOCHROME. The 114-kDa phytochrome was the source phytochrome preparation from which proteolytic fragments were prepared by limited proteolysis in this study. The phytochrome was obtained as a product of proteolysis by endogenous protease(s) existing in crude preparations of phytochrome, when the phytochrome was extracted and purified as Pr from 7-day-old etiolated pea (Pisum sativum cv. Alaska) seedlings (Yamamoto and Furuya, 1983).
The 114-kDa phytochrome was prepared by ammonium sulfate fractionation and brushite, DEAE agarose and size-exclusion chromatographies as described previously (Yamamoto and Furuya, 1983). The preparation was purified further with anion-exchange chromatography (Protein Pak G-DEAE, 8.2 x 75 mm, 10 um, 500 - 1000 Å pore size, Nihon-Waters, Tokyo, Japan) at ambient temperature, and reversed-phase chromatography (TSKgel Phenyl-5PW RP, 10 um, 1000Å pore size, 4.6 x 75 mm, Toyo Soda Mfg., Tokyo, Japan) at 40°C. In the anion-exchange chromatography, the phytochrome was eluted with a linear salt gradient from 0 to 0.5 M NaCl in 0.01 M Tris-HCl, pH 7.8. For re-

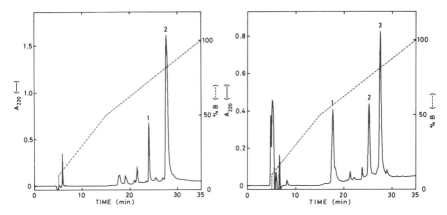

FIGURE 1. Separation of tryptic peptides present in the preparations of 62-kDa (left) and 39-kDa (right) chromopeptide by reversed-phase chromatography on a 4.6 x 250 mm Develosil C8 column (7 um, 300-Å pore size, Nomura Kagaku Co., Seto, Japan) equilibrated in 0.1% TFA containing 10 % isopropanol (solvent A) at 40°C. Polypeptides were eluted by increasing the concentration of solvent B (acetonitrile containing 10 % isopropanol and 0.1 % TFA) at a flow rate of 0.7 ml/min.

versed-phase chromatography phytochrome was eluted with a linear gradient of acetonitrile (0 – 100 %) with both solutions containing 0.02 % TFA and 10 % isopropanol.

2. 62-kDa CHROMOPEPTIDE. A sample of the 62-kDa chromopeptide was prepared by limited proteolysis of the 114-kDa phytochrome as Pr with 0.14 %(w/w)[4] TPCK-trypsin followed by size-exclusion chromatography with Sephacryl S-200 (Pharmacia, Uppsala, Sweden) (Yamamoto and Furuya, 1983). Subsequent chromatography of the sample with a reversed-phase column resolved two major polypeptides (Fig. 1, left). Fraction 1 had no absorption in the visible region and its apparent molecular mass was estimated as 7 kDa by SDS PAGE (Laemmli, 1970) using a 13.8 % polyacrylamide gel with an acrylamide:bisacrylamide ratio of 20:1 (Hashimoto et al., 1983). Fraction 2 had an absorption at 648 nm and an apparent molecular mass of 62 kDa. The peptide bands in the polyacrylamide gel were visualized by silver staining (Wray et al., 1981).

---

[4]The amount of trypsin added was reevaluated using the extinction coefficient of phytochrome reported by Litts et al. (1983). Therefore, the values in the present study are different from those appearing in Yamamoto and Furuya (1983).

3. 56-kDa CHROMOPEPTIDE. A sample of the 56-kDa chromopeptide was prepared by limited proteolysis of the 114-kDa
phytochrome as Pr with 0.3 %(w/w) thermolysin followed by
size-exclusion chromatography with Sephacryl S-200 (Yamamoto
and Furuya, 1983). The preparation was purified further with
anion-exchange chromatography (Mono Q, 5 x 50 mm, 10 um,
Pharmacia) at ambient temperature. The fraction which had
absorption at 666 nm was eluted with the linear gradient of 0
to 0.5 M NaCl in 0.01 M Tris-HCl, pH 7.8, and applied to a
reversed-phase column (Develosil C8) at 40°C. The peak which
had an absorption at 647 nm contained the 56-kDa chromopeptide.

4. 39-kDa CHROMOPEPTIDE. A preparation of the 39-kDa
chromopeptide was obtained by limited proteolysis of the 114-
kDa phytochrome as Pr with 1.2 %(w/w) TPCK-trypsin followed by
size-exclusion chromatography with Sephadex G-75SF (Pharmacia)
(Yamamoto and Furuya, 1983). The 39-kDa preparation was purified further with reversed-phase chromatography. The chroma-
togram in Fig. 1 (right) consisted of three fractions. Fraction 1 was the ovomucoid trypsin inhibitor which was used to
stop the trypsin digestion. Fraction 2 had an absorption at
646 nm but consisted of two polypeptides of apparent molecular
mass of 25 and 14 kDa based on SDS PAGE. Fraction 3 had also
an absorption at 648 nm and an apparent molecular mass of 39
kDa.
        Fraction 2 was diluted two-fold with water and rechroma-
tographed on the same column. The two polypeptides were separated from each other with a shallow linear gradient (1 % solvent B increase/min). The 14-kDa polypeptide, which eluted at
the lower concentration of solvent B, had no absorption in the
visible region, while the 25-kDa polypeptide had an absorption
at 646 nm.

B. Amino Acid Sequence Analysis of Phytochrome Chromopeptides

        Pea phytochrome fragments obtained as described above were
freeze-dried after the addition of SDS. The samples were reconstituted to 2 % SDS by the addition of water, and subjected
to amino-terminal sequence analysis using a gas-phase protein
sequencer with a PTH analyzer (model 470A and 120A, respectively; Applied Biosystems, Foster City, U. S. A.).
        A total of 7 amino-terminal sequences spanning 113 amino
acid residues were located within the amino acid sequence
deduced from the base sequence of oat phytochrome cDNA
(Hershey et al., 1985) (Table 2). The homologous sequence of
oat phytochrome is shown under the sequence of the pea poly-

TABLE II. Amino-terminal amino acid sequences of pea phytochrome fragments and their comparison with that deduced from the base sequence of oat phytochrome cDNA

Size (kDa)	Chromophore	N-Terminal amino acid sequence[a]	Homology[b] (%)

114	+	GSVD GDQQP XSNKV T
		      ** * **
		Oat: QRDGP PVQQG RSEKV I
		      52 60[c]
			36

62	+	VTTAYL	
56	+	AVPSV GDHPA LGIGT DIRTV FTAP ***** * * ***** +*++ *+ Oat: HAVPSV DDPPR LGIGT NVRSL FSDQ 111 120 130	58 (75)
39	+	VTTAYLN HIQRG KQIQP FGXLL ALDE * ***+ ***+* * ** ** ** **** Oat: KVI-AYLQ HIQKG KLIQT FGCLL ALDE 65 70 80	76 (84)
25	+	LQ SLASG SMEXL XDTMV QEVFE LT + ** * *** * *+* ***+ ** Oat: KIG SLPGG SMEVL CNTVV LEVFD LT 209 220 230	64 (77)
14	−	VTTAYLN HIQRG KQIQP FGXLL AL	
7	−	XVH LVXLE LSITH GGSGV PEAAL NQMFG NNVLE +* *+ ** * * * * ** * + *** * Oat: ENLH LIDLE LRIKH QGLGV PAELM AQMFF EDNKE 1048 1060 1070 1080	48 (58)

[a]X, unidentified residue; a deletion (−) is introduced to obtain maximal homology between pea and oat phytochromes; the symbols * and + denote identically placed and chemically similar residues, respectively, between the two phytochromes, where D=E, A=S=T, N=Q, F=Y=H, K=R, and V=I=L=M.
[b]Value in parentheses is % homology including chemically similar residues.
[c]Oat phytochrome numbering (Hershey et al., 1985).

peptides.

The 62- and 39-kDa chromopeptides and 14-kDa polypeptide shared a common amino terminus corresponding to Val-65 of oat phytochrome. The amino-terminal sequence of the 114-kDa chromopeptide overlapped that of the 62-, 39- and 14-kDa polypeptides. Both sequences formed a consecutive sequence corresponding to the oat phytochrome sequence Arg-52 to Glu-89.

The 56-kDa chromopeptide consisted of two different amino acid sequences in about equal quantities. The amino-terminal residue of one sequence corresponded to Ala-111 of oat phytochrome, while that of the other to Val-115.

C. Peptide Map of Pea Phytochrome

A peptide map of pea phytochrome was made based on the sequence data in Table 2 and the apparent molecular mass of the fragments (Fig. 2). Since the complete amino acid sequence of pea phytochrome remains unknown, the peptide map was composed by aligning pea fragments on the oat phytochrome sequence deduced from its cDNA base sequence (Hershey et al., 1985).

Recently pea phytochrome cDNA, corresponding to the amino acid sequence from Asp-28 to the carboxyl terminus of oat phy-

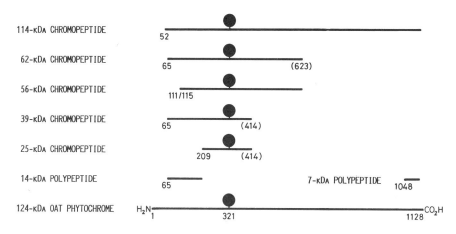

FIGURE 2. Relative alignment of the 114-kDa phytochrome and its tryptic and thermolytic fragments with the amino acid sequence deduced from the cDNA base sequence of oat phytochrome. A solid circle denotes phytochrome chromophore. Numbers indicate amino acid residue in oat phytochrome (Hershey et al., 1985). A number in parentheses is a tentative assignment.

tochrome, has been sequenced (Sato, N., personal communication). Surprisingly, the amino acid sequence deduced from the
pea cDNA included only 7 deletions and 2 insertions of amino
acid residues when compared with oat phytochrome, although the
difference between the apparent molecular mass of pea and oat
phytochromes (121 and 124 kDa, respectively (Vierstra et al.,
1984; Lumsden et al., 1985)) was as much as 3 kDa. Overall
homology in amino acid sequence between the two phytochromes
was 63 %.

Although the carboxyl-terminal sequences of the fragments
of pea phytochrome were not determined in the present study,
the position of their carboxyl terminus could be assigned
quite reasonably due to the overall similarity of the two phytochromes and narrow substrate specificity of trypsin.

1. 114-kDa PHYTOCHROME. The amino-terminal residue of the
114-kDa phytochrome corresponded to Arg-52 of oat phytochrome.
The amino-terminal peptide segment of pea phytochrome, corresponding to that from Ser-1 to Gln-51 in oat phytochrome, was
cleaved by endogenous protease(s) and lost during extraction
and purification of pea phytochrome.

The formula molecular mass from Arg-52 to the carboxyl
terminus of oat phytochrome (Gln-1128) is 119.3 kDa, significantly larger than the apparent molecular mass of the 114-kDa
phytochrome. On the other hand, the amino-terminal residue of
the 7-kDa polypeptide corresponded to Asn-1048 of oat phytochrome. The formula molecular mass from Asn-1048 to the carboxyl terminus of oat phytochrome is 9.0 kDa, which is also
larger than the apparent molecular mass of the 7-kDa peptide.
It is likely that the 7-kDa polypeptide is the carboxyl-terminal part of the 114-kDa phytochrome. The differences between
the apparent and formula molecular masses may indicate that
the 114-kDa phytochrome is missing a short segment of native
phytochrome at its carboxyl terminus as well as at its amino
terminus.

2. 62- AND 56-kDa CHROMOPEPTIDES. Since the amino terminus of the 62-kDa chromopeptide corresponded to Val-65 of oat
phytochrome, its carboxyl-terminal residue is tetatively assigned as Arg-623 in the present study. The formula molecular
mass of the fragment in oat phytochrome is 62.0 kDa.

The amino termini of two chromopeptides in the 56-kDa
chromopeptide preparation corresponded to Ala-111 and Val-115
of oat phytochrome. Since the formula molecular mass from
Val-115 to Arg-623 in oat phytochrome is 56.5 kDa, the carboxyl terminus of the 56-kDa fragment must be near that of the
62-kDa chromopeptide.

3. 39- AND 25-kDa CHROMOPEPTIDES. Since the amino terminus of the 39-kDa chromopeptide corresponded also to Val-65,
its carboxyl-terminal residue would correspond to Lys-414 or
Arg-425. In the former case, the formula molecular mass is
38.9 kDa, and in the latter, 40.2 kDa. The amino terminus of
the 25-kDa chromopeptide corresponded to Ile-209 of oat phytochrome. Formula molecular masses of the fragments from Ile-
209 to Lys-414 and from Ile-209 to Arg-425 are 23.6 and 24.9
kDa, respectively. Although it can not be determined at present which is the case, Lys-414 is taken as a tentative assignment of the carboxyl terminus of both the 39- and 25-kDa chromopeptides.

IV. STRUCTURE-FUNCTION RELATIONSHIP OF PEA PHYTOCHROME

A. Phototransformation

1. DOMAINS REQUIRED FOR Pr. The role of a few functional
domains in the photoreversible transformation of phytochrome
can be suggested on comparing the type of phototransformation
(Table 1) and a peptide map of phytochrome chromopeptides
(Fig. 2).
   The phototransformation of the 56-kDa chromopeptide indicates that all the functional domains required for the photoreversible transformation between Pr and Pfr(722) are located
within the polypeptide chain from Val-115 to Arg-623. Tryptic
digestion of the 62-kDa fragment to the 39-kDa fragment induced only a small blue shift of Pr, yielding P660. These
facts suggest that the chromophore conformation of Pr is maintained primarily by the functional domain composed of the region between Val-115 to Lys-414, and that the domain from Met-
415 to Arg-623 also contributes to the chromophore conformation secondarily. The formula molecular mass of the former
and the latter domain in oat phytochrome is 33.8 and 23.2 kDa,
respectively.
   P660 is a spectral form observed also when Pr is partially
denatured by urea (Butler et al., 1964), anilinonaphthalene
sulfonate (Hahn and Song, 1981; Eilfeld and Rudiger, 1984) or
saponin (Konomi et al., 1982). These reagents probably inter-
fere with an interaction of the 23-kDa domain with the chromophore and/or the other domains of phytochrome.

2. DOMAINS REQUIRED FOR Pfr.

a. Carboxyl-terminal half of the 62-kDa chromopeptide.
The phototransformation product of Pr caused by red-light
irradiation drastically changed from Pfr(722) to Pbl upon
tryptic degradation of the 62-kDa chromopeptide to the 39-kDa
chromopeptide, although the photoreversibility of the photoconversion remained unchanged. This clearly indicated that
the 23-kDa domain located in the carboxyl-terminal part of the
62- and 56-kDa fragments was essential for the chromophore
conformation of Pfr(722). The loss of interaction between the
23-kDa polypeptide and the Pfr chromophore and/or the other
parts of the phytochrome apoprotein produced Pbl instead of
Pfr.

Pbl is a spectral form observed also when phytochrome is
photoconverted from P660 in the presence of the denaturants.
The addition of the denaturants such as saponin (Konomi et
al., 1982) and anilinonaphthalene sulfonate (Eilfeld and
Rudiger, 1984) to Pfr did not produce Pbl as effectively as
did red-light irradiation of P660. This indicated that the
interaction between the 23-kDa domain and the chromophore in
Pfr was stronger than that in Pr.

b. The Amino-Terminal 6-kDa Segment of Native Phytochrome.
The important role of the amino-terminal segment of native
phytochrome for native spectral properties of Pfr has been
well established by studies with spectroscopy (Vierstra and
Quail, 1982, 1983; Litts et al., 1983; Jones et al., 1985),
monoclonal antibodies (Daniels and Quail, 1984; Cordonnier et
al., 1985; Lumsden et al., 1985; Chai et al., 1986), and
oxidizing reagents (Baron and Epel, 1983; Hahn et al., 1984;
Thummler et al., 1985). Furthermore, the segment was preferentially cleaved in Pr by endogenous proteases (Kerscher and
Nowitzki, 1982; Vierstra and Quail, 1982), trypsin (Lagarias
and Mercurio, 1985) and subtilisin (Jones and Quail, 1986b).

Essentially the same result was confirmed with pea phytochrome (Table 1). Namely, Pfr of the 114-kDa phytochrome
showed an absorption maximum at 722 nm (Yamamoto and Smith,
1981), while Pfr of the native 121-kDa phytochrome was at 730
nm (Lumsden et al., 1985). Photostationary photoequilibrium
was also different between the 114- and 121-kDa phytochromes
(Yamamoto and Smith, 1981; Tokutomi et al., 1986a). Native
121-kDa phytochrome was obtained when extracted as Pfr, while
the 114-kDa fragment was obtained when extracted as Pr
(Lumsden et al., 1985).

The peptide map in Fig. 2 demonstrates that the amino-
terminal region essential for Pfr(730) is located within the
segment from Ser-1 to Gln-51, of which the formula molecular

mass is 5.5 kDa in oat phytochrome. Although the effect of
the amino-terminal 6-kDa domain on the Pfr spectrum was evident, the effect was comparable to that of the 23-kDa domain
on the Pr spectrum. Therefore, the role of the 6-kDa domain
is probably secondary in the photoconversion of phytochrome,
correcting the interaction between the Pfr chromophore and the
23-kDa domain.

The site which corresponded to the amino-terminal side of
Arg-52 of oat phytochrome was cleaved in Pr by endogenous
protease(s) in pea extracts. In oat phytochrome the amino-
terminal side of Arg-52 (Jones and Quail, 1986b) and Asp-53
(Rudiger, W., communication in Discussion Session of the XVI
Yamada Conference) was also cleaved in Pr by subtilisin and
trypsin, respectively. It could be concluded that the residues around Arg-52 and Asp-53 are located in the region that
undergoes a phototransformation-induced reversible change in
conformation.

3. DOMAINS REQUIRED FOR PHOTOREVERSIBLE TRANSFORMATION.
Although the 39-kDa chromopeptide exhibited neither Pr nor
Pfr, it still maintained the structure necessary for photoreversible transformation. It is important to identify the
functional domain required for the photoreversible reaction.
The definite answer has not yet been obtained for this
problem.

The 25-kDa chromopeptide of pea phytochrome also showed

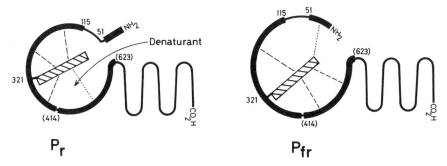

FIGURE 3. Schematic representation of possible interactions between phytochrome chromophore and its functional domains characterized in the present study. The hatched rectan-
gle is a chromophore. The bold line shows functional domains.
Numbers indicate amino acid residues in the oat phytochrome
protein (Hershey et al., 1985). Numbers in parentheses indicate a tentative assignment. Interactions between chromophore
and functional domains are indicated by broken lines, and weak
interactions by dotted lines.

photoreversible conversion between P658 and Pbl (Table 1). But it lost its photoreversibility more rapidly during repetitive photoconversion than the 39-kDa chromopeptide, and its absorption at 658 nm was less stable even in the dark (Yamamoto, K. T., unpublished). This suggested that the 14-kDa polypeptide from Val-65 to Lys-208 was necessary to maintain overall tertiary structure of the 39-kDa domain, even if it did not interact directly with the chromophore.

Jones and Quail (1986b) showed that the digest of native oat phytochrome with subtilisin, in which a polypeptide of 16 kDa was a chromophore-bearing peptide, displayed reversible phototransformation between P657 and Pbl. The chromopeptide was derived from a predominantly hydrophobic domain approximately between residues 190 and 330. Since the 16-kDa chromopeptide was not separated from the other polypeptides in the digest, possible interactions between the digested polypeptides and the 16-kDa chromopeptide might enable the digest to exhibit the reversible photoreaction.

4. Pbl AS AN INTERMEDIATE OF PHOTOTRANSFORMATION. We have suggested that Pbl of the 39-kDa chromopeptide could be the bleached intermediate of phytochrome phototransformation, Ibl (or Meta-Rb), detected in the phototransformation pathway from Pr to Pfr[5], since absorption and circular dichroism spectra of Pbl looked like those of Ibl (Yamamoto and Furuya, 1983). This hypothesis was supported by the study with low temperature spectroscopy (Reiff et al., 1985). If this is the case, absorption of light by the Pr chromophore must first disrupt the interaction between the chromophore and the 23-kDa domain (residues 415 - 623) resulting in Ibl, and a new interaction is formed thereafter between them in the dark, producing Pfr. The former process is rapid, while the latter is the slowest step in the photoconversion from Pr to Pfr (Kendrick and Spruit, 1977; Shimazaki et al., 1980).

In the present study I identified 3 functional domains with respect to phototransformation of phytochrome, namely, the amino-terminal 6-kDa domain (residues 1 - 51), the chromophore-bearing 34-kDa domain (residues 115 - 414), and the 23-kDa domain (residues 415 - 623). Fig. 3 schematically summarizes the interactions between the chromophore and the functional domains characterized and discussed here. Each of these domains might interact with the chromophore directly as

---

[5]Oat native 124-kDa phytochrome did not form Ibl in vitro in contrast to the 114/118-kDa phytochrome (Eilfeld and Rudiger, 1985). However, Ibl has been detected as a phytochrome intermediate in etiolated pea epicotyls (Kendrick and Spruit, 1973; Inoue and Furuya, 1985).

depicted in Fig. 3, and/or indirectly through the structural domain which provides a cavity for the chromophore. Quail et al. (1986) proposed that residues 150 - 300, a relatively extensive hydrophobic region, would be involved in forming the crevice for the chromophore.

B. Proton Uptake and Release

The 114-kDa pea phytochrome and its 62-kDa chromopeptide released and bound proton(s) in aqueous solutions in a photoreversible manner. Red-light irradiation of the sample caused proton uptake at pH's below 7.7 and release in the pH range above 7.7 (Tokutomi et al., 1982). Native 121-kDa pea phytochrome exhibited photoreversible proton uptake, but not release (Tokutomi et al., 1986b). The 39-kDa chromopeptide showed only photoreversible proton release (Tokutomi et al., manuscript in preparation). These facts suggest that phytochrome has a site for proton uptake in the 23-kDa domain and a site for proton release in the rest of the 62-kDa chromopeptide, and that the proton release from the latter site is inhibited by the amino-terminal 6-kDa segment of native phytochrome.

C. Dimerization

Native phytochrome (Jones and Quail, 1986a) as well as partially degraded phytochrome of 114-118 kDa (Hunt and Pratt, 1980) exists as a dimer in neutral buffer of low ionic strength. Jones and Quail (1986a) showed that the contact site(s) between the monomers was within 42 kDa from the carboxyl terminus. It was predicted from the hydropathy analysis (Hershey et al., 1985) and amino acid sequence homology between oat and zucchini phytochromes (Quail et al., 1986) that the monomers would stick to each other in an antiparallel manner at the contact site(s) located around residue 800 (Jones, A., communication in Discussion Session in the XVI Yamada Conference).

In this connection, it is interesting that the 7-kDa polypeptide, of which the amino-terminal residue corresponded to Asn-1048 of oat phytochrome, was contained in preparations of the 62-kDa chromopeptide of pea phytochrome (Fig. 1, left). Since the sample was prepared by size-exclusion chromatography as described previously, the 7-kDa polypeptide was probably attached to the 62-kDa chromopeptide and co-eluted in the chromatography. This suggested that the 7-kDa polypeptide might include one of the contact sites between the phytochrome monomers. Phytochrome monomers might stick to each other through the interaction between the 62-kDa chromopeptide of a monomer and the 7-kDa segment near the carboxyl terminus of the other monomer.

D. Structural Features as Suggested from Amino Acid Sequence Homology

The results shown in Table 2 are the first amino acid sequence data determined for dicotyledonous phytochrome. Since relatively low immunochemical cross-reactivity was observed between phytochrome from monocot and dicot plants (Cordonnier and Pratt, 1982; Saji et al., 1984; Cordonnier et al., 1986), comparison of the pea phytochrome sequence with that deduced from the cDNA sequence of oat phytochrome is interesting with respect to phytochrome structure. The amino acid sequence from Val-65 to Glu-89 shows the highest homology (76 %) among the 5 sequences determined in the present study; in contrast to that the segment located at the amino-terminal side of Val-65, namely, from Arg-52 to Lys-64, shows the lowest homology (36 %). Since it is generally assumed that the core of a globular protein would be relatively well conserved compared to external residues (Bajaj and Blundell, 1984), these results are consistent with the view that the 62-kDa chromopeptide assumes a globular, tightly packed structure, while the amino-terminal segment of native phytochrome is a flexible structure fully exposed to solvent in Pr (Cordonnier et al., 1985; Lagarias and Mercurio, 1985; Lumsden et al., 1985; Song, 1985; Quail et al., 1986). The joint between the two structural domains would be located around Val-65.

Among the 5 sequences measured in the present study, the sequence from Asn-1048 to Glu-1080 is the only sequence located in the carboxyl half of phytochrome. Its homology (48 %) is lower than that of sequences included in the 62-kDa chromopeptide. This is also consistent with the idea that the carboxyl half of phytochrome assumes a more extended form than the amino-terminal counterpart (Lagarias and Mercurio, 1985; Jones and Quail, 1986a).

A peptide map of oat phytochrome was also constructed from the amino-terminal sequence data of the tryptic peptides which were separated by SDS PAGE (Rudiger, W., communication in Discussion Session of the XVI Yamada Conference). Asp-53, Val-65, and Ile-209 were the amino-terminal residues of the tryptic fragments of oat phytochrome, indicating that the higher-order structure of pea and oat phytochrome is quite similar in the amino-terminal half of phytochrome.

V. CONCLUSION

During the last five years a lot of studies have been done on the structural organization of phytochrome (Lagarias, 1985; Song, 1985; Quail et al., 1986). Most of the contributions, however, have been made about the chromophore-bearing amino-terminal half of phytochrome, mainly using its chromophore as a reporter molecule. Study of the structural features in the carboxyl-terminal remainder of phytochrome has just started recently (Lagarias and Mercurio, 1985; Jones and Quail, 1986a).

There are several lines of evidence indicating interactions between the chromophore-bearing amino-terminal half (residues 1 - 624) and the carboxyl-terminal remainder of phytochrome. Photoequilibrium of phototransformation and its sensitivity to the ionic environment changed when the 62-kDa chromopeptide was cleaved from the 114-kDa pea phytochrome (Yamamoto and Smith, 1981; Yamamoto and Furuya, 1983). The 114-kDa pea phytochrome showed photoreversible self-aggregation in the presence of divalent metal ion, red-light-induced adsorption to a particulate fraction (Yamamoto et al., 1980), and a photoreversible change in hydrophobicity (Tokutomi et al., 1981). These changes were not observed with the 62-kDa chromopeptide. Photoconversion-induced conformational change was detected also in a 52-55 kDa carboxyl-terminal part of phytochrome by limited proteolysis (Lagarias and Mercurio, 1985). Differential reactivity of monoclonal antibodies against Pr and Pfr indicated light-induced conformational changes somewhere in the 118-kDa oat phytochrome (Thomas and Penn, 1986; Shimazaki, Y., personal communication).

In fact, relatively high amino acid sequence homology was observed from residue 70 to as far as residue 800, well outside of the chromophore-bearing domain, between oat and zucchini (Quail et al., 1986) and oat and pea phytochromes (Sato, N., personal communication). The highly conserved antigenic domain of a monoclonal antibody was also on the carboxyl-terminal half of phytochrome (Cordonnier et al., 1986).

These observations suggest the presence of functional domains also in the carboxyl-terminal half of phytochrome. The possible change in the interaction between the chromophore-bearing amino-terminal domain (residues 1 - 624) and the functional domains in the carboxyl-terminal remainder of phytochrome could explain the photoreversible phenomena described above. Much more work should be concentrated on the carboxyl-terminal half of phytochrome in order to fully understand the mechanism of phytochrome action.

ACKNOWLEDGMENTS

I thank Miss H. Kajiura and Mr. H. Hattori, Center for
Physical and Chemical Analysis, NIBB, for their helpful advice
and assistance in carrying out amino acid sequence analysis.
I also thank the Center for providing me a photodiode array
detector. I am grateful to Dr. N. Sato, Department of Biology, University of Tokyo, for showing me his unpublished data,
Miss Y. Tsuge for her technical assistance, and Prof. M.
Furuya for his encouragement and valuable discussion during
this study.

REFERENCES

Bajaj, M., and Blundell, T. (1984). Annu. Rev. Biophys.
    Bioeng. 13:453.
Baron, O., and Epel, B. L. (1983). Plant Physiol. 73:471.
Butler, W. L., Siegelman, H. W., and Miller, C. O. (1964).
    Biochemistry 3:851.
Chai, Y. G., Song, P. -S., Cordonnier, M. -M., and Pratt, L.
    H. (1986). Abstract XVI Yamada Conference, p. 66.
Cordonnier, M. -M., and Pratt, L. H. (1982). Plant Physiol.
    70:912.
Cordonnier, M. -M., Greppin, H., and Pratt, L. H. (1985).
    Biochemistry 24:3246.
Cordonnier, M. -M., Greppin, H., and Pratt, L. H. (1986).
    Plant Physiol. 80:982.
Daniels, S. M., and Quail, P. H. (1984). Plant Physiol. 76:
    622.
Eilfeld, P., and Rudiger, W. (1984). Z. Naturforsch. 39c:742.
Eilfeld, P., and Rudiger, W. (1985). Z. Naturforsch. 40c:109.
Hahn, T. -R., and Song, P. -S. (1981). Biochemistry 20:2602.
Hahn, T. -R., Song, P. -S., Quail, P. H., and Vierstra, R. D.
    (1984). Plant Physiol. 74:755.
Hashimoto, F., Horigome, T., Kanbayashi, M., Yoshida, K., and
    Sugano, H. (1983). Anal. Biochem. 129:192.
Hershey, H. P., Barker, R. F., Idler, K. B., Lissemore, J. L.,
    and Quail, P. H. (1985). Nucleic Acids Res. 13:8543.
Hunt, R. E., and Pratt, L. H. (1980). Biochemistry 19:390.
Inoue, Y., and Furuya, M. (1985). Plant Cell Physiol. 26:813.
Jones, A. M., and Quail, P. H. (1986a). Biochemistry 25:2987.
Jones, A. M., and Quail, P. H. (1986b). Abstract XVI Yamada
    Conference, p. 68.

Jones, A. M., Vierstra, R. D., Daniels, S. M., and Quail, P. H. (1985). Planta 164:501.

Kendrick, R. E., and Spruit, C. J. P. (1973). Photochem. Photobiol. 18:153.

Kendrick, R. E., and Spruit, C. J. P. (1977). Photochem. Photobiol. 26:201.

Kerscher, L., and Nowitzki, S. (1982). FEBS Lett. 146:173.

Konomi, K., Furuya, M., Yamamoto, K. T., Yokota, T., and Takahashi, N. (1982). Plant Physiol. 70:307.

Laemmli, U. K. (1970). Nature 227:680.

Lagarias, J. C. (1985). Photochem. Photobiol. 42:811.

Lagarias, J. C., and Mercurio, F. M. (1985). J. Biol. Chem. 260:2415

Litts, J. C., Kelley, J. M., and Lagarias, J. C. (1983). J. Biol. Chem. 258:11025.

Lumsden, P. J., Yamamoto, K. T., Nagatani, A., and Furuya, M. (1985). Plant Cell Physiol. 26:1313.

Nagatani, A., Yamamoto, K. T., Furuya, M., Fukumoto, T., and Yamashita, A. (1984). Plant Cell Physiol. 25:1059.

Quail, P. H., Colbert, J. T., Peters, N. K., Christensen, A. H., Sharrock, R. A., and Lissemore, J. L. (1986). Phil. Trans. Roy. Soc. Lond. B314:469.

Reiff, U., Eilfeld, P., and Rudiger, W. (1985). Z. Naturforsch. 40c:693.

Rice, H. V., and Briggs, W. R. (1973). Plant Physiol. 51:927.

Saji, H., Nagatani, A., Yamamoto, K. T., Furuya, M., Fukumoto, T., and Yamashita, A. (1984). Plant Sci. Lett. 37:57.

Shimazaki, Y., Inoue, Y., Yamamoto, K. T., and Furuya, M. (1980). Plant Cell Physiol. 21:1619.

Song, P. -S. (1985). In "Optical Properties and Structure of Tetrapyrroles" (G. Blauer, and H. Sund, ed.), p. 331. Walter de Gruyter, Berlin.

Stoker, B. M., McEntire, K., and Roux, S. J. (1978). Photochem. Photobiol. 27:597.

Thomas, B., and Penn, S. E. (1986). FEBS Lett. 195:174.

Thummler, F., and Rudiger, W. (1984). Physiol. Plant. 60:378.

Thummler, F., Brandlmeier, T., and Rudiger, W. (1981). Z. Naturforsch. 36c:440.

Thummler, F., Eilfeld, P., Rudiger, W., Moon, D. -K., and Song, P. -S. (1985). Z. Naturforsch. 40c:215.

Tokutomi, S., Yamamoto, K. T., and Furuya, M. (1981). FEBS Lett. 134:159.

Tokutomi, S., Yamamoto, K. T., Miyoshi, Y., and Furuya, M. (1982). Photochem. Photobiol. 35:431.

Tokutomi, S., Sato, N., Inoue, Y., Yamamoto, K. T., and Furuya, M. (1986a). Plant Cell Physiol. 27:765.

Tokutomi, S., Yamamoto, K. T., and Furuya, M. (1986b). Abstract XVI Yamada Conference, p. 73.

Vierstra, R. D., and Quail, P. H. (1982). Planta 156:158.

Vierstra, R. D., and Quail, P. H. (1983). Biochemistry
    22:2498.
Vierstra, R. D., Cordonnier, M. –M., Pratt, L. H., and Quail,
    P. H. (1984). Planta 160:521.
Weber, K., Pringle, J. R., and Osborn, M. (1972). Methods
    Enzymol. 26:3.
Wray, W., Boulikas, T., Wray, V. P., and Hancock, R. (1981).
    Anal. Biochem. 118:197.
Yamamoto, K. T., and Furuya, M. (1983). Plant Cell Physiol.
    24:713.
Yamamoto, K. T., and Smith, W. O., Jr. (1981) Plant Cell
    Physiol. 22:1149.
Yamamoto, K. T., Smith, W. O., Jr., and Furuya, M. (1980).
    Photochem. Photobiol. 32:233.

PHYTOCHROME FROM GREEN AVENA[1]

Lee H. Pratt

Department of Botany
University of Georgia
Athens, Georgia

Marie-Michèle Cordonnier

CIBA-GEIGY Biotechnology Research
Research Triangle Park, North Carolina

I. INTRODUCTION

Phytochrome was discovered largely as a result of investigations of photoperiodism in photosynthetically competent
plants grown under essentially natural environmental conditions (Hendricks, 1964).   Nevertheless, apart from a few
scattered reports, our present understanding of the physico-
chemical properties of phytochrome derives from the study of
this pigment as isolated from etiolated plant tissues.   There
are two principal reasons for this situation (Pratt, 1983).
First, chlorophyll in light-grown, photosynthetically competent tissues prevents spectrophotometric assay of phytochrome.
Second, light-grown plants contain one to two orders of magnitude less phytochrome than do etiolated, but otherwise comparable, plants.   The amount of this chromoprotein in crude
extracts of light-grown plants is thus near, or even below,
the limit of detection by spectral assay.

---

[1]Supported by DOE contract DE-AC-09-81SR10925 and NSF
grant PCM-8315882 to LHP and by Swiss National Funds grant
3:292-0:82 to MMC when she was at the University of Geneva,
where much of the work that is summarized here was done.

Phytochrome is a potent immunogen. Consequently, it has been possible to develop immunochemical assays that not only are unaffected by chlorophyll (Pratt, 1983), but are also sufficiently sensitive to permit phytochrome quantitation in crude, unfractionated extracts of green plants (Hunt and Pratt, 1980; Shimazaki *et al.*, 1983). These immunochemical assays have facilitated the recent characterization of phytochrome from green oat (*Avena sativa* L.) shoots. The results of these investigations will be summarized here.

II. DO GREEN PLANTS CONTAIN A UNIQUE POOL OF PHYTOCHROME?

Two observations, which Hillman (1967) referred to as paradoxes, led him to suggest that plants contain at least two pools of phytochrome: a bulk pool that is inactive, at least with respect to some photomorphogenic responses, and a quantitatively minor pool that is active. Phytochrome from etiolated plants would then constitute his hypothetical bulk pool, while the residual phytochrome that is present in light-grown tissues would be his active pool. More generally, Hillman also argued that we should expect to find several phytochromes, each "...differing only slightly in structure but significantly in biological activity..." from the others. This predicted heterogeneity in phytochrome could account for otherwise inexplicable photomorphogenic observations. At a minimum, therefore, one should consider seriously the possibility that phytochrome in a mature green plant might be different from that which has been characterized from etiolated seedlings.

Sporadic attempts to isolate phytochrome from light-grown plant tissues and to initiate its characterization, however, initially did little more than demonstrate that it was there (Lane *et al.*, 1963; Taylor and Bonner, 1967; Giles and von Maltzahn, 1968). Although differences in spectral properties were noted (Table I), these were attributed to the fact that the phytochrome was isolated from lower organisms, as opposed to angiosperms, rather than from light-grown, as opposed to etiolated, tissues (Taylor and Bonner, 1967).

Even though these direct attempts failed to determine whether phytochrome from green plants was different, indirect evidence that it was continued to accumulate. For example, Brockmann and Schäfer (1982) concluded, on the basis of an analysis of destruction kinetics in *Amaranthus* seedlings, that two pools of functionally distinct phytochrome exist within them. Independently, Shimazaki *et al.* (1983) and Thomas *et al.* (1984) developed enzyme-linked immunosorbent assays (ELISAs) with the intent to quantitate phytochrome in crude extracts of green oat shoots. Both groups found that they were unable to do so accurately, using antibodies directed to phytochrome from etiolated oat shoots. Both suggested that phytochrome from light-grown oats might be immunochemically distinct from that against which the antibodies had been made.

III.   PHYTOCHROMES FROM GREEN AND ETIOLATED *AVENA* ARE DIFFERENT

Tokuhisa and Quail (1983) presented the first direct evidence that phytochromes from green and etiolated oat shoots differed from one another. They found that antibodies directed to phytochrome from etiolated oats immunoprecipitated no more than about 30% of the photoreversibly detectable phytochrome in extracts of green oat shoots. Moreover, they observed that phytochrome from green and etiolated oat shoots differed spectrally. Because of the potential significance of this apparent confirmation of Hillman's (1967) suggestion that there may be multiple species of phytochrome, initial experiments were designed to determine (a) whether these multiple species of phytochrome arose from post-homogenization artifacts, or (b) reflected differences inherent to phytochrome from green and etiolated tissues (Tokuhisa *et al*., 1985; Shimazaki and Pratt, 1985).

TABLE I. Absorbance Properties of Phytochrome from Light-Grown Tissues[a]

Organism	Wavelength (nm)			Reference
	ΔA_{max}	Isosbestic	ΔA_{min}	
Avena				
etiolated[b]	666	688	732	Fig. 4
green	655	682	728	Fig. 4
bleached[c]	655	684	722	Jabben and Deitzer, 1978
Spinacia	661	685	722	Lane *et al*., 1963
Sphaerocarpus	655	680	720	Taylor and Bonner, 1967
Mesotaenium	649	670	710	Taylor and Bonner, 1967
Mnium hornum	658	685	721	Giles and von Maltzahn,
M. undulatum	657	677	724	1968

[a]Wavelengths are for Pr minus Pfr difference spectra. [b]Properties of phytochrome from etiolated oat shoots are given for comparison. [c]Difference spectrum measured for phytochrome *in vivo*, utilizing herbicide-bleached, achlorophyllous tissue.

Tokuhisa *et al*. (1985) demonstrated by mixing experiments that phytochrome from green oats was not a modified form of phytochrome of the type isolated from etiolated oats. Shimazaki and Pratt (1985) confirmed the results of Tokuhisa *et al*., demonstrating that the unique properties of phytochrome from green oats did not arise from interfering substances, including chlorophylls, present in extracts of green oats. Additional controls performed by Shimazaki and Pratt included (a) extraction of phytochrome from light-grown oat leaves that were bleached by treatment with norflurazon (Jabben and Deitzer, 1978), (b) repurification of hydroxyapatite-purified phytochrome from etiolated oats after its addition to the buffer used for homogenization of green oat leaves, and (c) purification of phytochrome from green oat leaves that had been kept for 48 h in darkness. In the latter case, the phytochrome that accumulated was like that isolated from fully etiolated oats, even though the leaves were green. Shimazaki and Pratt also presented preliminary evidence that an immuno-chemically distinct pool of phytochrome might exist in green pea shoots as well, an observation that has been confirmed and extended by Abe *et al*. (1985). All of these results are consistent with the suggestion that the differences that have been observed between phytochrome from etiolated and green oats are inherent to these two chromoproteins *in situ*. Nevertheless, they cannot exclude the possibility that the observed differences arise from differential lability of a phytochrome in light-grown oat leaves that differs only slightly from that found in etiolated oat shoots.

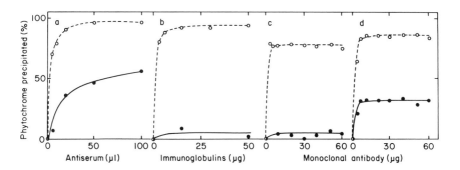

FIGURE 1. Immunoprecipitation of etiolated-oat (o) and green-oat (●) phytochrome by antibodies directed to etiolated-oat phytochrome. Antibodies were (a) present in whole anti-serum against approx. 120-kDalton phytochrome, (b) immunopuri-fied from this same antiserum, (c) monoclonal antibody oat-3, and (d) monoclonal antibody oat-9. Data are taken from Shima-zaki and Pratt (1985, 1986).

Given that there are two species of phytochrome in oats, appropriate terms are needed to distinguish between them. We suggest that phytochrome of the type that is most abundant in etiolated oats be referred to as etiolated-oat phytochrome. Conversely, that which is most abundant in green oats should be termed green-oat phytochrome. Assuming that a similar situation exists in other plants, for example peas (Shimazaki and Pratt, 1985; Abe *et al*., 1985; see Chapter 8), one would then have etiolated- and green-pea phytochrome. It must be emphasized that use of these terms is not meant to imply that there is no green-oat phytochrome in etiolated oats, nor that there is no etiolated-oat phytochrome in green oats. The use of these descriptive terms, instead of more abstract designations that give no hint as to the type of phytochrome in question (e.g., Abe *et al*., 1985), seems preferable.

IV. CHARACTERIZATION OF GREEN-OAT PHYTOCHROME

As anticipated from the initial report of Tokuhisa and Quail (1983), antiserum directed to etiolated-oat phytochrome fails to precipitate all photoreversibly detectable green-oat phytochrome (Fig. 1a). Polyclonal antibodies immunopurified by a column of immobilized etiolated-oat phytochrome from the same antiserum, however, precipitate little to no green-oat phytochrome (Fig. 1b). This difference between antiserum and immunopurified antibodies was observed with three independently prepared antisera (Shimazaki and Pratt, 1986). This inability to immunoprecipitate green-oat phytochrome with immunopurified polyclonal antibodies is in marked contrast to the observation of Tokuhisa *et al*. (1985), who could precipitate 30% of the photoreversibility from a green-oat phytochrome preparation with similarly immunopurified antibodies.

Tokuhisa *et al*. (1985) concluded that the phytochrome that was immunoprecipitated in their experiments was like etiolated-oat phytochrome. In contrast, immunoprecipitation data obtained with several monoclonal antibodies directed to etiolated-oat phytochrome (Fig. 1c and Shimazaki and Pratt, 1985) indicate that, at least for the green-oat phytochrome preparations examined, all photoreversibly detectable phytochrome from green oats is distinct from etiolated-oat phytochrome. The quantitative ELISA data of Shimazaki *et al*. (1983) are also inconsistent with the possibility that more than about 5% of the phytochrome in green oat leaves is like that from etiolated oats.

Shimazaki and Pratt (1985) did, however, find that one-third of the photoreversibility in a green-oat phytochrome preparation could be immunoprecipitated by monoclonal anti-

bodies oat-9 and oat-16 (Fig. 1d), both of which are directed
to etiolated-oat phytochrome. Since these antibodies apparently recognize the same epitope (Shimazaki et al., 1983), and
since they recognize etiolated- and green-oat phytochrome
equally well (Shimazaki and Pratt, 1985), it appears that
green-oat phytochrome does contain at least one antigenic
domain that is highly conserved with respect to etiolated-oat
phytochrome. Moreover, since only a portion of green-oat
phytochrome can be immunoprecipitated by oat-9, it is possible
that green-oat phytochrome itself may be heterogeneous.

For technical reasons, Tokuhisa et al. (1985) and Shimazaki and Pratt (1985) could not determine unambiguously the
native monomer size of green-oat phytochrome. With a monoclonal antibody (pea-25) directed to etiolated-pea phytochrome, however, Cordonnier et al. (1986b) demonstrated that
native green-oat phytochrome has a monomer size essentially
the same as that observed for etiolated-oat phytochrome (Fig.
2, lower panel). The inability of monoclonal antibody oat-22
to detect phytochrome in an extract of green oats, even though
it detects etiolated-oat phytochrome with comparable, if not
better, sensitivity than does pea-25 (Fig. 2, upper panel),
indicates again that green oat leaves contain at most a trace
quantity of etiolated-oat-like phytochrome.

Since oat-9 recognizes an epitope on the amino terminal
half of phytochrome (Shimazaki et al., 1986), while pea-25
binds to an epitope on the carboxyl terminal half (Cordonnier
et al., 1986a), green- and etiolated-oat phytochrome have at
least two domains in common. Furthermore, it appears that
oat-9 and pea-25 might be detecting different pools of green-
oat phytochrome (Cordonnier et al., 1986b), indicating again
that green-oat phytochrome might be heterogeneous.

Green-oat phytochrome is more susceptible to modification
in crude extracts than is etiolated-oat phytochrome (Fig. 3).
Since protease inhibitors retard this modification (Tokuhisa
and Quail, 1986), it can be attributed to proteolysis. Moreover, mixing experiments (Fig. 3 and Cordonnier et al., 1986b)
document that the difference in susceptibility to proteolysis
is due to inherent differences between etiolated- and green-
oat phytochrome, not to differences between protease levels in
etiolated- and green-oat extracts.

Tokuhisa et al. (1985) reported a monomer size for the
immunochemically distinct green-oat phytochrome of 118 kDaltons, significantly smaller than the 121-125 kDaltons reported
by Cordonnier et al. (1986b; see Fig. 2). It had initially
been speculated that the enhanced susceptibility of green-oat
phytochrome to protease activity (Fig. 3) might account for
this apparent size discrepancy (Cordonnier et al., 1986b).
Recent data reported by Tokuhisa and Quail (1986), however,
indicate that the difference in reported sizes is an artifact

arising from subtle differences in the methods by which they
were obtained. Nevertheless, this enhanced lability of green-
oat phytochrome reinforces the possibility that at least some
of the reported differences between etiolated- and green-oat
phytochrome might arise from post-homogenization modifications
of two phytochromes that differ very little from one another.
As a minimum, it is essential in future work that any characterization of green-oat phytochrome be accompanied by documentation that it has been unmodified.

As reported initially by Tokuhisa and Quail (1983), a Pfr
minus Pr difference spectrum for green-oat phytochrome devi-
ates markedly from that obtained for etiolated-oat phytochrome. Green-oat phytochrome exhibits a difference minimum
at 11 nm shorter wavelength and an isosbestic point that is
blue-shifted by 6 nm (Fig. 4, Table I). Immunoblot analysis
of the hydroxyapatite-purified, green-oat phytochrome that was

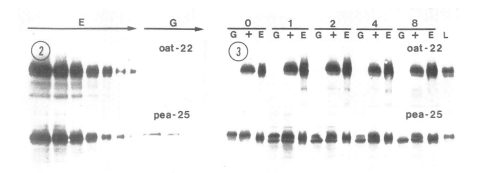

FIGURE 2. Immunoblots of sodium dodecyl sulfate polyacrylamide gels after electrophoresis of crude extracts of
rapidly frozen, lyophilized, etiolated (E) or green (G) oat
shoots. Sample loads from left to right were 20, 10, 5, 2, 1,
0.5 and 0.2 µl and 20, 10 and 5 µl of an etiolated- or green-
oat extract, respectively (estimated phytochrome content of 10
and 0.25 ng/µl, respectively). Top panel stained with oat-22,
bottom with pea-25. From Cordonnier *et al.* (1986b).

FIGURE 3. Blots of sodium dodecyl sulfate polyacrylamide
gels after electrophoresis of clarified extracts from green
(G) and etiolated (E) oats. Blots were immunostained with
oat-22, which reveals only etiolated-oat phytochrome, or
pea-25, which reveals both etiolated- and green-oat phytochrome. Extracts, either alone (lanes G, E) or after mixing
together (lanes +), were incubated for the indicated time in
hours prior to preparation for electrophoresis. An aliquot of
a crude, sodium dodecyl sulfate sample buffer extract of
lyophilized etiolated oat shoots (L) was included as a size
standard. From Cordonnier *et al.* (1986b).

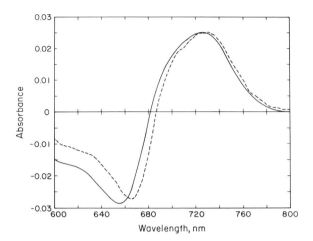

FIGURE 4. Pfr minus Pr absorbance difference spectra for hydroxyapatite-purified etiolated- and green-oat phytochrome (——— and – – –, respectively). The difference spectrum for etiolated-oat phytochrome was multiplied by 2.11 to facilitate comparison with that for green-oat phytochrome. Green-oat phytochrome was kindly provided by Dr. Y. Shimazaki. Spectra were measured with an Hitachi 320 scanning spectrophotometer interfaced with an IBM AT microcomputer as described elsewhere (Cordonnier *et al.*, 1986b).

used to obtain this difference spectrum indicated that it had been degraded to a minor extent, even though 10 mM iodoacetamide had been used throughout its preparation (Tokuhisa and Quail, 1986). Since upon such degradation the far-red difference maximum undergoes a blue-shift comparable to that observed with etiolated-oat phytochrome when it is similarly degraded (Cordonnier *et al.*, 1986b), it is likely that some or all of the minor blue-shift seen in the far-red peak (Fig. 4) arises from this partial degradation. It therefore appears that at least part of the reason for the earlier reported differences in spectral properties between phytochrome isolated from light- and dark-grown tissues (Table I) might reflect differences between etiolated- and green-plant phytochrome. Phylogenetic differences may not be as important as had originally been suspected.

Size exclusion chromatography through a TSK 3000 column indicates that green-oat phytochrome most likely exists in solution as a dimer (Cordonnier *et al.*, 1986b), as is already known to be the case for etiolated-oat phytochrome. Even though these two phytochrome pools are similar in monomer size and quaternary structure, however, they exhibit different peptide maps following partial proteolysis under denaturing conditions. This difference exists irrespective of whether peptide maps are probed with polyclonal antibodies (Tokuhisa *et al.*, 1985) or with monoclonal antibody pea-25 (Cordonnier *et al.*, 1986b). Since even pea-25 yields different maps, the difference cannot be ascribed solely to the detection of non-homologous peptides when immunostaining with polyclonal antibodies (Cordonnier *et al.*, 1986b).

V. WHAT IS THE BASIS FOR THE DIFFERENCES BETWEEN ETIOLATED- AND GREEN-OAT PHYTOCHROME?

Etiolated- and green-oat phytochrome can be either (a) differentially modified products of a single gene, or (b) products of different genes. While it is tempting, largely on the basis of differences in peptide maps following proteolysis under denaturing conditions (Tokuhisa *et al.*, 1985; Cordonnier *et al.*, 1986b), to conclude that the two proteins are products of different genes, the data summarized here do not permit this conclusion. Primary sequence data for green-oat phytochrome, or the demonstration that a unique gene exists for this form of phytochrome, will be needed to determine unambiguously how the two phytochromes differ.

It is already evident that etiolated- and green-oat phytochrome are differentially expressed. For example, immunochemically distinct, presumably green-oat, phytochrome is present in dry oat seed, while etiolated-oat phytochrome is apparently absent (Hilton and Thomas, 1985). Moreover, while etiolated-oat phytochrome accumulates during seedling growth in darkness, green-oat phytochrome appears to remain at an approximately constant level (Tokuhisa and Quail, 1986). Similarly, it appears that the phytochrome that accumulates during prolonged darkness, even in green oat leaves, is selectively of the etiolated-oat type (Shimazaki *et al.*, 1983). It thus appears that a plant accumulates etiolated-oat phytochrome in darkness, regardless of whether the plant is etiolated, and selectively degrades it in the light. The abundance of green-oat phytochrome, on the other hand, appears not to be markedly photoregulated.

VI. CONCLUSIONS AND FUTURE

While green-oat phytochrome is similar to etiolated-oat phytochrome in monomer size and quaternary structure, it differs in a number of respects. Its difference spectrum is blue-shifted in the red region, its immunochemical activity differs markedly, it is significantly more susceptible to an initial proteolytic cleavage, it yields a different peptide map upon proteolysis under denaturing conditions, and it displays differential photoregulation with respect to its own abundance.

Green-oat phytochrome does possess at least two epitopes in common with etiolated-oat phytochrome, one of which is the most highly conserved domain yet identified (Cordonnier *et al.*, 1986a). It is not clear, however, whether these epitopes are on the same or different pools of green-oat phytochrome. Whether this potential heterogeneity of green-oat phytochrome is real, or instead arises from differential post-homogenization modifications of a common green-oat phytochrome pool, remains to be determined. It may be that this possible heterogeneity within green-oat phytochrome is responsible for some of the differing observations that have been reported concerning its physicochemical properties.

The recent observation that etiolated oats contain a small amount of green-oat phytochrome, together with the observation that the two phytochromes co-purify (Shimazaki and Pratt, 1985), indicates that polyclonal antibodies directed to etiolated-oat phytochrome should be expected to be "contaminated" with antibodies to green-oat phytochrome. This "contamination" could then account, at least in part, for the ability of antisera directed to etiolated-oat phytochrome to immunoprecipitate as much as 80% of the photoreversibility from a green-oat phytochrome preparation, albeit with reduced avidity and with considerable variability from serum to serum (Shimazaki and Pratt, 1986). The inability of immunopurified anti-

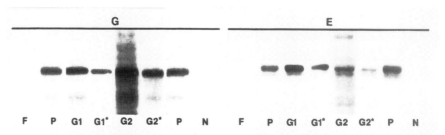

FIGURE 5. Immunoblots of sodium dodecyl sulfate polyacrylamide gels after electrophoresis of green-oat (G) or etiolated-oat (E) phytochrome and immunostaining with spent hybridoma medium from a cell line secreting antibody to formate dehydrogenase from *Clostridium thermoaceticum* (F), pea-25 (P), spent hybridoma medium from two cell lines secreting antibody to green-oat phytochrome before (G1, G2) and after (G1*, G2*) 10-fold dilution, or non-immune mouse immunoglobulins. Data obtained with the assistance of Dr. Y. Shimazaki.

bodies to precipitate a significant quantity of green-oat phytochrome, at least in our own hands, indicates that the frequency with which immunoprecipitating antibodies are made to common epitopes is low.

A discrepancy between our observations (Shimazaki and Pratt, 1985, 1986; Cordonnier *et al.*, 1986b) and those of Tokuhisa *et al.* (1985) that is difficult to explain at present is the following. Tokuhisa *et al.* indicated that green oat leaves contain two kinds of phytochrome, with about one-third being like that from etiolated oats and two-thirds being the immunochemically distinct, green-oat type. In contrast, we have never obtained evidence for a second pool of etiolated-oat-like phytochrome in green oats, even though the assays used were sufficiently sensitive to detect it if it were there in the abundance reported by Tokuhisa *et al.* While the discrepancy might derive from the fact that we used tissues grown under different conditions and harvested at slightly different ages, both the data of Shimazaki *et al.* (1983) and recent unpublished experiments designed to test these possibilities indicate that they are not responsible.

Resolution of this discrepancy, as well as answers to the questions already posed, will require the development of new tools. For the moment it must remain unresolved. Perhaps the most important tool will be the development of monoclonal antibodies directed to, and specific for, green-oat phytochrome. They will be the probes needed to determine whether green-oat phytochrome is itself heterogeneous, as preliminary data indicate might be the case, and to determine whether a unique gene for green-oat phytochrome exists. Although antibodies directed to green-oat phytochrome have recently been obtained. (Fig. 5), they are not sufficiently specific for green-oat phytochrome to serve as the needed probes. One antibody (tentatively designated G2), however, clearly immunostains green- better than etiolated-oat phytochrome.

As originally argued by Hillman (1967), it should not be surprising to learn that phytochrome is heterogeneous. Phytochrome performs many different functions. It may be that there is a different form of phytochrome responsible for performing each type of function. For example, photoregulation of de-etiolation is primarily a red/far-red, photoreversible, inductive phenomenon. In contrast, photoregulation of morphogenesis in daylight results from continuing modulation by phytochrome of growth and development. Since these two types of phytochrome-regulated behavior are phenomenologically so different, it should not be surprising, at least in retrospect, that there might be two different photoreceptors, one for each response type.

REFERENCES

Abe, H., Yamamoto, K. T., Nagatani, A., and Furuya, M. (1985). *Plant Cell Physiol.* 26:1387.

Brockmann, J., and Schäfer, E. (1982). *Photochem. Photobiol.* 35:555.

Cordonnier, M.-M., Greppin, H., and Pratt, L. H. (1986a). *Plant Physiol.* 80:982.

Cordonnier, M.-M., Greppin, H., and Pratt, L. H. (1986b). *Biochemistry* 25:7657.

Giles, K. L., and von Maltzahn, K. E. (1968). *Can. J. Bot.* 46:305.

Hendricks, S. B. (1964). In "Photophysiology," Vol. I. (A. C. Giese, ed.), p. 305. Academic Press, New York.

Hillman, W. S. (1967). *Ann. Rev. Plant Physiol.* 18:301.

Hilton, J. R., and Thomas, B. (1985). *J. Exp. Bot.* 36:1937.

Hunt, R. E., and Pratt, L. H. (1980). *Plant Cell Environ.* 3:91.

Jabben, M., and Deitzer, G. F. (1978). *Planta* 143:309.

Lane, H. C., Siegelman, H. W., Butler, W. L., and Firer, E. M. (1963). *Plant Physiol.* 38:414.

Pratt, L. H. (1983). *Encycl. Plant Physiol.* 16:152.

Shimazaki, Y., and Pratt, L. H. (1985). *Planta* 164:333.

Shimazaki, Y., and Pratt, L. H. (1986). *Planta* 168:512.

Shimazaki, Y., Cordonnier, M.-M., and Pratt, L. H. (1983). *Planta* 159:534.

Shimazaki, Y., Cordonnier, M.-M., and Pratt, L. H. (1986). *Plant Physiol.* 82:109.

Taylor, A. O., and Bonner, B. A. (1967). *Plant Physiol.* 42:762.

Thomas, B., Crook, N. E., and Penn, S. E. (1984). *Physiol. Plant.* 60:409.

Tokuhisa, J. G., and Quail, P. H. (1983). *Plant Physiol.* 72:S85.

Tokuhisa, J. G., and Quail, P. H. (1986). In "Proceedings of XVI Yamada Conference on Phytochrome and Plant Photomorphogenesis" (M. Furuya, ed.), p. 81. Okazaki, Japan.

Tokuhisa, J. G., Daniels, S. M., and Quail, P. H. (1985). *Planta* 164:321.

APPLICATION OF MONOCLONAL ANTIBODIES
TO PHYTOCHROME STUDIES

Akira Nagatani[1,2]
Peter J. Lumsden
Koji Konomi[2]
Hiroshi Abe[2]

Division of Biological Regulation
National Institute for Basic Biology
Myodaijicho, Okazaki, Japan

I.   INTRODUCTION

Although phytochrome is a well-known plant photoreceptor, the molecular mechanism of its action remains unclear. Specific probes for the molecule are essential to elucidate its structure and function; specific antibodies, which are now readily available and applicable to most proteins, are one of the choices for such probes.

Immunochemical studies of phytochrome, which has been proven to be a good antigen, started some fifteen years ago with conventional polyclonal antibodies (Hopkins and Butler, 1970). This was followed by various applications of the technique such as immunochemical analysis of the phytochrome

---

[1]Supported by Grant-in-Aid (60790110) from the Ministry of Education, Science and Culture.
[2]Present address: Frontier Research Programs, The RIKEN Institute, Hirosawa, Wako, Japan.

Abbreviations: CD, circular dichroism; cDNA, complementary DNA; C-terminus, carboxyl terminus; ELISA, enzyme-linked immunosorbent assay; IgG, immunoglobulin G; N-terminus, amino terminus; PAGE, polyacrylamide gel electrophoresis; RIA, radioimmunoassay

molecule, radioimmunoassay, immunocytochemistry and immunoaffinity purification of phytochrome (reviewed in Pratt,
1982).

The development of monoclonal antibodies in 1975 (Köhler
and Milstein) enhanced the potential scope of the application
of immunochemical methods to the study of phytochrome, since
the technique provides a specific probe for a single epitope
whose size is in the order of several amino acids (Atassi,
1975). In addition, homogeneous molecular species of antibody
can easily be produced in large quantities by inoculating
antibody producing cells into the peritoneal cavity of mice
(Goding, 1980).

In this chapter, we review new findings made in this field
since 1983, when anti-phytochrome monoclonal antibodies were
first reported (Nagatani et al., 1983; Cordonnier et al.,
1983), especially findings which would be impossible or very
difficult to achieve with polyclonal antibodies.

II.   PRODUCTION AND INITIAL CHARACTERIZATION
OF MONOCLONAL ANTI-PHYTOCHROME ANTIBODIES

A.   Production of Monoclonal Antibodies

Monoclonal antibodies against phytochrome were obtained
rather easily without any modification of the well-established
technique using mice (Nagatani et al., 1983, 1984,; Cordonnier
et al., 1983) or rats (Thomas et al., 1984b) as spleen cells
donor. Various screening methods such as RIA with phyto-
chrome-coated red blood cells (Nagatani et al., 1983),   ELISA
(Cordonnier et al., 1983; Thomas et al., 1984b; Daniels and
Quail, 1984; Whitelam et al., 1985), and dot-immunoblotting
(Lumsden et al., 1985) have yielded a range of monoclonal
antibodies large enough for most purposes.

However, the monoclonal antibodies obtained by different
screening methods seemed to be of somewhat different types.
We observed that all of our six monoclonal antibodies obtained
using a sandwich ELISA (Shimazaki et al., 1983) as the screening method failed to recognize phytochrome in western
blotting. These monoclonal antibodies seemed to correspond to
"Type 4" monoclonal anti-oat phytochrome antibodies reported
by Daniels and Quail (1984), which reacted with phytochrome in
ELISA but not in western blotting. In contrast, at least 15
of 16 monoclonal antibodies screened by dot-immunoblotting
(Hawkes et al., 1982), in which antigen protein was SDS-
denatured and adsorbed onto nitrocellulose sheet, reacted with
phytochrome in sandwich ELISA as well as in western blotting
(Nagatani, unpublished data). Thus, the correct screening

method should be chosen according to the purpose.

B. Mapping of Epitopes

As already mentioned, recognition of a small and restricted region of the antigen protein characterises monoclonal
antibodies. Obviously, then, it is desirable to map the
epitopes for the obtained antibodies on the antigen molecule
before use. Conventionally, this has been achieved by testing
the cross-reaction of the antibody with various proteolytic
fragments of phytochrome (Daniels and Quail, 1984; Nagatani et
al., 1984; Thomas et al., 1986).

Figure 1 shows the possible arrangement of epitopes for
the 25 monoclonal antibodies (Nagatani et al., 1984; Lumsden
et al., 1985; Abe et al., 1985; Nagatani, unpublished data)
against pea or rye phytochrome obtained in our laboratory, and
Table 1 summarizes the results of the mapping of anti-phytochrome monoclonal antibodies reported up to now.

It is clear that a small domain at the N-terminus of

TABLE 1. Frequency of Domain Recognition by Monoclonal
Antibodies

Anti-gen	Screen-ing Method	Frequency			References
		N-terminal[a]	chromo-phoric[b]	non-chromo-phoric[c]	
Oat (124kDa)	ELISA	65%	25%	10%	Daniels & Quail (1984)
Pea (121kDa)	Dot-immuno	5/8	2/8	1/8	Lumsden et al.(1985) & Abe et al.(1985)
Oat (124kDa)	ELISA	3/3	0/3	0/3	Cordonnier et al. (1985)
Oat (124kDa)	ELISA	4/6	1/6	1/6	Holdsworth & Whitelam (1986)
Pea (114kDa)	SRBC-RIA	–	4/7	3/7	Nagatani et al. (1984)
Oat (114/118kDa)	ELISA	–	16/16	0/16	Cordonnier et al. (1984)
Pea (114kDa)	Dot-immuno	–	3/8	5/8	Nagatani unpublished

[a]N-terminal domain which can easily removed from Pr by
proteolysis.
[b]chromophoric domain of about 60 kDa.
[c]non-chromophoric domain of between 60 and 50 kDa which is
assumed to be the C-terminal half of the phytochrome molecule.

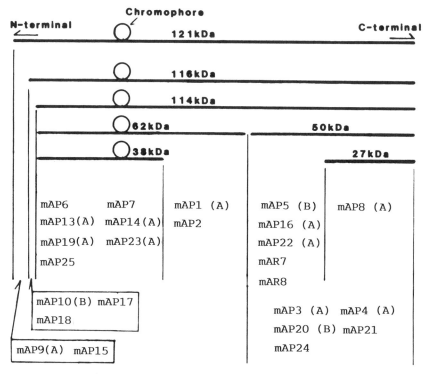

FIGURE 1. Possible arrangement of the epitopes for 25
monoclonal anti-phytochrome antibodies. Locations of the
epitopes were indicated by the cross-reaction of the antibody
with phytochrome fragments. The fragments used were prepared
as follows: 121kDa, see Lumsden et al. (1985). 116 kDa, see
Abe et al. (1985). 114, 62, 38 kDa, see Yamamoto and Furuya
(1983). N-terminal amino acid sequences were also determined
to locate the position of the fragment along the phytochrome
polypeptide (Yamamoto, K. T., personal communication). 50
kDa, phytochrome was treated with 0.2 % trypsin for 15 min at
25°C as described in Yamamoto and Furuya (1983). 27 kDa, see
Tomizawa et al. (1986). The monoclonal antibodies were
produced as follows: mAP1 to 9, see Nagatani et al. (1984).
mAP9, 10, and mAP15 to 18, see Lumsden et al. (1985) and also
Abe et al. (1985). mAP13, 14, mAP19 to 22, and mAP23 to 25,
hybridomas were obtained using mice immunized with 114 kDa pea
phytochrome and positive clones were screened by dot
immunoblotting. mAR 7 and 8, monoclonal antibodies were
prepared from a mouse immunized with rye phytochrome using dot
immunoassay as screening method. (A) indicates that the
antibody reacted with phytochrome I but not with phytochrome
II, while (B) indicates that the antibody reacted with both
phytochrome I and II (for details, see Section VI.).

native phytochrome is highly immunogenic, as first reported by Daniels and Quail (1984). This supports the idea that the N-terminus of phytochrome is exposed, since loops and/or protruding regions of protein are supposed to be the 'best' antigenic domain (Barlow et al., 1986). However, it is still unclear whether the chromophoric or non-chromophoric half of the molecule is the more immunogenic (Table 1).

III. CROSS-REACTION OF MONOCLONAL ANTIBODIES AGAINST PHYTOCHROMES OF VARIOUS PLANT SPECIES

Since the phytochrome molecule must contain many antigenic domains, it would be expected that among the antibodies raised to various phytochrome species would be some that bind to the molecule regardless of its source. The search for such antibodies has two main objectives. First to identify conserved domains, which by inference may be related to the molecular function of the molecule. Secondly, such antibodies may be used in studies of organisms from which much physiological information about phytochrome has been obtained, but from which it is very difficult to purify phytochrome.

A. Antigenic Diversity of Phytochromes from Various Plant Species

There are now several reports from different laboratories describing the cross-reactivity of monoclonal antibodies to phytochrome in crude or partially purified extracts from a number of plants. Saji et al. (1984), using a radioimmuno-assay with sheep red blood cells, found some cross-reactivity between rye monoclonal antibodies and oat phytochrome, and between pea monoclonal antibodies and dicots, but none between rye monoclonals and dicots or between pea monoclonals and monocots. Using an ELISA screening method with partially purified phytochrome for 16 monoclonal antibodies raised against etiolated oat and 9 against pea phytochrome, Cordonnier et al. (1984) identified a range of monoclonal antibodies, including some which were more specific for mono-cots than dicots, and some which were highly specific for the antigen used. Two monoclonal antibodies to pea and six to oat bound strongly to phytochrome from other species.

More recently the use of western blotting of SDS sample buffer extracts enables the native molecular size of the molecule to be retained. Using this method, Whitelam et al. (1985) found that three out of six monoclonal antibodies raised against etiolated oat phytochrome and originally

screened by ELISA, immunostained phytochrome from both monocots and dicots. Cordonnier et al. (1986a) found that an
antibody raised against etiolated pea, which on western blots
immunostained extracts from etiolated and light-grown oats and
pea, also stained extracts from the moss Physcomitrella and
the algae Mougeotia, Mesotaenium and Chlamydomonas. Using
proteolytically degraded fragments of phytochrome, the epitope
of this antibody was mapped to the non-chromophoric, C-
terminal half of the molecule.

These results confirm the diversity of phytochrome in
terms of its immunogenicity, as previously indicated by studies with polyclonal antibodies (Cordonnier et al., 1982).

To get antibodies that show wide cross-reactivity, it may
be better to screen hybridomas by dot-immunoblotting or
western blotting than to screen by ELISA or RIA, since all
cross-reacting antibodies reported were positive in western
blotting. However, when hybridomas were screened by ELISA or
RIA, not all the final monoclonal antibodies were positive in
western blotting (for detail see previous part of this
chapter). Thus the frequency of cross-eacting antibodies may
be greater when screened by western blot.

B. Application to Physiological Studies
of Monoclonal Antibodies Which Recognize
Phytochromes Regardless of Its Source

One of the most important developmental changes under
phytochrome control in green plants is the photoperiodic induction of flowering, which we have studied in the short day
plants Pharbitis nil and Lemna paucicostata 441. Three separate reactions involving phytochrome have clearly been demonstrated (Saji et al., 1982; Lumsden and Furuya, 1986; Lumsden
et al., 1987). To understand the mechanism of phytochrome
action, it is necessary to be able to quantify and localise
the phytochrome. We have therefore tested a wide range of
monoclonal antibodies raised against both pea and rye, against
crude extracts of Pharbitis and Lemna, and found one which
immunostained extracts from etiolated Pharbitis on western
blots, and which was mapped to the C-terminal half of the
molecule (unpublished).

In a preliminary experiment using this antibody we examined the staining of extracts following transfer to light.
Plants were harvested after 12 and 24 h, frozen in liquid
nitrogen, powdered and extracted with boiling SDS buffer.
Extracts were then subjected to PAGE, transferred electro-
phoretically to nitrocellulose, and immunostained. There was
clear staining of a band corresponding to phytochrome in the
etiolated tissue, but no bands could be detected in the

greening tissue, confirming the similar observations in oat seedlings (Tokuhisa et al., 1985; Cordonnier et al., 1986b) and pea epicotyl (Abe et al., in preparation). Thus the physiological responses observed in those plants may be regulated mainly via phytochrome II which was recently reported to be antigenically distinct from well-characterised phytochrome I (for detail, see Section VI of this Chapter).

We also determined whether radiolabelling of the second antibody could increase the sensitivity of detection. Using this method we were unable to detect phytochrome in greened Pharbitis extracts, but found an increase in activity from extracts of Lemna which had been returned to darkness for 10 or 20 days (unpublished).

IV. STRUCTURAL ANALYSIS OF PHYTOCHROME PHOTOTRANSFORMATION USING MONOCLONAL ANTIBODIES

One of the potential uses of monoclonal antibodies is as probes of molecular structure or structure/function relationships. Applied to phytochrome, the main objective is to identify regions specific for the biologically active Pfr form of the molecule, since these may represent a functional region of the molecule. Two approaches have to date been employed; to identify epitopes where preferential binding occurs, and to identify regions where binding affects function.

A. Monoclonal Antibodies Which React Differently with Pr and Pfr

Of eight monoclonal antibodies raised against 114 kDa pea phytochrome, Nagatani et al. (1984), using a RIA with sheep red-blood cells, did not find any which showed preferential binding to Pr or Pfr, and Cordonnier et al. (1984), using an ELISA in which phytochrome was first bound directly to the plate, found none out of 23 monoclonal antibodies to pea and oat.

Subsequently, using an ELISA in which assay plates are first coated with antibody and then phytochrome (sandwich ELISA), at least three different epitopes have been identified which show preferential binding between Pr and Pfr.

Cordonnier et al. (1985) screened monoclonal antibodies raised against 124 kDa oat phytochrome by ELISA with phytochrome as either Pr or Pfr. Three, which showed greater affinity for Pr than for Pfr were mapped to the 6 kDa N-terminus. Using a western blotting screening method to identify monoclonal antibodies, we also obtained anti-pea phyto-

chrome monoclonal antibodies which stained the 121 kDa molecule but not the 114 kDa (Lumsden et al., 1985). On an immunoaffinity column this antibody did not bind to native pea
phytochrome in the Pfr form but did bind Pr (unpublished).

More recently, Thomas and Penn (1986) confirmed that one
of their antibodies originally reported to discriminate between Pr and Pfr bound more strongly to Pfr and was not associated with the N-terminal region. Shimazaki et al. (1986)
also found an antibody which bound preferentially to Pfr and
which was mapped on the chromophore half of the molecule but
away from the N-terminus.

These results agree with proteolytic studies (Daniels and
Quail, 1984) and phosphorylation studies (Wong et al., 1986)
that conformational changes occur in a number of regions of
the molecule, and offer the possibility of precisely locating
the parts of the molecule with regulatory functions.

B. Monoclonal Antibodies Which Affect the
Photochemical Properties of Phytochrome

Another approach which may be used is to examine the
effect of monoclonal antibodies on photochemical properties of
phytochrome. Cordonnier et al. (1985) reported that the antibody which was mapped to the N-terminus of etiolated oat
phytochrome, and which showed a 4-5 fold greater affinity for
Pr than Pfr, induced a blue-shift in the absorption maximum of
Pfr in vitro, and induced dark reversion of Pfr to Pr. We too
carried out a detailed study of the effect of several monoclonal antibodies on the spectral properties of immunopurified
pea phytochrome in vitro (Lumsden et al., 1985). At 10°C dark
reversion from Pfr was observed. In the presence of monoclonal anti-pea phytochrome antibodies mAP1, 3 or 5, which
bind away from the chromophore, and mAP7, which binds near the
chromophore (Nagatani et al., 1984), the rate of the reversion was reduced. None of these antibodies affected the
absorption spectra of phytochrome. In the presence of mAP9,
which binds near the N-terminus, the absorption at the red-
light-induced photostationary state was reduced and the rate
of dark reversion was increased (Fig. 2), resembling partially
degraded phytochrome of 114 kDa, but with no evidence of
proteolysis.

More recently it has been reported that the antibody that
bound to the N-terminus of oat phytochrome (Cordonnier et al.,
1985) also suppressed the photoreversible CD spectral change
in the far UV which normally indicates an approximately 3 %
increase in α-helical folding of the apoprotein. Other antibodies binding away from the N-terminus had no effect. Hence,
it was suggested that a Pfr-chromophore / N-terminus inter-

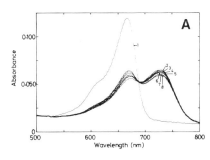
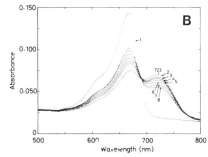

FIGURE 2. Pfr dark reversion of purified pea phytochrome in the presence (B) or absence (A) of mAP9. Measurement was done in phosphate-buffered saline at 10°C. The spectra were recorded, first as Pr (1), and then in the following sequence; 0 (2), 5 (3), 10 (4), 20 (5), 40 (6), 80 (7) and 120 min (8) in the dark after saturating red-light irradiation (for detail, see Lumsden et al., 1985).

action resulted in additional α-helical folding of the Pfr (Chai et al., 1986).

These results support the observation that the N-terminus play an important role in phytochrome phototransformation (Vierstra and Quail, 1982), and also indicate that dark reversion of phytochrome in vivo may either be increased or decreased due to binding of phytochrome with receptor sites within a cell. This raises the possibility that reversion may occur at different rates in different cellular locations.

V. ESTIMATION OF CONTENT OF NON-CHROMOPHORIC PHYTOCHROME
    PROTEIN MOIETY (APOPHYTOCHROME) BY COMPARING
    IMMUNOCHEMICAL AND SPECTROPHOTOMETRIC ASSAYS

Since the discovery of phytochrome (Butler et al., 1959), biologically functional holophytochrome, consisting of apoprotein and chromophore, has been detected spectrophotometrically by its unique photoreversible absorbance changes. To understand the mechanism of assembly of the phytochrome chromophore and its protein moiety it is necessary to estimate the content of non-chromophoric phytochrome protein moiety. Since an anti-phytochrome antibody recognizes a restricted region of protein moiety of the phytochrome molecule, and the spectrophotometric signal reflects the active holophytochrome, up to now, the only way to determine the fraction of non-chromophoric phytochrome has been to compare the contents of spectrophotometrically active phytochrome and immunochemically

active phytochrome. In this section, we will describe a
refinement of the immunoassay following the introduction of
monoclonal anti-phytochrome antibodies, and a refinement of
the spectrophotometric assay of phytochrome in crude plant
extracts. Its application to the measurement of phytochrome
in embryonic axes of pea seed in our laboratory will also be
described.

A. Detection of Phytochrome by ELISA

For the immunochemical quantitation of phytochrome protein
moiety using polyclonal antibody, RIA was first developed by
Hunt and Pratt (1979, 1980). Although the detection limit was
improved to about 1 ng of phytochrome, this method was
troublesome and expensive. These limitations could be over-
come by ELISA.

Following the production of mouse monoclonal anti-phytochrome antibodies, the double antibody sandwich method of
ELISA has been applied (Shimazaki et al., 1983; Thomas et al.,
1984a, 1984b; Jordan et al., 1984; Hilton and Thomas, 1985).
Shimazaki et al. (1983) reported that this method had a higher
sensitivity, down to as little as 100 pg. The assay consists
of sequential reactions in plastic microplates. In brief,
wells of polystyrene microplate are coated with rabbit anti-
phytochrome antibody. Test solution containing phytochrome,
mouse anti-phytochrome antibody, and enzyme-conjugated anti-
mouse IgG antibody are then reacted sequentially with the
complex on the surface of the well, and the "Sandwich", that
is the resulted complex on the plate, is detected by the
enzyme reaction.

Using this method, the content of phytochrome protein,
regardless of adding chromophore in crude plant homogenates,
could be determined in terms of protein weight.

B. Estimation of Phytochrome Content
in Crude Extracts by Spectrophotometry

To compare the spectrophotometric data and immunochemical
data, it is essential to estimate the content of spectrophotometrically detectable phytochrome in terms of protein weight.

To date it has not been possible to determine the exact
quantity of phytochrome detected spectrophotometrically in
crude plant homogenates as well as in tissue, because of the
scattering factors (Pratt, 1978). We have attempted to solve
this problem by adding highly purified phytochrome as an
internal standard to crude homogenates. The results obtained
showed that the apparent content of phytochrome could be

compensated for by the estimated scattering factor of the crude extract (Konomi et al., 1985).

C. Synthesis of Holophytochrome in Embryonic Axes During Imbibition

The content of phytochrome in pea embryonic axes was determined during imbibition in the dark or in the light using both immunochemical and spectrophotometric assays (Konomi et al., 1985).

In the homogenates from non-imbibed fragments, no significant level of phytochrome could be detected by either assays, indicating that neither inactivated holophytochrome nor its protein moiety exist in the embryonic axes of dormant pea seed.

Fragments of embryonic axes quickly absorbed water during the first 1-2 h after the start of imbibition on agar medium, after which the fresh weight stayed at a constant level for a further 10 h. In the dark at $25^{O}C$, both holophytochrome and antigenically active phytochrome became detectable at least 3 h after the start of imbibition, after which each the content increased during incubation.

In the light-grown axes, spectrophotometrically detectable phytochrome did not increase significantly during incubation. Immunochemically detectable phytochrome increased, but its content was lower than in the dark.

These data indicate that, the appearance and increase of holophytochrome in imbibed embryonic axes is affected by light, and that in the light there seemed to be a population of non-chromophoric phytochrome.

D. Synthesis of Phytochrome Apoprotein in the Presence of Gabaculine, an Inhibitor of Chromophore Synthesis

Recently, Gardner and Gorton (1985) reported that gabaculine inhibited the synthesis of spectrophotometrically detectable phytochrome in developing etiolated pea seedlings. Therefore, we investigated whether gabaculine inhibits the synthesis of holophytochrome and whether this influences the synthesis of phytochrome apoprotein in the pea embryonic axes (Konomi and Furuya, 1986).

In dark-imbibed axes, the content of spectrophotometrically detectable phytochrome was reduced in the presence of gabaculine at concentrations of 0.002 mM or higher, while little inhibitory effect of gabaculine on the synthesis of immunochemically detectable phytochrome was observed at concentrations of less than 1 mM. In the presence of 0.5 mM

gabaculine, the time course of the increase of immunochemically detectable phytochrome was unaffected (Fig. 3).   In
pea grown tissue treated with gabaculine, Jones et al. (1986)
obtained similar results   by western blotting.

Thus, it was concluded that there might be no coordination
between phytochrome chromophore synthesis and its apoprotein
synthesis and that the latter could exist stably in the
embryonic axes, though it is still unclear whether the phytochrome chromophore by itself exists in the cell.   Comparing
the results from both spectrophotometric and immunochemical
assays, as   described above, may be a good approach to
understanding the regulation of phytochrome synthesis.

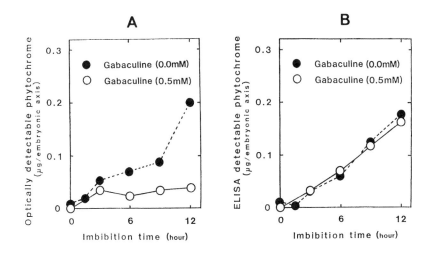

FIGURE 3.   Effect of gabaculine on the time course of the
synthesis of spectrophotometrically detectable (A) and
immunochemically (B) detectable phytochrome.   Embryonic pea
axes were imbibed on 0.2% agar with 0.5mM gabaculine in the
dark at 25°C.   The content of phytochrome protein in the crude
homogenates was determined by spectrophotometric internal
standard assay and ELISA. (after Konomi et al., 1985; Konomi
and Furuya, 1986).

VI.   APPLICATION OF MONOCLONAL ANTIBODIES TO A STUDY
      OF A PHYTOCHROME SPECIES (PHYTOCHROME II) WHICH
      IS ANTIGENICALLY DISTINCT FROM THE PHYTOCHROME
      SPECIES (PHYTOCHROME I) PREDOMINANT
      IN ETIOLATED TISSUE

A.   Purification and Initial Characterization
            of Phytochrome II

Initial efforts to isolate phytochrome from light-grown,
green plants were only made by a few workers, without the aid
of specific antibodies (Lane et al., 1963; Taylor and Bonner,
1967; Giles and von Maltzahn, 1968; Shimazaki et al., 1981),
but the results were far from satisfactory.

We have succeeded in purifying phytochrome from light-
grown pea seedlings by a combination of conventional column
chromatography and immunoaffinity purification (Abe et al.,
1985). In brief, the procedure consists of extraction in
potassium phosphate buffer, ammonium sulfate fractionation,
brushite adsorption and batch elution, DEAE-agarose and monoclonal anti-pea phytochrome antibody (mAP)(Nagatani et al.,
1984) conjugated agarose (mAP-agarose) chromatography. Phytochrome from light-grown tissue partially purified by brushite
and DEAE-agarose chromatography was separated immunochemically
into two fractions; one was phytochrome which bound to mAP3
(phytochrome I) and the other was phytochrome which did not
bind to mAP3 (phytochrome II). This observation seemed to
confirm the preliminary reports on antigenically distinct
phytochrome in crude extracts of light-grown oat tissue
(Shimazaki et al., 1983; Thomas et al., 1984a). Furthermore,
phytochrome II, which did not bind to mAP3, could be purified
by mAP10-agarose chromatography to 65% purity (Fig. 4).

Proteolytic fragmentation of phytochrome I purified by
mAP3 from both etiolated and light-grown pea tissue was
carried out using Staphylococcus aureus V8 protease under
denatured conditions (Cleveland et al., 1977). Phytochrome
fragments generated by the protease during SDS-PAGE were
detected by western blotting with polyclonal anti-pea phytochrome antibodies against etiolated pea phytochrome (Abe et
al., 1985). The digestion pattern of phytochrome I purified
by mAP3 from both etiolated and light-grown pea tissue was
essentially the same. However, the digestion pattern of
phytochrome II from light-grown pea tissue, which did not bind
to mAP3 but bound to mAP10, was different from that of phytochrome I purified by mAP3 (Abe et al., 1985). Similar results
obtained from comparative peptide maps of green- and
etiolated-oat phytochrome were reported by Tokuhisa et al.
(1985) and Cordonnier et al. (1986b).

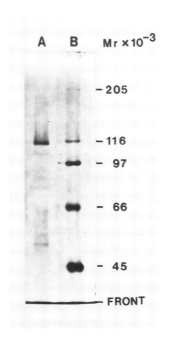

FIGURE 4.   SDS-PAGE of immunoaffinity purified phytochrome II from light-grown pea tissue. About 100 ng of phytochrome II (lane A) purified by mAP10-conjugated agarose column was electrophoresed on 7.5% acrylamide gel and stained by silver staining.   Lane B contains molecular weight standards, the sizes of which are shown on the right.

Next, reactivities of the mAPs with phytochrome II were checked by western blotting or monoclonal antibody-conjugated column chromatography. As a result, those mAPs were grouped into two types; Type A reacts only with phytochrome I and Type B reacts with both. Among 14 mAPs raised against phytochrome I, only three were Type B antibodies (Fig. 1). Two of those, namely mAP5 and mAP20, were mapped on the non-chromophoric C-terminal half of phytochrome I and another one, mAP10, was mapped on the 5 kDa N-terminus of phytochrome I. In oat, a monoclonal antibody which was specific to the non-chromophoric C-terminal half of etiolated-oat phytochrome also recognized green-oat phytochrome (Tokuhisa et al., 1985; Cordonnier et al., 1986b).

These findings suggest that light-grown green pea and oat tissue contain at least two molecular species of phytochrome, phytochrome I and II, which are distinct from each other in terms of primary structure of the phytochrome peptide. A few common epitopes exist on the phytochrome I and II molecules, which indicates that the two phytochromes share a few common amino acid sequences.

B. Detection of Phytochrome II in Crude Extracts of Tissue

As described above, antigenically distinct phytochromes were found to exist in light-grown oat and pea tissue. However, the content of each phytochrome in crude extracts of the tissues has not been estimated. We first attempted to establish an immunochemical method to distinguish phytochrome I and II by using mAPs raised against phytochrome I, because monoclonal antibodies against phytochrome II had not then been produced.

In brief, to separate phytochrome II from a crude plant extract in which both molecules presumably exist, the crude extract was passed through an mAP(Type A)-conjugated agarose column to exclude phytochrome I. Subsequently, the non-bound fraction was applied to a Type B-agarose column to collect phytochrome II. Each of the phytochrome molecules bound to the relevant mAP-agarose was eluted by SDS-containing buffer and detected by western blotting. To detect phytochrome I by ELISA, the test sample was assayed by mAP(Type A) which did not react with phytochrome II, while to detect phytochrome II the non-bound fraction from the Type A-agarose column was assayed by mAP(Type B).

Using the methods described above, phytochrome I and II were examined in pea epicotyl tissue (Abe et al., in preparation) and pea embryonic axis (Konomi et al., in preparation).

In pea epicotyl tissue grown for 5 days in the dark, the changes in the content of both phytochrome I and II content in crude extract were roughly estimated as a function of time before and after the onset of continuous white light irradiation. The existence of phytochrome II was recognized not only in light-grown tissue but also in etiolated tissue. The level of phytochrome II was almost the same before and after the light irradiation, but the level of phytochrome I which accumulated in the etiolated tissue, was drastically decreased by the light irradiation.

In the embryonic axes imbibed in the dark, phytochrome II was also detected by ELISA. Its content increased during imbibition, and reached one quarter of that of phytochrome I 12 hr after the start of imbibition. In the light, the phytochrome I increase was reduced, while the content of phytochrome II was unaffected.

These results indicate that in pea epicotyl and embryonic axis synthesis of phytochrome II is unlikely to be under the control of light, whereas phytochrome I might be light-regulated (Otto et al., 1984).

C. Phytochrome II Specific Monoclonal Antibodies

Specific probes for the molecule are essential to study
its molecular properties and function. Some laboratories have
already succeeded in producing monoclonal antibodies against
phytochrome purified from several species of etiolated plants
(Cordonnier et al., 1983; Daniels and Quail 1984; Nagatani et
al., 1983, 1984; Thomas et al., 1984b; Whitelam et al., 1985).
We have recently succeeded in producing five monoclonal
antibodies to pea phytochrome II (Abe et al., in preparation).
These monoclonal antibodies were obtained by fusing NS-1
myeloma cells with spleen cells from a BALB/c mouse immunized
with purified pea phytochrome II described above. Screening
of antibody producing cells was performed by a dot-immunoblotting and western blotting. All the antibodies recognised
pea phytochrome II by western blotting but did not recognise
pea phytochrome I.
To elucidate the nature of phytochrome in light-grown
tissue, monoclonal antibodies are a useful and necessary
specific tool. Phytochrome II specific monoclonal antibodies
will be available for detection, quantification and purification of the phytochrome II molecule, isolation of cDNA for
phytochrome II and so on.

VII.   APPLICATION OF ANTIBODIES IN THE MOLECULAR
       BIOLOGY OF PHYTOCHROME

It has been well-established that molecular biological
techniques are one of the most potent tools for various kinds
of biological studies. In fact, some important findings have
already been made in the field of phytochrome study, although
the application of molecular biological techniques has only
recently started (for details, see Chapter 2b). The importance of specific antibodies as a probe in this field cannot
be overestimated, since at various stages of molecular
biological methods most protein can only be detected or
identified easily immunochemically.
In this section we describe applications of anti-phytochrome antibodies in this field, and also refer to some recent
findings.

A. Detection of Translatable mRNA for Phytochrome

Until recently, the appearance and accumulation of phytochrome in plant tissue had been measured exclusively by
spectrophotometry. However, the recent introduction of

immunochemistry and molecular biology to phytochrome studies has made it possible to determine the content not only of spectrophotometrically active phytochrome but also of mRNA for phytochrome.

With the aid of specific antibody to the phytochrome, Gottmann and Schäfer (1982) reported detection of translatable phytochrome mRNA. Two groups then reported that the level of translatable mRNA for phytochrome was regulated by phytochrome itself (Gottmann and Schafer, 1983; Colbert et al., 1983). Both extracted and purified RNA from etiolated oat seedlings, performed in vitro translation with the extracted RNA and analysed the translated products by immunoprecipitation with polyclonal anti-phytochrome antibodies. Their results clearly suggested that brief irradiation of etiolated oat seedlings with red light lowered the level of translatable mRNA for phytochrome. This was also confirmed in etiolated pea epicotyls using monoclonal antibody to detect the translation product (Otto et al., 1984)

Inhibition of accumulation of the mRNA by light was also reported in embryonic axes of pea seeds using similar procedures (Sato and Furuya, 1985). By comparing their data with phytochrome protein accumulation (Konomi et al., 1985), the authors suggested that, in embryonic axes, light regulates the level of phytochrome not only via accumulation of the mRNA but also at steps involved in the synthesis and turnover of the protein.

B.   Isolation of cDNA Using an Expression Vector

One of the most promising ways of studying the primary structure of phytochrome and for quantifying phytochrome mRNA is the use of cDNA. Furthermore, to elucidate the regulation of phytochrome synthesis itself, it is necessary to analyze genomic sequences for phytochrome.

The first report of the isolation of cDNA for phytochrome was by Hershey et al. (1984). The isolation procedures included two steps which could not have been done without specific antibody. First, the authors prepared a RNA fraction enriched in oat phytochrome mRNA. This was done by separating poly(A) RNA according to size and testing the fractions by in vitro translation and immunoprecipitation. Next, the clones which preferentially hybridized to phytochrome-enriched cDNA probes were tested by hybridization-selection and in vitro translation. Specific antibody plays a key role in such techniques for identifying the translation product.

More recently, Tomizawa et al. (1986) reported the isolation of cDNA for pea phytochrome, using specific antibodies more extensively. The authors immunoprecipitated polysomes to enrich phytochrome mRNA and screened a cDNA library, which was constructed using an E. coli expression vector, by colony immunological assay. They also immunoprecipitated in vitro translation products in a hybridization arrest translation assay.

As demonstrated above, specific antibody is vital at some stages in cDNA cloning, since it is often the one and only probe for identifying the sequences. When a specific antibody is available, recent progress in the field indicates that cDNA cloning for most proteins can be achieved.

VIII. CONCLUSION

This chapter has been concerned with recent applications of immunological techniques to the study of phytochrome. Specific antibodies have played and will continue to play an important role in a wide range of phytochrome studies from very biochemical work such as pinpointing active sites on the molecule to very physiological approaches such as quantitation of phytochrome content in tissue. Further, techniques which have not yet been fully applied to the study of phytochrome but which seem very promising include differential staining of Pr and Pfr phytochrome by immunocytochemistry, introduction and expression of monoclonal antibody genes in plants by gene technology, and so on.

ACKNOWLEDGEMENTS

We thank Miss M. Kato for preparing monoclonal antibodies, and also Miss Y. Tsuge for her technical assistance. We thank professor Masaki Furuya for his valuable discussion. A.N. and P.J.L. were supported by postdoctoral fellowships from the Japanese Society for the Promotion of Science.

REFERENCES

Abe, H., Yamamoto, K. T., Nagatani, A., and Furuya, M. (1985). Plant Cell Physiol. 26:1387.
Abe. H., Konomi, K., and Furuya, M. in preparation.
Atassi, M. Z. (1975). Immunochemistry 12:423.
Barlow, D. J., Edwarks, M. S., and Thornton, J. M. (1986). Nature 322:747.

Butler, W. L., Norris, K. H., Siegelman, H. W., and Hendricks, S. B. (1959). Proc. Natl. Acad. Sci. U.S.A. 45:1703.

Chai, Y. G., Song, P.-S., Cordonnier, M.-M., and Pratt, L. H. (1986). Abstract of the XVI Yamada Conference p. 66.

Cleveland, D. W., Fischer, S. G., Kirschner, M. W., and Laemmli, U. K. (1977). J. Biol. Chem. 252:1102.

Colbert, J. T., Hershey, H. P., and Quail., P. H. (1983). Proc. Natl. Acad. Sci. U.S.A. 80:2248.

Cordonnier, M.-M., and Pratt, L. H. (1982). Plant Physiol. 70:912.

Cordonnier, M.-M., Smith, C., Greppin, H., and Pratt, L. H. (1983). Planta 158:369.

Cordonnier, M.-M., Greppin, H., and Pratt, L. H. (1984). Plant Physiol. 74:123.

Cordonnier, M.-M., Greppin, H., and Pratt, L. H. (1985). Biochem. 24:3246.

Cordonnier, M.-M., Greppin, H., and Pratt, L. H. (1986a). Plant Physiol. 80:982.

Cordonnier, M.-M., Greppin, H., and Pratt, L. H. (1986b). Biochem. 25:7657.

Daniels, S. M., and Quail, P. H. (1984). Plant Physiol. 76:622.

Gardner, G., and Gorton, H. L. (1985). Plant Physiol. 77:540.

Giles, K. L., and von Maltzahn, K. E. (1968). Can. J. Bot. 46:305.

Goding, J. W. (1980). J. Immunol. Methods 39:285.

Gottmann, K., and Schafer, E. (1982). Photochem. Photobiol. 35:521.

Gottmann, K., and Schäfer, E. (1983). Planta 157:392.

Hawkes, R., Niday, E., and Gordon, J. (1982). Analytical Biochem. 119:142.

Hershey, H. P., Colbert, J. T., Lissemore, J. L., Baker, R. F., and Quail., P. H. (1984). Proc. Natl. Acad. Sci. U.S.A. 81:2332.

Hilton, J. R., and Thomas, B. (1985). J. Experimental Bot. 36:1937.

Holdsworth, M. L., and Whitelam, G. C. (1986). Abstract of the XVI Yamada Conference p. 65.

Hopkins, B. W., and Butler, W. L. (1970). Plant Physiol. 45:567.

Hunt, R. E., and Pratt, L. H. (1979). Plant Physiol. 64:327.

Hunt, R. E., and Pratt, L. H. (1980). Plant Cell Environ. 3:91.

Jordan, B. R., Partis, M. D., and Thomas, B. (1984). Physiol. Planta. 60:416.

Jones, A. N., Allen, C. D., Gardner, G., and Quail, P. H. (1986). Plant Physiol. 81:1014.

Köhler, G. and Milstein, C. (1975). Nature 257:495.

Konomi, K., Nagatani, A., Furuya, M. (1985). Photochem.
    Photobiol. 42:649.
Konomi, K., and Furuya, M. (1986). Plant Cell Physiol.
    27:1507.
Konomi, K., Abe, H., and Furuya, M. in preparation.
Lane, H. C., Siegelman, H. W., Butler, W. L., and Firer, E. M.
    (1963). Plant Physiol. 38:414.
Lumsden, P. J., Yamamoto, K. T., Nagatani, A., and Furuya, M.
    (1985). Plant Cell Physiol. 26:1313.
Lumsden, P. J., and Furuya, M. (1986). Plant Cell Physiol.
    27:1541.
Lumsden, P. J., Saji, H., and Furuya, M. (1987). Plant Cell
    Physiol. submitted.
Nagatani, A., Yamamoto, K. T., Furuya, M., Fukumoto, T., and
    Yamashita, A. (1983). Plant Cell Physiol. 24:1143.
Nagatani, A., Yamamoto, K. T., Furuya, M., Fukumoto, T., and
    Yamashita, A. (1984). Plant Cell Physiol. 25:1059.
Otto, V., Schäfer, E., Nagatani, A., Yamamoto, K. T., and
    Furuya, M. (1984). Plant Cell Physiol. 25:1579.
Pratt, L. H. (1978). Photochem. Photobiol. 27:81.
Pratt, L. H. (1982). Ann. Rev. Plant Physiol. 33:557.
Sato, N., and Furuya, M (1985). Plant Cell Physiol. 26:1511.
Saji, H., Furuya, M., and Takimoto, A. (1982). Plant Cell
    Physiol. 23:623.
Saji, H., Nagatani, A., Yamamoto, K. T., Furuya, M., Fukumoto,
    T., and Yamashita, A. (1984). Plant Sci. Lett. 37:57.
Shimazaki, Y., Moriyasu, Y., Pratt, L. H., and Furuya, M.
    (1981). Plant Cell Physiol. 22:1165.
Shimazaki, Y., Cordonnier, M.-M., and Pratt, L. H. (1983).
    Planta 159:534.
Shimazaki, Y., Cordonnier, M.-M., and Pratt, L. H. (1986).
    Plant Physiol. 82:109.
Taylor, A. O. and Bonner, B. A. (1967). Plant Physiol. 42:762.
Thomas, B., Crook, N. E., and Penn, S. E. (1984a). Physiol.
    Planta. 60:409.
Thomas, B., Penn, S. E., Butcher, G. W., and Galfre, G.
    (1984b). Planta 160:382.
Thomas, B. and Penn, S. E. (1986). Planta 159:534.
Tokuhisa, J. G., Daniels, S. M., and Quail, P. H. (1985).
    Planta 164:321.
Tomizawa, K., Komeda, Y., Sato, N., Nagatani, A., Iino, T.,
    and Furuya, M. (1986). Plant Cell Physiol. 27:1101.
Vierstra, R. D., and Quail, P. H. (1982). Planta 156:158.
Whitelam, G. C., Anderson, M. L., Billett, E. E., and Smith
    (1985). Photochem. Photobiol. 42:793.
Wong, Y.-S., Cheng, H.-C., Walsh, D. A., and Lagarias, J. C.
    (1986). J. Biol. Chem. 261:12089.
Yamamoto, K. T., and Furuya, M. (1983). Plant Cell Physiol.
    24:713.

III. PROBLEMS AND PROSPECTS IN SPECTROPHOTO-
METRICAL AND BIOPHYSICAL APPROACHES

Phototransformation pathway
of "native" pea phytochrome

Yasunori Inoue

Department of Botany
Faculty of Science
University of Tokyo
Hongo, Tokyo, Japan

INTRODUCTION

The phototransformation pathway of phytochrome at a physiological temperature was determined in isolated "small" oat
(Linschitz at al. 1966, Linschitz and Kasche 1967, Pratt and
Butler 1970, Braslavsky et al. 1980), "large" oat (Cordonnier
at al. 1981, Pratt et al. 1982, Pratt et al. 1984) and "large"
pea (Shimazaki et al. 1980, Cordonnier et al. 1981, Inoue et
al. 1982) phytochromes using the flash photolysis technique.
Recent works have shown that many physico-chemical characteristics of "native" phytochrome differ from those of the
"large" phytochrome (see Quail et al. 1983). The phototransformation pathway of "native" phytochrome has not been completely clarified. Ruzsicska and his co-workers (1985) determined decay of the first intermediate from Pr to Pfr in both
"small" and "native" oat phytochrome by flash photolysis and
reported that the kinetic parameters of both species were
similar. On the other hand, Eilfeld and Rüdiger(1985) reported that a bleached intermediate (I_{bl}) observed in the
phototransformation pathway of "small" (Linschitz et al. 1966)
and "large" (Pratt et al. 1982) oat phytochrome was normaly
not formed in the phototransformation of "native" oat phytochrome determined by low temperature spectroscopy. Therefore,
the entire phototransformation pathway of "native" phytochrome
should be analyzed at physiological temperature.

The phototransformation pathway of pea phytochrome has
also been determined *in vivo*(Inoue and Furuya 1985). It
should be interesting to compare of the phototransformation
pathway of isolated "native" pea phytochrome with those of

PHYTOCHROME AND
PHOTOREGULATION IN PLANTS

117

isolated "large" pea phytochrome and phytochrome *in vivo* to
clearly identify the controlling factor of the phototransformation pathway. The monomer molecular weight of isolated
"native" phytochrome is equal to that of phytochrome *in vivo*
but the micro-environment of isolated "native" phytochrome is
rather similar to that of isolated "large" phytochrome.

In the present study, the phototransformation pathway of
isolated "native" pea phytochrome was determined at 2°C using
the flash photolysis technique, and the kinetic parameters
were compared with those of isolated "large" pea phytochrome
and phytochrome *in vivo* at the same temperature. Finally, the
effects of pH on the kinetics of phototransformation were
determined.

A. Phototransformation pathway of isolated "native" pea phytochrome.

The phototransformation pathway of isolated "native" pea
phytochrome from Pr to Pfr was determined with a transient
spectrum analyzer (Furuya et al. 1984) and compared with those
found for isolated "large" pea phytochrome (Shimazaki et al.
1980, Cordonnier et al. 1981) and phytochrome in pea epicotyl
tissue (Inoue and Furuya 1985).

Phytochrome was isolated from the dark-grown pea tissue as
Pfr (Vierstra and Quail 1983). Procedures for isolating
phytochrome have been described in the previous paper (Lumsden
et al. 1985). Immunoblot analysis of the phytochrome sample
confirmed the nativeness of the polypeptide size (see Lumsden
et al. 1985). The absorption spectrum of photoequilibrium
mixture of Pr and Pfr under the red light showed absorption
maximum at 730 nm. Absorption change ratio between red and
far-red region reached 1.07. Procedures for measuring the
transient spectra changes have been described in previous
papers (Shimazaki et al. 1980, Inoue et al. 1982, Pratt et al.
1982, 1984, Inoue and Furuya 1985). The samples were always
kept at 2°C.

***Relationship between incident energy of a laser flash and
the amount of produced Pfr***   The fluence response curve of a
single red laser flash at 640 or 650 nm on the amount of flash
induced Pfr showed a plateau at about 30 mJ/pulse irrespective
of the flash wavelength. This fluence level is equal to that
of isolated "large" pea phytochrome (Shimazaki et al. 1980)
but about two times higher than that determined in phytochrome
in vivo (Inoue and Furuya 1985). The amount of Pfr induced by
a saturation level of the red flash was about half of that
attained at the photostationary equilibrium with continuous
red light irradiation. This level is equal to that in pea
epicotyl (Inoue and Furuya 1985), but ca. 1.6 times higher

than that in isolated "large" pea phytochrome (Shimazaki et al. 1980). This result suggested that in "native" phytochrome, the quantum efficiency of phototransformation from Pr to the first intermediate (I_{692}) or from I_{692} to Pr is different from that of large phytochrome.

Flash-induced transient spectra The difference spectrum obtained by subtracting the original absorption spectrum of Pr from the spectrum measured at 5 µs after a red flash irradiation on Pr showed an absorbance increase at about 693 nm (Fig. 1). The difference spectrum determined at 2.5 ms after the flash had a low positive peak at about 700 nm and showed an eminent absorption decrease at about 670 nm (Fig. 1). These flash-induced absorption spectra changes were principally equal to those determined in "large" pea phytochrome (Shimazaki et al. 1980) and phytochrome in pea epicotyl (Inoue and Furuya 1985). The difference spectrum at 2.5 ms after the flash irradiation clearly suggested the formation of an intermediate that had a relatively low extinction coefficient (I_{bl}).

Kinetics of the first intermediate decay Absorbance in the red region indicated by the difference spectrum at 5 µs after a flash irradiation in Fig. 1 decreased rapidly. The

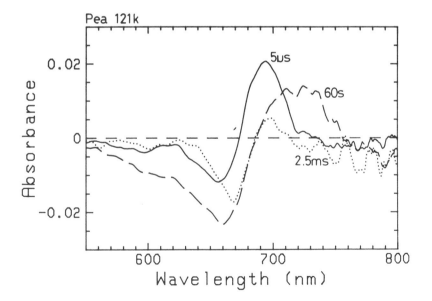

FIGURE 1. Difference spectra between flash-induced intermediates of isolated "native" pea phytochrome determined at 5 µs(——), 2.5 ms(······) and 60 s(— — —) after a red laser flash irradiation on Pr and the original Pr.

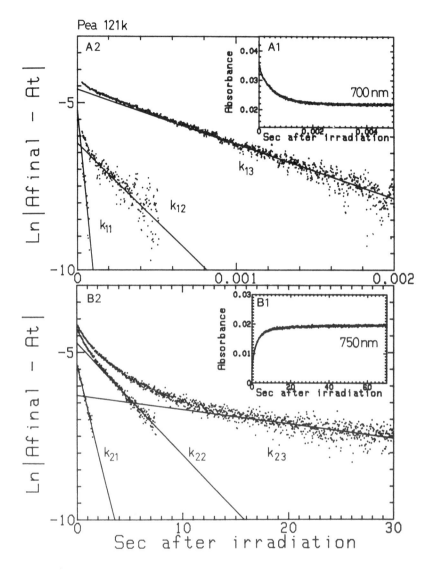

FIGURE 2. Laser-flash-induced absorbance change in
isolated "native" pea phytochrome. (A1): Rapid absorbance
decrease at 700 nm. (A2): A natural logarithm of the absolute
value of the difference between absorbance at 4 ms after a
flash and each data point in A1(upper dotted line). (k_{11-13}):
Resolved three first-order reaction components. (B1): Slow
absorbance increase at 750 nm. (B2): Natural logarithm of
the absolute value of the difference between absorbance at 60 s
after a flash and each data point in B1(upper dotted line).
(k_{21-23}): Resolved three first-order reaction components.

rapid absorbance decay ceased within about 2 ms after the
flash irradiation (Fig. 2A1). The natural-log plot obtained
by subtracting the absorbance value at completion of the
reaction from that at each time point in the data in Fig. 2A1
showed a complex curve (dotted line close to a solid line k_{13}
in Fig. 2A2). Assuming that this curve was the sum of independent parallel reactions, it was resolved into three first-
order components (Solid lines, k_{11}, k_{12} and k_{13} in Fig. 2A2).
Rate constants for the fast, moderate and slow processes were
42000, 4100 and 1500 s^{-1}, respectively. Extrapolation of
solid lines k_{11}, k_{12} and k_{13} in Fig. 2A2 to zero time showed
about 26, 20 and 54% of the absorbance changes at 700 nm were
due to the fast, moderate and slow processes, respectively
(Table 1).

   Kinetics of Pfr appearance Slow absorbance increase in
the far-red region which probably corresponded to the formation of Pfr continued about 60 s after the excitation of Pr
with a red flash (Fig. 2B1). A natural-log plot obtained by
subtracting the absorbance value at each time point in the
data in Fig. 2B1 from that at completion of the reaction
showed also a complex curve (dotted line closed to a solid
line k_{23} in Fig. 2B2). Assuming that this curve was the sum
of independent parallel reactions, it was resolved into three
first-order components (Solid lines k_{21}, k_{22} and k_{23} in

TABLE 1. Comparison of kinetic data for the reaction
monitored during the phototransformation of Pr to Pfr at ca.
2°C in pea phytochrome.

Reaction	Rate constants(% amount) (s^{-1})			
	"large"[a]	"large"[b]	"native"[c]	Tissue[d]
Decay of the first intermediate				
k_{11}	46000(25)	46000(7)	42000(26)	-----
k_{12}	2500(75)	2500(59)	4100(20)	2570(100)
k_{13}	---[e]	430(34)	1500(54)	----
Decay of the second intermediate				
k_{21}	2.3 (30)	4.1 (31)	2.4 (14)	5.1 (37)
k_{22}	0.4 (35)	0.9 (69)	0.4 (69)	1.8 (50)
k_{23}	0.07(34)	---[e]	0.08(17)	0.36(13)

[a]Shimazaki et al. 1980 & Shimazaki, Y., Doctoral thesis.
[b]Cordnnier et al. 1981.
[c]Present work.
[d]Inoue and Furuya 1985.
[e]not determined.

Fig. 2B2). Rate constants for the fast, moderate and slow
processes were 2.4, 0.4 and 0.08 s^{-1}, respectively (Table 1).
Extrapolation of solid lines k_{21}, k_{22} and k_{23} in Fig. 2B2 to
zero time showed that about 14, 69 and 17% of the absorbance
change at 750 nm were due to the fast, moderate and slow
processes, respectively (Table 1).

*Comparison of kinetic data between "large", "native" and
in vivo phytochrome* Currently determined kinetic data for
"native" pea phytochrome were compared with those determined
in "large" pea phytochrome (Shimazaki et al. 1980, Shimazaki
Doctral Thesis 1981, Cordonnier et al. 1981) and phytochrome
in pea epicotyl (Inoue & Furuya 1985).

Rate constants of the decay of the first and the second
intermediates determined for "native" phytochrome were similar
to those determined for "large" phytochrome (Table 1). In
particular, rate constants of the second intermediate decay to
Pfr in "native" phytochrome showed good coincidence with those
determined for the "large" phytochrome by Shimazaki et al.
1980. On the other hand, the rate constants determined for
phytochrome in pea epicotyl differed slightly from those observed for isolated phytochrome in spite of the molecualr
weights of the monomers being the same.

B. Effect of pH on the phototransformation
of phytochrome.

Comparison of kinetic data from isolated "large" and
"native" pea phytochrome and phytochrome in pea tissue suggested that the micro-environment of phytochrome molecules had
some effects on the phototransformation of phytochrome. Buffer conditions of isolated phytochrome samples used in previous flash photolysis studies were similar; the pH was always
7.8 (Linschitz et al. 1966, Linschitz and Kasche 1967, Pratt
and Butler 1970, Shimazaki et al. 1980, Braslavsky et al.
1980, Cordonnier et al. 1981, Inoue et al. 1982, Pratt et al.
1982 1984, Ruzsicska et al. 1985). Recently, we have found
that absorption spectra of isolated "native" pea phytochrome
measured under continuous red-light irradiation and 55 s after
the irradiation were affected by the pH condition (Tokutomi et
al. 1986). Therefore, we tried to determine the effect of pH
on the kinetics of phototransformation of isolated "native"
pea phytochrome from Pr to Pfr.

Phytochrome samples were dissolved in 10 mM potassium
phosphate(pH 6.8 or 7.8) or potassium borate (pH 8.8) buffer
containing 0.1 mM Na_2EDTA. The sample temperature was increased from 2°C to 10°C. At basic pH, the reaction speed of
the second intermediate decay at 2°C decreased too slowly to
observe the entire reaction within the limit of the instrument

TABLE 2.   Effect of pH on kinetic data for the reaction monitored during the phototransformation of Pr to Pfr at $10^{\circ}C$ in isolated "native" pea phytochrome.

pH	Rate constants(% amount) (s^{-1})		
	k_1	k_2	k_3
Decay of the first intermediate			
6.8	------	13600(6)	4200(94)
7.8	------	13000(14)	3900(86)
8.8	------	9000(26)	3700(74)
Decay of the second intermediate			
6.8	40(11)	5.7(56)	1.5 (33)
7.8	32(19)	5.5(33)	1.4 (48)
8.8	12(45)	1.7(8)	0.38(47)

used(100 s).

Kinetics of the first intermediate decay   Higher temperatures induced higher rate constants, with the fast component decaying too fast to determine the correct rate constant at any pH. Rate constants of the moderate and the slow components were about three times higher than the corresponding rate constants determined at $2^{\circ}C$. Rate constants of both the moderate and slow components were higher under acidic condition, but the difference was not very high (Table 2 upper row). In other words, difference in pH had little effect on the decay of the first intermediate.

Kinetics of the second intermediate decay   Rate constants of the second intermediate decay were strongly affected by pH. At neutral and acidic pH, rate constants of the corresponding components were similar. At basic pH, rate constants of all three components were reduced to ca. 1/3 of the corresponding value found at neutral pH (Table 2, lower row). The amount of the moderate speed reaction component decreased when the pH shifted from acidic to basic, and the amount of the fast speed component increased, and vice versa.

DISCUSSION

Oat phytochromes, have many physicochemical characteristics which differ between "native" and "large" molecules (Quail et al. 1983). In the case of phototransformation, kinetic data determined in "small" (Linschitz et al. 1966,

Linschitz and Kasche 1967, Braslavsky et al. 1980), "large"
(Cordonnier et al. 1981, Pratt et al. 1982) and "native"
(Ruzsicska et al. 1985) isolated oat phytochromes suggested a
parallel reaction pathway from Pr to Pfr. Rate constants of
the first intermediate decay were similar irrespective of the
differences in monomer molecular weights. Eilfeld and
Rüdiger (1985) recently reported that in "native" oat phytochrome, the intermediate having a relatively low extinction
coefficient was not formed when the phototransformation pathway was determined by low-temperature spectroscopy. But the
bleached intermediate was clearly formed at physiological
temperature, as reported in the Round Table Discussion I of
this conference by Inoue and Rüdiger.

In pea phytochrome, kinetic data for "large" (Shimazaki et
al. 1980, Cordonnier at al. 1981), "native" (Fig. 2) phytochrome and phytochrome *in vivo* (Inoue and Furuya 1985) also
suggested a parallel reaction pathway from Pr to Pfr. Comparison of kinetic data for "large" and "native" isolated pea
phytochromes in Table 1 showed good coincidence in the rate
constants of not only the first but also the second intermediate decay. Absorption spectra changes induced by a red
flash irradiation on Pr were also similar between "large"
(Shimazaki et al. 1980) and "native" (Fig. 1) isolated phytochrome. These results suggested that the photochromic nature
of phytochrome molecules was principally determined by the
apoprotein moiety of the chromophore domain, and the difference in monomer molecular weight had little effect on the
phototransformation of the phytochrome.

Kinetic data of phytochrome phototransformation for pea
epicotyl tissue (Inoue and Furuya 1985) suggested that some
differences in the micro-environment of phytochrome molecules
were rather influential on the phototransformation of the
phytochrome. This assumption was confirmed by data determined
at different pH (Table 2),which showed that pH differences had
little effect on the kinetics of the first intermediate decay,
but had an obvious effect on that of the second intermediate
decay. Proton uptake and release occurred during phototransformation of "large" pea phytochrome (Tokutomi et al. 1982).
At basic pH, absorption of "native" pea phytochrome in the
far-red region was kept at a relatively low level under the
continuous red light (Tokutomi et al. 1986). These results
suggested that under the continuous red light, the second
phototransformation intermediate having a relatively low extinction coefficient accumulated at basic pH. Slow rate constants of the second intermediate decay at basic pH in Table 2
could explain the accumulation of the second intermediate
under the continuous red light. These data suggested that the
proton uptake reported in "large" pea phytochrome (Tokutomi et
al. 1982) would occur in the decay of the second intermediate

to Pfr.

Dichroic orientation of the phytochrome molecule in fern protonema also changed in the final step of phototransformation (Kadota et al. 1986). These results suggested that physiologically important events occurred in the final step of phototransformation, namely, in the decay of I_{b1} to Pfr.

At basic pH, the rate constants of all three reaction components in the decay of the second intermediate were reduced to ca. 1/3 of the corresponding rate constants at neutral pH irrespective of the kind of reaction component (Table 2). These results also supported the parallel reaction hypothesis. In this case, the amount of the moderate speed reaction component decreased when the pH shifted to basic. This finding suggested that the parallel reaction pathway of phytochrome phototransformation was not brought by a rigid difference in the phytochrome molecule, such as about a difference of monomer molecular weight, but rather by some flexible difference, such as the state of association.

ACKNOWLEDGMENTS

This work was made possible by collaboration with Dr. K. T. Yamamoto, Dr. S. Tokutomi and Professor M. Furuya. The author would like to thank them for their thorough and painstaking efforts and fruitful advice during this work.

REFERENCES

Braslavsky, S. E., Matthews,J. I., Herbert, H. J., de Kok, J., Spruit, C. J. P. and Schaffner, K. (1980). Photochem. Photobiol. 31:417.
Cordonnier, M. -M., Mathis, P. and Pratt, L. H. (1981). Photochem. Photobiol. 34:733.
Eilfeld, P. and Rüdiger, W. (1985). Z. Naturforsch. 40c:109.
Furuya, M., Inoue, Y. and Maeda, Y. (1984). Photochem. Photobiol. 40:771.
Inoue, Y. and Furuya, M. (1985). Plant Cell Physiol. 26:813.
Inoue, Y., Konomi, K. and Furuya, M. (1982). Plant Cell Physiol. 23:731.
Kadota, A., Inoue, Y. and Furuya, M. (1986). Plant Cell Physiol. 27:867.
Linschitz, H. and Kasche, V. (1967). Proc. Natl. Acad. Sci. USA 58:1059.
Linschitz, H., Kasche, V., Butler, W. L. and Siegelman, H. W. (1966). J. Biol. Chem. 241:3395.

Lumsden, P. J., Yamamoto, K. T., Nagatani, A. and Furuya, M. (1985). Plant Cell Physiol. 26:1313.

Pratt, L. H. and Butler, W. L. (1970). Photochem. Photobiol. 11:361.

Pratt, L. H., Inoue, Y. and Furuya, M. (1984). Photochem. Photobiol. 39:241.

Pratt, L. H., Shimazaki, Y., Inoue, Y. and Furuya, M. (1982) Photochem. Photobiol. 36:471.

Quail, P. H., Colbert, J. T., Hershey, H. P. and Vierstra, R. D., (1983). Phil. Trans. R. Soc. Lond. B303:387.

Ruzsicska, B. P., Braslavsky, S. E. and Schaffner, K. (1985). Photochem. Photobiol. 41:681.

Tokutomi, S., Yamamoto, K. T., Miyoshi, Y. and Furuya, M. (1982). Photochem. Photobiol. 35:431.

Tokutomi, S., Inoue, Y., Sato, N., Yamamoto, K. T. and Furuya, M. (1986). Plant Cell Physiol. 27:765.

Vierstra, R. D. and Quail, P. H. (1983). Biochemistry 22:2498.

BIOCHEMISTRY OF THE PHYTOCHROME CHROMOPHORE

Wolfhart Rüdiger

Botanisches Institut, Universität München

I. INTRODUCTION

The transduction chain from light absorption to physiological responses in photomorphogenesis starts at the chromophore of the photoreceptor. In the case of phytochrome, the
tetrapyrrole chromophore is responsible for light absorption.
Physical and chemical events which occur in the chromophore
after light absorption can be considered as first steps of
the transduction chain. Subsequent steps also involve the
protein part of phytochrome. Chromophore-protein interactions
are therefore essential for the function of phytochrome.

II. CHROMOPHORE STRUCTURE IN SMALL CHROMOPEPTIDES

Properties of the tetrapyrrole chromophore can be modified
by the native protein in manifold ways. In order to recognize
the properties of the undisturbed chromophore, non-covalent
interactions between chromophore and protein have to be removed e.g. by denaturation or proteolysis. The former method
is fast but leaves still the full-length of the peptide chain
which is covalently bound to the chromophore. The second method is somewhat slower but allows removal of most of the
peptide chain. Digestion of phytochrome with pepsin thus
leads to small chromopeptides which contain only a short peptide sequence of 8-11 amino acid covalently linked with the
chromophore.

The structure of the chromophore in such small chromopeptides was investigated by chemical and physical methods as
summarized by Rüdiger and Scheer (1983) and Rüdiger et al.

(1985). The methods include UV-Vis and NMR spectroscopy, chemical degradation into small fragments, cleavage of the covalent linkage and comparison of the protein-free chromophore
with phytochromobilin obtained by total synthesis. The structures deduced from these investigations are given in Fig. 1.
The Pr chromophore is the 15(Z) isomer whereas the Pfr chromophore is the 15(E) isomer. This difference in the chromophore structure was not detected in earlier investigations of
small chromopeptides (Fry and Mumford, 1971) because of the
light sensitivity of the Pfr chromophore: irradiation transforms the Pfr chromophore into the Pr chromophore. This reaction and other properties of the 15(E) (= Pfr) chromophore
were investigated with model chromophores the most important
of which was phycocyanobilin. This chromophore of phycocyanin
is closely related with phytochromobilin including the covalent binding to the peptide chain. The only difference consists in the side chain at C-18: it is a vinyl group in phytochromobilin but an ethyl group in phycocyanobilin (see Fig.
1). The natural 15(Z) phycocyanobilin can be converted into
the 15(E) isomer by a chemical reaction sequence. The 15(E)
isomer was used as model for the Pr chromophore whereas the
15(Z) isomer served as model for the Pr chromophore (Thümmler
and Rüdiger, 1983; Thümmler et al., 1983).

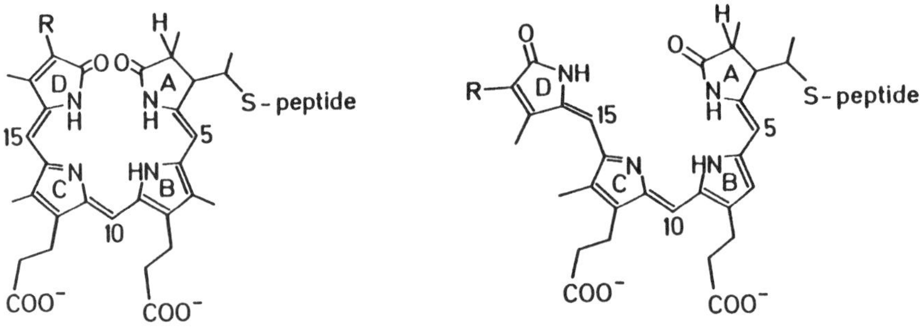

FIGURE 1. Chromophore structure as determined in small
chromopeptides. R = Ethyl: Phycocyanin, R = Vinyl: Phytochrome. Left: 15(Z) configuration = Pr, right: 15(E) configuration = Pfr.

It is well known that free bilins adopt cyclic conformations which have the shape of a flat helix (Scheer, 1981).
This is also true for the chromophores of small chromopeptides as indicated in Fig. 1. However, this does not reflect
the conformation in native phycocyanin or native phytochrome.
This question will be treated in section III.

III. CHROMOPHORE-PROTEIN INTERACTIONS

Whereas investigations on the phytochrome chromophore can
be performed with enriched phytochrome fractions, investigations on the protein part require entirely purified phytochrome preparations. We therefore developed a simple and
rapid, large-scale isolation procedure which results in 25%
overall yield of pure 124 kDa phytochrome starting with 3 kg
etiolated oat tissue (Grimm and Rüdiger, 1986). The purity
index A_{667}/A_{280} is 0.99, i.e. higher than in most previous
reports. It is sufficiently high for spectral measurements in
the UV region.

1. Chromophore Conformation in Native Pr and Phycocyanin

The exact absorption in the UV-A region (350-400 nm) is
needed for calculation of the oscillator strength of the
second chromophore absorption band. The oscillator strength
ratio f_2/f_1 varies with the conformation of bilins: it is
high for cyclic-helical conformations but low for extended
conformations.

This criterion was used in order to conclude that the
chromophores in native phycocyanin have extended conformations whereas they have predominantly cyclic-helical conformations in the denaturated state or in small chromopeptides
(Scheer, 1981). As seen in Fig. 2, this difference exists
also between Pr and the small chromopeptides derived there-
from. The native chromophore conformation has been elucidated by X-ray analysis for crystalline phycocyanin from Masti-
gocladus laminosus and from Agmenellum quadruplicatum
(Schirmer et al., 1985; 1986). It is indeed an extended but
slightly distorted conformation. It should be noted that such
distortion is the way for transformation of one conformation
into another, e.g. extended into cyclic and vice versa. With
increasing torsion angel, extended conformations approach
more and more the cyclic-helical conformations.

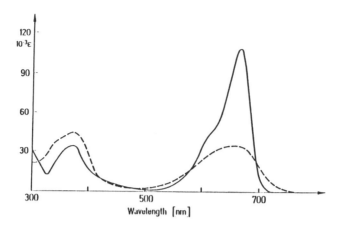

FIGURE 2. Absorption spectrum of Pr (———) and Pr
chromopeptides (---) (Eilfeld and Rüdiger, unpublished).

From the oscillator strength ratio of Pr, a semicircular
conformation for the native chromophore was deduced (Song et
al., 1979; Song, 1984). The conclusion, that internal hydrogen bonding occurs between rings A and C and leads to a
proton transfer, is however not confirmed by recent results
on the primary reaction (see below).

2. Reactivity of Pr and Phycocyanin

As indicated in section II, cis-trans (or Z-E) isomerization of the chromophore must occur in the course of the
photoreaction from Pr to Pfr. The absolute quantum yield for
the phototransformation Pr - Pfr, determined with light
pulses of seconds or minutes, is 17%. This photochemical
reactivity of Pr is not shared by small chromopeptides from
Pr (or phycocyanin) or by native phycocyanin. A high fluorescence yield (absolute quantum yield 80%) is obtained with
native phycocyanin but neither with Pr nor small chromopeptides. In the latter case, "free" chromophores (i.e. chromophores which are not perturbed or sterically constrained by any
native protein) can easily transform the excitation energy
into rotational energy. Because many ways of rotation are
possible, radiationless deexcitation must be very effective.
This is not the case in the native biliproteins. We can
assume that the phycocyanobilin chromophores are strongly
fixed in native phycocyanin; the consequence is that deexcitation occurs nearly exclusively via fluorescence. If cis-
trans isomerization is the primary photoreaction in phyto-

chrome (see section IV), the chromophore must have enough
space for this reaction. Therefore, it cannot be as strongly
fixed as in phycocyanin. On the other hand, it must be more
strongly fixed than in small chromopeptides in order to avoid
arbitrary rotations. The question whether or not bilin chromophores undergo photochemical (Z,E) isomerization is a
matter of degree of chromophore fixation according to this
view. This proposal is supported by experiments in which reversible photochemical (Z,E) isomerization of certain chromophores was observed in partially unfolded phycocyanin
(Fischer, Siebzehnrübl and Scheer, unpublished results).

3. Long-wavelength absorption of Pfr

The first absorption band of native Pfr is bathochromically shifted versus the absorption band of the small Pfr
chromopeptide (see Fig. 3). Such a dramatic shift cannot be
explained by merely conformational changes of the chromophore; it must be due to a type of chromophore-protein interaction which is specific for Pfr. When considering the nature
of this chromophore-protein interaction, the following
findings have to be considered: (1) It is not present in Pr
since the Pfr absorption band is bathochromically shifted
also if compared with that of Pr. (2) It appears only at the
last intermediate step of Pfr formation (for phytochrome
intermediates see Fig. 4). The shift is not observed in meta
Ra or meta Rb, it appears partly in meta Rc and only in full
in Pfr. The shift disappears with formation of lumi F. (3)

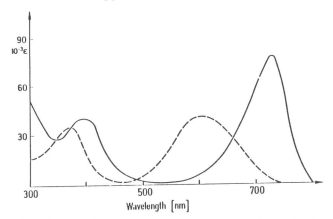

FIGURE 3. Absorption spectrum of Pfr (———), calculated
from photoequilibrium with 85% Pfr/15% Pr, and Pfr chromopeptides (---) (Eilfeld and Rüdiger, unpublished).

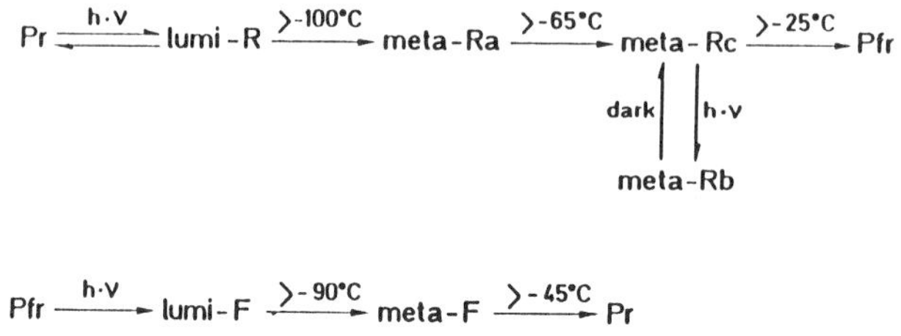

$$Pr \xrightleftharpoons{h \cdot v} lumi-R \xrightarrow{>-100°C} meta-Ra \xrightarrow{>-65°C} meta-Rc \xrightarrow{>-25°C} Pfr$$

$$dark \Big\uparrow \quad \Big\downarrow h \cdot v$$

$$meta-Rb$$

$$Pfr \xrightarrow{h \cdot v} lumi-F \xrightarrow{>-90°C} meta-F \xrightarrow{>-45°C} Pr$$

FIGURE 4. Intermediates of phytochrome photoconversions as detected by low temperature spectroscopy (after Eilfeld and Rüdiger, 1985).

The shift can be abolished by certain chemical reagents, e.g. 8-anilinonaphthalene-1-sulfonate (Eilfeld and Rüdiger, 1984). The resulting absorption spectrum of modified Pfr is very similar to that of the intermediate meta Rb. One conclusion of these observations is that the shift is not directly related to (Z,E) isomerization. It has been demonstrated, that the native intermediate meta Rb has the 15(E) configuration (Rüdiger, 1983) as well as Pfr which had been bleached by anilinonaphthalene-1-sulfonate or by other reagents (Thümmler and Rüdiger, 1984). Thus the shift can be abolished without change of the configuration. Another conclusion, namely that a certain part of the peptide chain must interact with the chromophore (Eilfeld and Rüdiger, 1984), is supported by partial proteolysis of phytochrome (Reiff et al., 1985; Grimm et al., 1986). A 39 kDa fragment which is fully photoreversible has a "normal" absorption spectrum in the Pr form but is bleached after phototransformation (see Fig. 5). Since also the normal intermediates on the pathway Pr- Pfr are detectable with this 39 kDa fragment, we conclude that the photoconversion proceeds in the normal way except for the last step, appearance of the long wavelength absorption of Pfr. Sequence analysis showed that the 39 kDa fragment extends from valine 66 to arginine-426 of the original phytochrome sequence (Grimm et al., 1986). Since a 59 kDa fragment which contains the sequence from valine-66 to arginine-596 possesses a normal bathochromically shifted absorption band in the Pfr form, the specific interaction site for the Pfr chromophore must be localized within amino acid residues 426 and 596. It is presently not known whether the interaction with the chromophore involves only one or more amino acid residue(s). Theoretical calculations predict a significant bathochromic shift in the absorption band if a point charge approaches

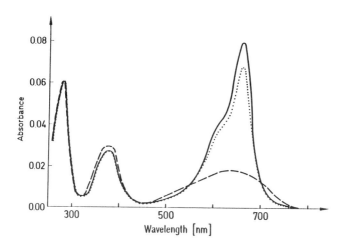

FIGURE 5.   Absorption spectrum of 39 kDa fragment of
phytochrome. ( ) Pr form, (--) bleached form after red
irradiation, (- -) photoequilibrium between Pr and bleached
form, obtained by far-red irradiation (after Reiff et al.,
1985).

certain parts of phytochromobilin, especially the 18-vinyl
group (Scharnagl, 1986). A model was proposed in which (Z,E)
isomerization leads to a movement of ring D which causes the
vinyl group to approach a point charge of the peptide chain
(Rüdiger et al., 1985). The identity of the charged group has
still to be elucidated.

    Interestingly, a chromophore-containing 23.5 kDa fragment
of phytochrome which extends from residue isoleucin-210 to
arginine-426 shows only some residual stabilization of the
"native" chromophore absorption - and hence the corresponding
conformation - in the Pr form but no stabilization whatsoever
in the Pfr form: it behaves like the small chromopeptides
(see section II) after irradiation (Grimm et al., 1986).
Since A. Jones described a 16 kDa fragment with the spectral
properties of the 39 kDa fragment during the Yamada Conference, it has to be checked which part of the peptide chain
below isoleucin-210 is contained in that fragment because
this should be the most important part for stabilization of
the Pr chromophore.

IV NATURE OF THE PRIMARY PHOTOREACTION ON
PHYTOCHROME

For effective signal transduction in photoperception, the
energy of the absorbed photon has to be chaneled effectively
to formation of the primary photoproduct. This means that the
step from the excited state of the photoreceptor to the primary photoproduct must have a low thermal activation barrier.
On the other hand, the back reaction from the primary photoproduct to the ground state of the photoreceptor must involve
to a high thermal activation barrier. It must at least be
higher than the activation barrier for relaxation to subsequent products, including the final stable photoproduct.
Otherwise the primary photoproduct would easily relax to the
ground state of the photoreceptor and in this way vaste the
energy of the absorbed photon.

We consider here Pr as the photoreceptor, lumi R as the
primary photoproduct and Pfr as the final photoproduct. The
above mentioned conditions are fulfilled here. The activation
energy for formation of lumi R from excited Pr (= Pr *) is
only 3.6 kJ/mol (Eilfeld et al., 1986), that for the next
step lumi R - meta Ra about 56 kJ/mol (Ruzsicska et al.,
1985). Although the absolute value for the thermal activation
energy of the back reaction lumi R - Pr is not known, it must
be higher than the above given values: at 2°C or higher
temperature, all lumi R decays to Pfr (Pratt et al., 1984).

An important finding was the lack of a deuterium isotope
effect for the activation energy of the reaction Pr* - lumi R
(Eilfeld et al., 1986). This is a strong argument against the
earlier proposal that the primary photoreaction of phytochrome should be a proton transfer. This proposal had been
made for two reasons (Sarkar and Song, 1981): 1. The yield of
Pfr upon red irradiation of (partially degraded) phytochrome
under steady-state conditions is lower in D_2O than in H_2O. 2.
The quantum yield of fluorescence of Pr is increased in D_2O
as compared to H_2O.

The first effect turned out to be a bleaching effect by
the solvent D_2O which is found with partially degraded, but
not with 124 kDa phytochrome (Eilfeld and Rüdiger, unpublished
results). Also quantum yield of phototransformation and
photoequilibrium of 124 kDa phytochrome are not changed upon
exchange of H_2O by D_2O (Moon et al., 1984; 1985). There is
also no deuterium isotope effect for the yield of lumi R at
77 K (Song, 1985).

The second effect which has repeatedly been discussed as argument for intramolecular proton tunneling between the tetrapyrrolic nitrogens in the excited state of the chromophore (Song, 1985) is true only for the whole integral fluorescence which consists of three kinetic components (Ruzcicska et al., 1985). The main component with the shortest life time (48 ps) which makes up 90% of total phytochrome and which shows photoreversibility (Ruzcicska et al., 1985) does not exhibit any deuterium isotope effect; the isotope effect is only found for an "impurity" with a long lifetime ($\approx$ 1 ns) which does not show photoreversibility (Brock et al., 1987). In summary, the observed deuterium isotope effects are not related to the primary photoreaction. At present, no experimental evidence for proton transfer as the primary photoreaction is available.

Another proposal for the primary photoreaction is Z,E isomerization of the chromophore (Rüdiger, 1983; Rüdiger et al., 1985). Photochemical isomerization of C,C double bands is well known. It meets the requirements state above: In the excited state, uncoupling of Π-electrons results in a facile torsion around the quasi-single bond. This step has a low activation barrier. In the ground state, torsion around the double band of either isomer has a high thermal activation barrer. It is certainly higher than the thermal activation barrier for a mere proton transfer even if Z,E isomerization in the case of a bilatriene chromophore can be facilitated by a series of tautomerization reactions (Song, 1985).

The positive evidence for Z,E isomerization so far available is the difference between the Pr chromophore (15 Z configuration) and the Pfr chromophore (15 E configuration) as analyzed in small chromopeptides (see section II). The argument that such reaction could have occurred during proteolysis can be ruled out: Proteolysis experiments of Pr and Pfr were performed under exactly identical conditions. The thermal activation barrier for this step is high (see above) but can be catalyzed by acidification. Under these conditions, only the Pfr chromophore can be transformed into the Pr chromophore but not vice versa. Therefore, the Pfr chromophore in the chromopeptide cannot be an artifact of the method of preparation.

REFERENCES

Brock, H., Ruzsicska, B.P., Arai, T., Schlamann, W.,
    Holzwarth, A.R., Braslavsky, S.E., and Schaffner, K.
    (1987). Biochemistry 26: in press.
Eilfeld, P. and Rüdiger, W. (1984). Z.Naturforsch. 39c:
    742-745.
Eilfeld, P. and Rüdiger W. (1985). Z.Naturforsch. 40c:
    109-114.
Eilfeld, P., Eilfeld, Petra, and Rüdiger, W. (1986).
    Photochem. Photobiol. 44: 761-770
Fry, K.T. and Mumford, F.E. (1971). Biochem. Biophys. Res.
    Commun. 45: 1466-1473.
Grimm, R. and Rüdiger, W. (1986). Z. Naturforsch. 41c:
    988-992..
Grimm, R., Lottspeich,F., Schneider, Hj.A.W. and Rüdiger, W.
    (1986). Z. Naturforsch. 41c: 993-1000.
Moon, D.K., Jeen, G.S. and Song, P.S. (1984). Photochem.
    Photobiol. 39: 849.
Moon, D.K., Jeen, G.S., and Song, P.S. (1985). Europ. Symp.
    Photomorphogenesis in Plants, Wageningen
Pratt, L.H., Inoue,Y., and Furuya, M. (1984). Photochem.
    Photobiol. 39: 241-246.
Reiff, U., Eilfeld, P., and Rüdiger, W. (1985). Z.
    Naturforsch. 40c: 693-698.
Rüdiger, W. (1983). Phil. Trans. Roy. Soc. London B 303:
    377-386.
Rüdiger, W. and Scheer, H. (1983). In "Photomorphogenesis"
    (W. Shropshire and H. Mohr, eds.), Encyclopedia of Plant
    Physiology New Series, Vol. 16 A, pp. 116-151, Springer,
    Berlin,Heidelberg, New York.
Rüdiger, W., Eilfeld, P., and Thümmler, F. (1985). In:
    Optical properties and structure of tetrapyrroles
    (G.Blauer, H. Sund, eds.)W. de Gruyter & Co. Berlin, pp.
    349-366
Ruzcicska, B.P., Braslavsky, S.E., and Schaffner, K.(1985).
    Photochem. Photobiol. 41: 681-688
Sarkar H.K. and Song, P.S. (1981). Biochemistrz 20: 4315-4320
Scharnagl, Chr. (1986). Dissertation Technical University
    Munich.
Scheer, H. (1981). Angew. Chem. 93: 230-250.
Schirmer, T., Bode, W., Huber, R., Sidler, W., and Zuber, H.
    (1985). J. Mol.Biol. 184: 257-277.
Schirmer, T., Huber, R., Schneider, M., Bode, W., Müller, M.,
    and Hackert, M.L. (1986). J. Mol. Biol. 188: 651-676.
Song, P.S. (1984). In "Optical Properties and Structure of
    Tetrapyrroles" (G. Blauer and H. Sund, eds.), pp.

Song, P.S.(1985). In: Sensory Perception and Transduction in
    Aneural Organisms (G. Colombetti, F. Lenci and P.S. Song,
    eds.) Plenum Press New York pp. 47-59.
Song, P.S., Chae,Q., and Gardner, J.G. (1979). Biochim.
    Biophys. Acta 476: 479-495.
Thümmler, F. and Rüdiger, W.(1983). Tetrahedron 39: 1943-1951
Thümmler, F. and Rüdiger, W. (1984). Physiol. Plant 60:
    378-382.
Thümmler, F., Rüdiger, W., Cmiel, E., and Schneider,S.
    (1983). Z. Naturforsch. 38c: 359-368.

STRUCTURE-FUNCTION RELATIONSHIP OF THE PHYTOCHROME CHROMOPHORE

Pill-Soon Song

Department of Chemistry
University of Nebraska
Lincoln, Nebraska 68588
U.S.A.

Iwao Yamazaki

Institute for Molecular Science
Okazaki 444, Japan

I. INTRODUCTION

In this chapter, we examine the structural changes, and thus their functional implication, in the phototransformation of phytochrome as simplified in the following scheme:

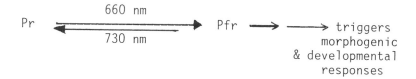

Our knowledge of the phytochrome molecule itself is not complete, but we know great deal more about it now than a few years ago, as the reader can appreciate this from reading other chapters in this volume. However, we know little about the mechanism of the phytochrome phototransformation and its structural implication in the physiological activity of the active form, Pfr, of the chromoprotein. For a detailed review of the phytochrome molecule, the reader is referred to Pratt

(1982), Furuya (1983), Smith (1983), Song (1984), Lagarias (1985), and this book (Furuya, 1987).

In this chapter, the mechanism of the phytochrome phototransformation and the structural basis of the biological reactivity of the Pfr form are discussed based on information currently available, but due to page space limitation, no attempt is made to cover all the relevant literature. In particular, we will be mainly concerned with the role of the chromophore in the phototransformation and its structure-function relationship in the active form of oat phytochrome(Pfr). Other aspects of the phytochrome protein and its phototransformation are covered in several chapters in this book.

II. CHROMOPHORE CONFIGURATION/CONFORMATION

Accompanying the phototransformation of oat phytochrome from its Pr to Pfr form, there is a red shift of the absorption band maximum from 666 to 730 nm (Figure 1). This shift corresponds to a relatively small energy difference of 15.6 kJ/mol. The red shift is even smaller in going from Pr to its first intermediate (for example, absorbing at 696 nm, 77 K; Song et al., 1981), yielding an energy difference of 7.9 kJ/mol. These small energy differences contrast with the large spectral shifts found in the photoisomerization of rhodopsin. It is likely that the energy differences arise dominantly from the ground state stabilities of the Pr vs. Pfr or intermediate (I-700; see for example, Pratt et al., 1984; Ruzsicska et al., 1985). Thus, we suggest that structural and conformational changes of the chromophore and its apoprotein environment involved in the phototransformation of phytochrome are relatively small and subtle. This is also predictable from the magnitude of the spectral shifts noted, although the energy differences alone cannot tell us whether they mainly arise from the ground or excited state stabilities.

What then is the structural change of the chromophore in the phototransformation of phytochrome ? Based on a high resolution proton NMR study of phytochrome chromopeptides which exhibit markedly blue-shifted absorption spectra relative to the visible absorption (Qy) bands of the native Pr and Pfr forms, photoisomerization of the tetrapyrrolic chromophore has been proposed, as discussed in the previous chapter by Ruediger.

There have been a number of proposals on the possible mechanisms of the phototransformation. However, experimental

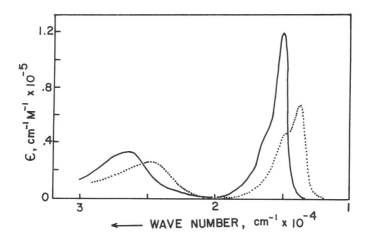

FIGURE 1. The absorption spectra of 124 kDa oat phytochrome at room temperature, expressed as a function of wavenumber for f number (oscillator strengths) calculations. The Soret bands were divided into two parts with respect to the absorbance peak, and the integrated area of the longer wavelength part was then doubled for the near UV oscillator strength. Solid line, Pr: dotted line, 86% Pfr.

supports for these speculative proposals are lacking. Based on steady state fluorescence measurements in water and heavy water and in analogy to the relaxation behaviors of free base porphyrins, proton transfer within the chromophore pore was suggested as a primary photoprocess of phytochrome leading to the formation of intermediate I-700 (Sarkar and Song, 1981). However, evidence now suggests that proton transfer is not likely in the primary photoprocess of native, 124 kdalton oat phytochrome, since solvent deuterium isotope effect on the fluorescence yield of phytochrome is small and markedly lower in 124 kdalton phytochrome than in degraded phytochrome used in the previous study (Moon et al., 1985). It remains to be seen at what stage of the phototransformation an intramolecular, i.e., within the tetrapyrrole pore, or intermolecular, i.e., between chromophore and apoprotein, proton transfer occurs along the thermal pathway leading to the formation of the Pfr form of phytochrome and is responsible for proton uptake/release in solution (Tokutomi et al., 1982, 1986). Although it remains to be established as to the tautomeric/or ionic structure and conformation of the Pfr form of phytochrome in its native state, photoisomerization of the chromophore during the phototransformation of phytochrome is supported by the work on phytochrome chromopeptides, as reviewed in the previous chapter. In this connection, we have examined the primary photoprocess of phytochrome as a function of viscosity.

Fluorescence decays of phytochrome in phosphate buffer were measured as a function of viscosity (glycerol) by using a synchronously pumped, cavity-dumped dye laser and a time-correlated, single-photon counting apparatus (Yamazaki et al., 1985). All the fluorescence decay curves can be fitted with two exponential decays with the lifetimes of 50-70 ps and 1.1-1.2 ns. Decay curve analyses based on the non-linear, least squares method showed that fitting is better with three components than with two components (Holzwarth et al., 1984; Song et al., 1986). However, the two component analysis appears to be adequate for interpreting the decay data, because the relative contribution of a third component to the overall fluorescence intensity is small. Although a complicated model can be proposed to account for the three component decays observed, a model based on the two component decays is satisfactory for a qualitative understanding of the viscosity dependence investigated by Song et al. (1986).

The short lifetime component (τ_1) of the fluorescence decay of phytochrome is slighly viscosity-dependent, and that of the longer component (τ_2) is viscosity-independent. However, their amplitudes are strongly viscosity-dependent. Several examples from the published work (Song et al., 1986) suffice to illustrate the viscosity dependences, as shown in Table 1.

The two exponential decays of the phytochrome fluorescence can be interpreted by the following scheme:

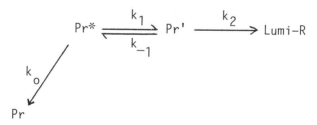

where Pr* and Pr' represent the excited singlet state of phytochrome and its primary intermediate, respectively, with the corresponding rate constants k_o (sum of both radiative and radiationless rate constants) and k_1/k_{-1}.

Rate equations for the above kinetic scheme can be written as follows:

$$d(Pr^*)/dt = - (k_o + k_1)(Pr^*) + k_{-1}(Pr') \qquad (1)$$

$$d(Pr')/dt = k_1(Pr^*) - (k_{-1} + k_2)(Pr') \qquad (2)$$

Solutions of these equations in terms of the concentrations of Pr* and Pr' can be expressed as follows:

$$
\begin{aligned}
(Pr^*) &= (q_2 - (k_o + k_1))\exp(-q_1 t)/(q_2 - q_1) \\
&\quad + ((k_o + k_1) - q_1)\exp(-q_2 t)/(q_2 - q_1)
\end{aligned} \qquad (3)
$$

$$(Pr') = k_1(\exp(-q_1 t) - \exp(-q_2 t))/(q_2 - q_1) \qquad (4)$$

where

$$2q_1, 2q_2 = (k_o + k_1 + k_{-1} + k_2) \pm \sqrt{(k_o + k_1 - k_{-1} - k_2)^2 + 4k_1 k_{-1}} \qquad (5)$$

with the reasonable assumption for an efficient photoreceptor that k_1 is greater than the absolute difference $|k_o - k_2|$, the following expressions can be derived:

$$
\begin{aligned}
1/\tau_1 &= q_1 = k_1 + k_{-1} \\
1/\tau_2 &= q_2 = k_2 + (k_o k_{-1})/(k_1 + k_{-1})
\end{aligned} \qquad (6)
$$

The relative amplitudes of the two lifetime components are given by:

$$A_2/A_1 = k_{-1}/k_1 \qquad (7)$$

with

$$k_1 = (1/\tau_1)/(1+(A_2/A_1)), \quad k_{-1} = (A_2/A_1)k_1 \qquad (8)$$

Values of these rate constants are shown in Table 1. The rate constant k_1 decreases somewhat with increasing viscosity, whereas k_{-1} increases steeply with viscosity. These viscosity dependences are explicable in terms of Eq.(6), with two rate constants having opposing dependencies on viscosity and with k_1 being much greater than k_{-1}. The viscosity independence of τ_2 is attributable to k_2 which contributes dominantly to the long fluorescence decay component, as can be seen from Eq.(6). The viscosity dependences of rate constants calculated, k_1 and k_{-1}, (Table 1) may reflect conformational/configurational isomerization of the Pr chromophore in its primary photoprocess. However, k_1 depends only moderately on the medium viscosity, whereas k_{-1} increases strongly with increasing viscosity. These differential viscosity dependences of the forward and reverse rates probably arise from a restricted conformational flexibility of the chromophore in the primary intermediate and/or its proximal amino acid residues under higher viscosity conditions. In less viscous media, the chromophore may assume an equilibrium distribution of conformations within a glycerol-accessible pocket or solvent cage from which the reverse reaction proceeds with the reorganization

of peptide chains and/or the solvent cage. In highly viscous
solution, the reorganization is less extensive. This
qualitatively explains the differential dependences of the
reversible photoprocess in the excited state of phytochrome.

TABLE 1. Fluorescence lifetimes of phytochrome(Pr) as a
          function of viscosity. For rate constants listed,
          see the text for calculations.

Viscosity*	Temp, K	τ_1(ps)	A(%)	τ_2(ns)	A(%)	k_1**	k_{-1}**
0.98	294	53	98	1.14	2	1.84	4.4
3.1	294	62	93	1.13	7	1.49	11.9
19.0	292	64	78	1.17	22	1.29	26.1

* Relative viscosity. 20 mM phosphate buffer, pH 7.8, with 1
mM EDTA and 2.8 mM 2-mercaptoethanol. ** $s^{-1} \times 10^{10}$ and $s^{-1} \times 10^{8}$, respectively.

     The viscosity dependence of the primary photoprocess
described above suggests that the chromophore moves, possibly
with a concerted motion of the peptide chain(s), during the
relaxation of the excited state as well as its
photoequilibrium with an early intermediate. Conformations of
the configurational isomers shown in Figure 1, Chapter 10,
are not known in the native state of phytochrome, but it is
possible that the configurational isomerization within the
chromophore pocket is microviscosity-dependent because of the
relatively small molecular size of glycerol used as a viscous
medium. Since the visible:near-ultraviolet oscillator
strength ratio $(f(Qy,x)/f(Bx,y))$ is a sensitive function of
the chromophore configuration and/or conformation (Song et
al., 1979), this ratio has been evaluated for the native oat
phytochrome from the absorption spectra expressed as a
function of wavenumber, yielding values of 1.24 and 1.19 for
the Pr and Pfr species, respectively, as shown in Figure 1.
The closeness of the two ratios means that the chromophore
conformations of the two spectral forms are not significantly
different. A semi-extended or semi-circular conformation is
consistent with the observed oscillator strength ratios for
Pr and Pfr (Song et al., 1979; Song and Chae, 1979; Sugimoto
et al., 1984). Since the ratio is significantly higher for
phycocyanins than for phytochrome, we propose that the
chromophore conformation of the latter is somewhat less
extended than that of the former as revealed for crystalline

phycocyanin from M. laminosus and A. quadruplicatum by X-ray
analysis (Schirmer et al., 1985, 1986), either by assuming a
coplanar conformation or by minimizing the steric hindrance
with rotation about the single bond(s). However, the
structural and conformational identity of the Pfr chromophore
in its native protein must still be worked out in order to
understand the mechanism of the phototransformation of
phytochrome.

III. CHROMOPHORE TOPOGRAPHY

We now discuss the chromophore topography and its
variation during the phototransformation of phytochrome. In
particular, the differential accessibility of the chromo-

TABLE 2. Relative accessibilities of the phytochrome
chromophores as expressed in terms of rates of
oxidation with TNM, oxygen/bilirubin oxidase,
reduction with sodium borohydride. Experimental
conditions can be found in the respective
references cited in the text.

Phytochrome	Treatment	Relative Rate		Ratio(Pfr/Pr)
		Pr	Pfr	
124 kDa	TNM	1.00	8.41	8.41
	NaBH₄	1.00	2.34	2.34
	O₂/BO	1.00	10.19	10.19
	" +ANS	54.94	140.26	2.55
118 kDa*	TNM	0.97	39.14	40.35
	NaBH₄	1.99	5.04	2.53
60 kDa	TNM	5.71	88.97	15.58
Chromo-peptide**	TNM	827.59	879.31	1.06

*This may contain a minor contaminant phytochrome of
114 kDa molecular mass, although our preparations
usually showed little or no 114 kDa band on SDS
gel (Hahn et al., 1984). ** For neutral solution
(Thuemmler et al., 1985).

phores of the Pr and Pfr species of phytochrome has been studied by various techniques such as spectral bleaching of the phytochrome chromophore with denaturing agents (see reviews by Pratt, 1978; Furuya, 1983).

Tetranitromethane (TNM) has been used as a chemical probe for the chromophore accessibilities in phytochrome because this oxidant preferentially reacts with the tetrapyrrolic chromophore, although it also reacts with tyrosyl residues, but at a considerably slower rate (Hahn et al.,1984). Table 2 summarizes results of TNM oxidation and borohydride (BH_4) reduction of phytochrome. From this table, we can conclude that the Pfr chromophore is more accessible to these agents than the Pr chromophore. However, the differential accessibilities of the chromophores are modulated by the integrity of apoprotein in phytochrome, as shown in Table 2.

Bilirubin oxidase bleaches the phytochrome color, i.e., visible absorbance band disappears, as the enzyme catalyzes the oxidation of the tetrapyrrolic chromophore in the presence of molecular oxygen (Kwon et al., 1987). Results are shown in Table 2. Consistent with other data shown in Table 2, again the Pfr chromophore is more readily oxidized by the enzyme than is the Pr chromophore. These results suggest that the former is more accessible to the oxidants than the latter. However, alternative interpretations cannot be ruled out based on these data alone. For example, the chromophores in the native proteins, Pr and Pfr, might possess intrinsically different chemical reactivities which could then determine the differential catalytic rates of the chromophore oxidation observed.

Whether the differential catalytic oxidation rates of the Pr and Pfr chromophores of phytochrome as substrates for bilirubin oxidase are predominantly determined by their topography or intrinsic chemical reactivity can be partially resolved by examining the effect of ANS on the enzyme-catalyzed oxidation of the phytochrome chromophores. From Table 2, it can be seen that 8-anilinonaphthalene 1-sulfonate (ANS) accelerates the enzyme-catalyzed oxidation of the chromophores and that the ratio of the rate for Pfr to that for Pr is markedly reduced, as compared to the ratio observed in the absence of ANS. Since ANS forces exposure of the tetrapyrrole chromophore by presumably binding to the chromophore crevice and/or elsewhere in the apoprotein (Hahn and Song, 1981), the reduced reaction rate ratio observed in the presence of ANS is, thus, attributable to the similarity in reactivity between the Pr and Pfr chromophores. In other word, the differential reaction rates for Pr and Pfr observed with bilirubin oxidase in the absence of ANS are largely due to their topographic dispositions. ANS also dissociates dimeric phytochrome species into monomers in solution (Choi et al., 1987). To what extent the monomerization contributes to the exposure of the phytochrome chromophore to an oxidant species and/or to bilirubin oxidase remains to be elucidated.

Figure 2 shows the spectral bleaching of phytochrome with relatively high concentrations of ANS as an indirect probe for the chromophore topography of phytochrome (Hahn and Song, 1981). Tetrapyrroles in their free form resume a cyclic conformation, as described later (see also Falk and Mueller, 1983). ANS binding to phytochrome exerts apparently the same kind of effect, causing a marked loss of absorbance in the visible spectral region of phytochrome. The Pfr chromophore is more susceptible to ANS bleaching than the Pr chromophore, but the ANS susceptibility strongly depends on the apoprotein integrity, particularly to the 6/10 kdalton amino terminal peptide segment (Figure 2). This observation suggests that the amino terminus segment interacts and/or masks the Pfr chromophore of phytochrome, thus making it less susceptible to cyclize in the presence of ANS. Similar results have been reported by Eilfeld and Ruediger (1984). Borohydride reduction of the Pr and Pfr chromophores of phytochrome yielded results closely resembling those obtained with ANS (Chai et al., 1987). In the next section, we explore further the aspect of interactions between the chromophore and apoprotein, especially the amino terminus segment.

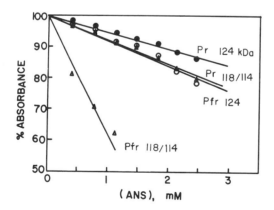

FIGURE 2.   Bleaching   of the main   absorbance bands of phytochrome (ca. 1 μM) with ANS in 20 mM K-phosphate buffer, pH 7.8, 1 mM EDTA.

IV. INTERACTIONS BETWEEN CHROMOPHORE AND APOPROTEIN

To probe a conformational change induced by the chromophore phototransformation of phytochrome, the number of exchangeable protons in phytochrome in tritiated water was measured for the Pr and Pfr forms (Hahn et al., 1984). There is a significant decrease in the number of exchangeable protons in going from Pr to Pfr (this conclusion contrasts with the one in the original paper, because a lower molar extinction coefficient (73 000 $M^{-1}cm^{-1}$; Tobin and Briggs, 1973) was used in the original paper. With more recent value of 121,000 $M^{-1}cm^{-1}$ (Cha et al., 1983), the Pfr form is seen to possess less exchangeable protons than the Pr form, suggesting that the molecular topography of the phytochrome apoprotein changes with the chromophore accessibility or movement.

An examination of the data shown in Table 2 as well as in Figure 2 suggests that the Pfr chromophore is more exposed than the Pr counterpart, as a result of the phytochrome phototransformation. Furthermore, there is indirect evidence that the Pfr chromophore interacts with the N-terminus segment of apoprotein, as mentioned above. Recently, additional evidence for the possible interaction between the chromophore and the N-terminus peptide chain has been obtained by circular dichroism (CD) spectroscopy (Vierstra et al., 1987), showing that a photoreversible CD change occurs during the phototransformation of phytochrome and that, only in 124 kdalton phytochrome, 3 % increase in alpha-helical folding is observable in the Pfr form of phytochrome (Figure 3). The photoreversible CD change, and the increased alpha-helicity in the Pfr form, are suppressed by a monoclonal antibody that binds near the N-terminus sequence, as shown in Figure 4 (Chai et al., 1987). The photoreversible CD spectral change can also be inhibited by sodium borohydride, which bleaches the chromophore by reducing it, and by TNM, which oxidizes the chromophore of phytochrome. Although explanations of these results based on indirect interactions between the chromophore and the N-terminus segment are possible, it is more likely that an additional alpha-helical folding of the Pfr form of phytochrome results from a photoreversible interaction between the Pfr form of the phytochrome chromophore and the N-terminus segment.

It is interesting that of 52 amino acid residues from the N-terminus shown below (Hershey et al., 1985), 34 amino acid residues make up three sequences where alpha-helical folding is favorable, according to the Chou-Fasman method (Chou and Fasman, 1978) of predicting secondary structures of proteins:

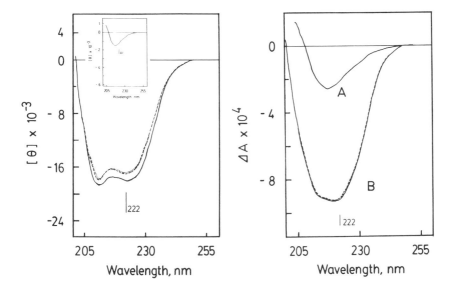

FIGURE 3. Far–UV CD spectra of 124–kDa Pr(broken line), Pfr (solid), and cycled Pr (dotted)in 20 mM phosphate buffer, pH 7.8, 277 K. See Chai et al. (1987).
Inset: Oat–25 monoclonal antibody.

FIGURE 4. Far–UV CD spectra of Pr(broken line), Pfr (solid) and cycled Pr (broken line) mixed with oat–25 monoclonal antibody (A), under the same conditions as in Figure 3. See Chai et al. (1987) for details.

N–ser–ser–ser–arg–pro–<u>ala–ser–ser–ser–ser–ser–arg</u>–asn

<u>–arg–gln–ser–ser–gln–ala–arg–val–leu–</u>
<u>ala–gln–thr–thr–leu–asp–ala–glu–leu–</u>

asn–ala–glu–tyr–glu–glu–ser–gly–asp–ser–phe–asp–tyr–

<u>ser–lys–leu–val–glu–ala–gln–arg–asp</u>––––––

These sequences (underlined) represent 3% of the total amino acid residues of 124 kdalton oat phytochrome. This may be merely coincidental, but it is tempting to assign locations of the additional 3% alpha–helices observed to the above sequences. It is well known that prosthetic groups such as the tetrapyrrolic heme of myoglobin (Beychok, 1966) exhibit propensity to facilitate an alpha–helical conformation of peptide chains, and perhaps, this type of interaction is responsible for the photoreversible alpha–helical folding in phytochrome. Although the mode of interactions, if exists, between the Pfr-chromophore and the N-terminus sequences is not known, it is noted that the N-terminus residues (eg., ser and thr) and the chromophore possess hydrogen bonding functionalities such as the hydroxyl group on the former and the pyrrolic-NH- and -N=, carbonyl and carboxylate groups on the latter.

V. FUNCTIONAL IMPLICATIONS

The possible functional implication of the proposal that the Pr $\longrightarrow$ Pfr phototransformation involves (a) reorientation and/or topographic change of the chromophore and (b) alpha-helical folding of the N-terminus sequences as a result of an interaction with the chromophore in the Pfr form is speculated here.

First, what is the possible functional implication of the chromophore reorientation and/or topographic change in Pr $\longrightarrow$ Pfr phototransformation ? To discuss this question, polarotropism of the chloroplast movement in Mougeotia (Chapter 18) can be used as an example for the possible role of chromophore reorientation in the phototransformation of phytochrome (see also Chapter 19 for polarotropism in fern). For chloroplast rotation, photoreceptor molecules in this organism, believed to be phytochrome, are apparently fixed dichroically in plasmalemma or other organellar matrix. To explain mutually perpendicular polarizations of the red and far red irradiations for a polarotropic movement of chloroplast, i.e., Pr and Pfr transition dipoles are perpendicular to each other with respect to the cell surface plane, several possibilities can be envisioned. Among them are that (i) the Qy transition dipoles of the red and far-red absorption bands in Pr and Pfr, respectively, are mutually orthogonal, with chromophore and apoprotein rigidly fixed to their binding sites, (ii) the directions of the transition dipoles of both spectral forms are unaltered but the apoprotein precisely rotates during phytochrome phototransformation, by an angle sufficient to achieve the orthogonality and reversibility of red and far-red polarized excitations, (iii) the apoprotein remains fixed, but the chromophore reorientation occurs upon Pr $\longrightarrow$ Pfr phototransformation, along with a subtle conformational change in and around the chromophore binding pocket; this would result in a greater angle of orientation between the transition dipoles of the Pr and Pfr forms, and (iv) both the chromophore and apoprotein reorient or rotate in concert to achieve a mutually orthogonal direction of the transition dipoles in the Pr and Pfr forms of phytochrome.

The model (i) above is least likely of the four possibilities since the chromophore conformations, even with a configurational isomerization, of the Pr and Pfr species are similar as far as the pi-electron framework is concerned, vide supra (also see Chapter 12 by Suzuki et al.). Figure 5

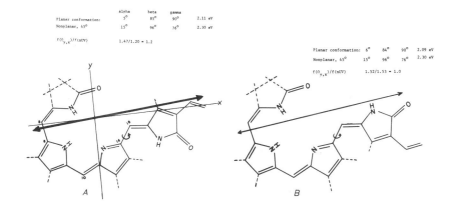

FIGURE 5. Orientation of the Qy transition dipole in phytochrome calculated by the semi-empirical self-consistent field MO/CI method using the Pariser-Parr-Pople-Mataga-Nishimoto approximations(see Song and Chae, 1979 for details).

(A) Pr with a semi-extended/semi-circular, planar conformation consistent with the spectral shape shown in Figure 1. The calculated f ratio(visible/Soret) is 1.2 (inset).

(B) Pfr(?), a semi-extended conformation, with 15,16-E configuration. The calculated f ratio is 1.0 (inset) in agreement with the spectral shape shown in Figure 1.

The effect of a 45 rotation around the 14, 15-s-bond, to relieve steric requirements, is to blue shift the Qy visible absorption band (Inset;calculated blue shifts (eV)corresponds to 49 and 54 nm for (A) and (B),respectively. Angles alpha, beta, and gamma for the dipoles are with respect to cartesian coordinates, x, y and z axes, respectively.

shows the calculated directions of the Q_y transition dipoles for Pr and Pfr in their planar conformations, predicting that the polarization direction does not change markedly upon Pr ⟶ Pfr phototransformation. This prediction is still valid even if the conformation of ring D (or ring IV) is out-of-plane by some angle either due to isomerization or rotation about the 14-15 single bond (see Figure 5: inset, for a spectral effect of the s-bond rotation).

Model (ii), i.e. rotation of protein, is possible, although the phytochrome molecule is large and possibly dimeric, as in solution. No experimental evidence is currently available to evaluate the applicability of this

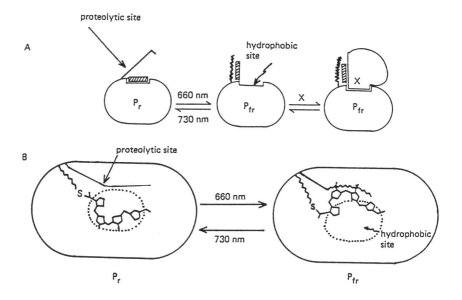

FIGURE 6. A schematic model for the topographic changes involved in phototransformation of phytochrome from Pr to Pfr form. Top panel, side view; X, undefined membrane or putative receptor/binding site. Lower panel, top view; Proteolytic site on the N-terminus segment of Pr is shown. The N-terminus segment assumes an alpha-helical conformation in the Pfr form (Modified from Song, 1983).

model in chloroplast polarotropism. It is possible that the

phototransformation of Pr to Pfr generates a new binding surface on the Pfr protein approximately perpendicular to the binding surface plane of the Pr protein. Assuming that phytochrome serves as a peripheral membrane protein, one can then postulate that Pr rotates upon its phototransformation to the Pfr form. The latter then binds to the lipid bilayer core via the newly generated hydrophobic surface. Since both electrostatic and hydrophobic forces contribute to the binding of phytochrome to membrane (Kim and Song, 1981; Hong et al., 1984; Furuya et al., 1981), an anisotropic rotation of the protein molecule can be envisioned. An anisotropic rotation of the protein molecule due to the relocation of the binding site of the Pr and Pfr forms, for example, the Pfr chromophore binding to the N-terminus segment preferentially, can be reasonably accommodated in the chromophore reorientation/hydrophobic model of Pfr-phytochrome (Figure 6), if the newly generated hydrophobic surface on the Pfr protein results from the exposure and reorientation of the chromophore.

   Models (iii) and (iv) are most appealing, since some degree of chromophore movement is implicated from the currently available information on the chromophore topography of phytochrome. Linear dichroic measurements of the immobilized phytochrome suggest that the chromophore indeed reorient by 32 or 148 degrees upon Pr ⟶ Pfr phototransformation (See Chapter 14 for details and references). A recent analysis has indicated that the angle of chromophore reorientation is 50 degrees (Y. Inoue, this Conference). Further work is warranted to resolve the above models. As it stands now, these remain speculative.

   What is the possible, functional implication of alpha-helical conformation of the N-terminus sequences in the Pfr form of phytochrome ? In general terms, one can speculate that the following circumstances, on the assumption that the N-terminus in fact is the locus of light-induced alpha-helical folding in the Pfr form of phytochrome, vide supra, are involved in the primary event of phytochrome action:
(i) The alpha-helical sequence acts as an anchor to a membraneous site ("X" in Figure 6) for phytochrome, in analogy to membrane proteins in which the role of alpha-helical domain is to span the membrane layer,
(ii) the alpha-helical domain serves to recognize a specific DNA sequence or site either promoter/operator, enhancer or some other types of regulatory sequences within plant chromatin assembly; alternatively, it may interact with RNA, particularly at duplex-structured RNA sites, and
(iii) folding of alpha-helical segments is thermodynamically favorable; an additional alpha-helix along the N-terminus can

serve to interact with a putative phytochrome receptor protein ("X" in Figure 6).

Nothing is as yet known about the putative phytochrome receptors. Since phytochrome mediates diverse responses to red light in plants, some involving and some apparently not involving gene expression, it is possible that there are more than one common type of receptors in a plant cell. We need much more knowledge on the molecular biology of phytochrome action before some of the speculative models discussed here and elsewhere by others can be ascertained for their validity, if any.

ACKNOWLEDGEMENTS

This work was supported by the U.S.PHS NIH Grant (GM36956) and the Robert A. Welch Foundation (D-182) to PSS. Part of the work and writing of this review at Institute for Molecular Science, Okazaki, were supported by a visiting professorship from the National Research Institutes to PSS. PSS also wishes to thank his current and former students, Y. G. Chai, J. K. Choi, T. I. Kwon, B. R. Singh, T. R. Hahn and H. K. Sarkar for their valuable contributions. We also thank Professor W. Ruediger for making his chapter available to PSS.

REFERENCES

Beychok, S.(1966) Science 154:1288.
Cha, T.A., Maki, A.H., and Lagarias, J.C. (1983) Biochemistry 22:2846.
Chai, Y.G., Singh, B.R., and Song, P.S. (1987) Anal. Biochem., in press.
Chai, Y.G., Song, P.S., Cordonnier, M.M., and Pratt, L.H.(1987) Biochemistry, in press.
Choi, J.K., Kwon, T.I., and Song, P.S. (1987) Photochem. Photobiol. 45: in press.
Chou, P.Y., and Fasman, G.D. (1978) Annu. Rev. Biochem. 47:251.
Eilfeld, P., and Ruediger, W. (1984) Z. Naturforsch. 39c:742.
Falk, H., and Mueller, N. (1983) Tetrahedron 39:1875.
Furuya, M. (1983) Phil. Trans. R. Soc. (Lond.) B303:361.
Furuya, M., Freer, J.H., Ellis, A., and Yamamoto, K.T. (1981) Plant Cell Physiol. 22:135.
Hahn, T.R., and Song, P.S. (1981) Biochemistry 20:2602.
Hahn, T.R., and Song, P.S. (1984) Biochemistry 23:1219.

Hahn, T.R., Song, P.S., Quail, P.H., and Vierstra, R.D. (1984) Plant Physiol. 74:755.

Hershey, H.P., Baker, R.F., Idler, K.B., Lissemore, J.L., and Quail, P.H. (1985) Nucleic Acids Res. 13:8543.

Holzwarth, A.R., Wendler, J., Ruzsicska, B.P., Braslavsky, S.E., and Schaffner, K. (1984) Biochim. Biophys. Acta 791:265.

Hong, C.B., Hahn, T.R., and Song, P.S. (1984) Photochem. Photobiol. 39:23S.

Kim, I.S., and Song, P.S. (1981) Biochemistry 20:5482.

Kwon, T.I., Singh, B.R., Choi, J.K., Chai, Y.G., and Song, P.S.(1987) To be submitted.

Lagarias, J.C. (1985) Photochem. Photobiol. 42:821.

Moon, D.K., Jeen, G.S., and Song, P.S.(1985) Photochem. Photobiol. 42:633.

Pratt, L.H. (1978) Photochem. Photobiol. 27:81.

Pratt, L.H. (1982) Annu. Rev. Plant Physiol. 33:557.

Pratt, L.H., Inoue, Y., and Furuya, M. (1984) Photochem. Photobiol. 39:241.

Ruzsicska, B.P., Braslavsky, S.E., and Schaffner, K. (1985) Photochem. Photobiol. 41:681.

Sarkar, H.K., and Song, P.S. (1981) Biochemistry 20:4315.

Schirmer, T., Bode, W., Huber, R., Sidler, W., and Zuber, H. (1985) J. Mol. Biol. 184:257.

Schirmer, T., Huber, R. Schneider, M., Bode, W., Mueller, M., and Hackert, M.L. (1986) J. Mol. Biol. 188:651.

Smith, W.O., Jr.(1983) In "Encyclopedia of Plant Physiology: Vol. 16A, Photomorphogenesis" (W. Shropshire, Jr. and H. Mohr, eds.), p.96. Springer-Verlag, Berlin.

Song, P.S. (1983) Annu. Rev. Biophys. Bioeng. 12:35.

Song, P.S. (1984) In "Advanced Plant Physiology" (M.B. Wilkins, ed.), p. 354, Pitman Books, London.

Song, P.S., and Chae, Q. (1979) Photochem. Photobiol. 30:117.

Song, P.S., Chae, Q., and Gardner, J.G. (1979) Biochim. Biophys. Acta 576:479.

Song, P.S., Sarkar, H.K., Kim, I.S., and Poff, K.L. (1981) Biochim. Biophys. Acta 635:369.

Song, P. S., Tamai, N., and Yamazaki, I. (1986) Biophys. J. 49:645.

Sugimoto, T., Inoue, Y., Suzuki, H., and Furuya, M. (1984) Photochem. Photobiol. 39:697.

Thuemmler, F., Eilfeld, P., Ruediger, W., Moon, D.K., and Song, P.S. (1985) Z. Naturforsch. 40c:215.

Tobin, E.M., and Briggs, W.R. (1973) Photochem. Photobiol. 18:487.

Tokutomi, S., Inoue, Y., Sato, N., Yamamoto, K.T., and Furuya, M. (1986) Plant Cell Physiol. 27:765.

Tokutomi, S., Yamamoto, K.T., Miyoshi, Y., and Furuya, M. (1982) Photochem. Photobiol. 35:431.

Vierstra, R.D., Quail, P.H., Hahn, T.R., and Song, P.S.
    (1987) Photochem. Photobiol. 45:429.
Yamazaki, I., Tamai, N., Kume, H., Tsuchiya, H., and Oba, K.
    (1985) Rev. Sci. Instrum. 56:1187.

THEORETICAL APPROACHES
TO PHOTOREVERSIBLE SPECTRAL CHANGES
OF PHYTOCHROME CHROMOPHORE

Hideo Suzuki, [†]Tohru Sugimoto and Etsuro Ito

Department of Physics
Waseda University
Tokyo, Japan

[†]Biophysics Laboratory
Kanto Gakuin University
Yokohama, Japan

I. INTRODUCTION

As the first step to disclose the origin of the reversible spectral changes of phytochrome, we have theoretically been investigating the light absorption property of phytochrome chromophore with inclusion of the interaction between the chromophore and its microenvironment. And we have found that the reversible spectral changes of phytochrome can be reproduced by the *cis-trans* isomerization of pyrrole ring D around the double bond of its neighboring methyne bridge (Suzuki *et al.*, 1975; Sugimoto *et al.*, 1977, 1984, 1985).

The next step of the problem is to examine theoretically the possibility of the *cis-trans* isomerization of pyrrole ring D from the viewpoint of the dynamical phototransformation of phytochrome chromophore, of which the process is expected to consist of the following steps.

When phytochrome chromophore absorbs an incident light, it undergoes a transition from its ground to its excited state. The excited chromophore remains in the excited state for its lifetime T, which can be estimated as follows. (i) The radiative lifetime of the excited chromophore is of the order of 10^{-9} s. If the excited chromophore did not change its conformation within the radiative lifetime, the excitation energy would insignificantly be exhausted without giving any informa-

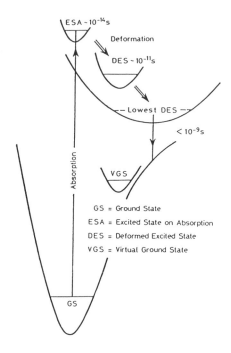

Fig. 1. A representation of the transient excited states
in a whole process from absorption to emission of light.

tion about light stimulas to the protein moiety of phytochrome. For this reason, the upper limit of T seems to be of
the order of 10^{-9} s. (ii) On the other hand, any conformational change of the excited chromophore can not take place
for 10^{-14} s after the light absorption by the chromophore,
since the conformational change of the excited chromophore is
first caused by the change in bond lengths for stabilization
due to the stretching vibrations of bonds whose periods are of
the order of 10^{-13} s. For this reason, the lower limit of T
seems to be of the order of 10^{-13} s. (iii) Moreover, to give
rise to a significant change in bond lengths, every bond of
the excited chromophore must undergo over hundreds of stretching vibrations. For these reasons, it would seem that the
lifetime T of the excited state of the chromophore is of the
order of 10^{-11} s (Fig. 1).
    The excited chromophore changes its conformation during T
to give some information about light stimulas to the protein
moiety of phytochrome, leading to its conformational change.
Then, the excited state of the chromophore will have to be
sorted into the following two transient states. One is the
excited state formed immediately after the absorption of

light, of which the lifetime is of the order of 10^{-14} s, being shorter than the periods of stretching vibrations of bonds. Needless to say, this transient excited state has the same bond lengths as those in the ground state, which we have called "the excited state on absorption". The other is the excited state which comes out after the excited state on absorption, being accompanied by the energy relaxation of stretching vibrations of bonds (i.e. by the change in bond lengths for stabilization). This transient excited state has different bond lengths and conformation from those in the excited state on absorption, which we have called "the deformed excited state". According to this definition, the excited state from which the spontaneous emission takes place can be regarded as the lowest deformed excited state. In addition, we have also defined such low-lying state that has the same bond lengths and conformation as in the deformed excited state, which we have called "the virtual ground state" to discriminate it from the ground state.

During the lifetime T of the excited state, it is expected that the excited chromophore undergoes a non-adiabatic transition from some deformed excited state to a certain virtual ground state without emitting radiant energy. And after this transition, phytochrome chromophore would gradually change its state from the virtual ground state to the most stable state (i.e. the ground state of the phototransformed chromophore), being accompanied by the conformational change of the protein moiety of phytochrome. If this change in conformation is restricted within a small region around the chromophore, it will be completed during the period of microseconds (it is thus expected that some intermediate has a lifetime of the order of 10^{-6} s).

We thus suppose that P_r is phototransformed from the ground state of P_r to that of P_{fr} through the excited state on absorption, the deformed excited state and the virtual ground state, and that the excited states in the process of $P_r \rightarrow P_{fr}$ phototransformation are not identical to those of $P_{fr} \rightarrow P_r$ one.

Under these considerations, we have been developing a method of calculating the deformed excited states of π-electron systems in consistence with their own bond lengths, and we have been studying the possibility of the cis-trans isomerization of phytochrome chromophore through its deformed excited state.

In the present paper we would like to mention about the following three problems: (a) the role of the change in the bond lengths of phytochrome chromophore in its deformed excited state; (b) the change in total energy of the chromophore in the cis-trans isomerization of pyrrole ring D; (c) a theoretical model for the change in chromophore structure corresponding to the photoreversible spectral change of phytochrome.

II. METHODS OF CALCULATION

The chemical structure proposed by Lagarias and Rapoport (1980) for phytochrome chromophore is adopted in the present work. Molecular orbital calculations are performed according to the previously reported formulation of the ZDO approximation of LCAO-ASMO-SCF-CI theory for π-electrons (Suzuki *et al.*, 1973, 1975; Nakachi *et al.*, 1981).

The essential points of the methods of calculation are as follows. (1) Bond length $R(g)$ in the ground state is assumed to vary with bond order $P(g)$ in the ground state by the relation

$$R(g) = R_s - (R_s - R_d)/[1 + K(1 - P(g))/P(g)].$$

Here R_s (R_d) is the length of normal single-bond (double-bond); K is a constant. (2) Bond length $R(e)$ in the deformed excited state is assumed to vary with bond order $P(e)$ in that state by the relation

$$R(e) = \bar{R} - (R_s - \bar{R})[(P(e) - \bar{P})/\bar{P}] \exp(-\gamma P(e)).$$

Here $\bar{R}$ $(\bar{P})$ is the mean value of $R(e)$'s $(P(e)$'s) in the deformed excited state; γ is a constant. (3) All the bond angles of the pyrrole rings with a cyclic π-system are improved according to the previous methods (Suzuki *et al.*, 1973, 1975). (4) Coulomb integral α (corresponding to the minus of ionization potential I) is assumed to vary with π-electron density $P(g)$. For example, as for carbon atom, α_C is determined as

$$\alpha_C = 12.95 - 4.056\times(3.60 - 0.35P_C)$$

$$- 1.035\times(3.60 - 0.35P_C)^2$$

in eV units, according to the variable electronegativity method along the valence-state isoelectric series of (B^-, C, N^+). (5) A sufficient number of singly- and doubly-excited configurations of π-electrons are included in the CI calculation. (6) The interaction of the π-electron system of phytochrome chromophore with a point-charge and a point-dipole is regarded as a perturbation, and its effect on the optical absorption of the π-electron system is examined according to the stationary perturbation theory for degenerate states. This calculation is performed by changing the position of a point-charge and the position and moment of a point-dipole.

III. THE CHANGE IN TOTAL ENERGY
      OF PHYTOCHROME CHROMOPHORE CORRESPONDING TO
      CIS-TRANS ISOMERIZATION OF PYRROLE RING D
      ABOUT C(15)-C(16) DOUBLE BOND

As the first step to examine theoretically the deformed
excited state of phytochrome chromophore, the protonated nitrogen in pyrrole ring C in the chromophore is replaced by a
pyrrolenine nitrogen in the present section, and the conformation of the chromophore is assumed to be of *trans-trans-trans*
linear form, for simplicity.

First, the ground and the deformed excited state of π-
electrons in phytochrome chromophore are calculated. One of
the most remarkable difference between the two states can be
seen in the bond lengths of the chromophore. They are shown
in Fig. 2, from which it is found that the bond lengths in the
deformed excited state are longer than those in the ground
state, and that the rates of elongation of double bonds are
higher than those of single bonds. The extent of the π-electron system of the deformed excited state is larger than that
of the ground state. Although C(14)-C(15) or C(15)-C(16) bond
in the ground state shows a single- or a double-bond character respectively, it is noticed that in the deformed excited
state both bonds show a similar character.

Next, the change in total energy of phytochrome chromo-

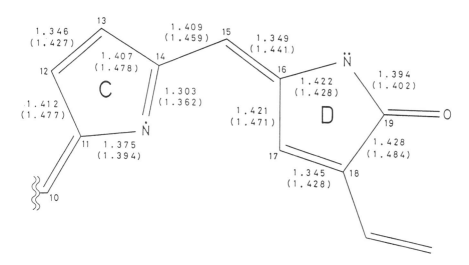

Fig. 2. Bond lengths (in A units) of phytochrome chromophore, calculated for the ground state and the deformed excited state (enclosed with parentheses).

phore in the *cis-trans* isomerization of pyrrole ring D about
C(15)-C(16) double bond is examined. The π-electron system of
the chromophore consists of even π-electrons, atomic cores and
σ-bonds. By rotating pyrrole ring D around C(15)-C(16) double
bond, the total energy of the whole system is calculated for
the ground state, the excited state on absorption, the deformed excited state and the virtual ground state. In the calculation, the initial conformation of the chromophore is taken
to be of *trans-trans-trans* linear form.

The results are shown in Fig. 3, from which it is found
that both the total energies of the ground state and the excited state on absorption increase as the rotation angle θ approaches 90 degrees. It is also found that the total energy
of the virtual ground state increases as θ approaches 90 degrees whereas that of the deformed excited state decreases,
and that a close approach of the total energies between the

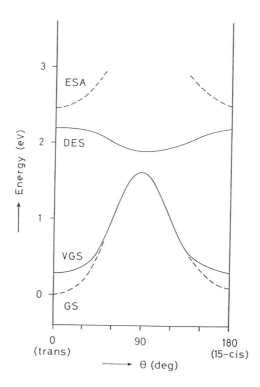

Fig. 3. The change in total energy of phytochrome chromophore in the *cis-trans* isomerization of pyrrole ring D about
C(15)-C(16) double bond, calculated by rotating pyrrole ring D
from *trans-* to *cis-*conformation. Abbreviations are defined in
Fig. 1.

virtual ground and the deformed excited state occurs in the region of θ = 90 degrees. This close approach of the total energy is ascribable to the change in bond lengths of phytochrome chromophore in the deformed excited state shown in Fig. 2.

When a barrierless and downward concave curve of excited-state total energy approaches closely an upward convex curve of ground-state total energy, it is expected according to the Landau-Zener theory (Zener, 1932) that a non-adiabatic transition occurs between these two states. On the basis of this theory, it is deduced from Fig. 3 that a non-adiabatic transition between the deformed excited state and the virtual ground state takes place to give rise to the *cis-trans* isomerization of pyrrole ring D about C(15)-C(16) double bond.

We have thus arrived at the conclusion that the *cis-trans* isomerization of pyrrole ring D about C(15)-C(16) double bond is probable as the initial reaction of phytochrome phototransformation, and that the change in bond lengths in the deformed excited state of phytochrome chromophore plays an important role in its conformational change corresponding to phytochrome phototransformation.

IV. MODEL FOR THE CHANGE IN CHROMOPHORE STRUCTURE OF PHYTOCHROME CORRESPONDING TO ITS PHOTOREVERSIBLE SPECTRAL CHANGE

An attempt was made to provide the simplest probable model for the chromophore structure of phytochrome, with the fewest number of bonds concerned with the twisting of the π-electron system (Sugimoto *et al.*, 1984). The model can reproduce the absorption spectra of phytochrome and its phototransformation intermediates (Shimazaki *et al.*, 1980; Inoue *et al.*, 1982).

The conformational change in the chromophore model (Fig. 4) corresponding to P_r-P_{fr} phototransformation may be summarized as follows: (1) pyrrole ring D rotates through 90 degrees around C(15)-C(16) double bond within 10 µs after light absorption, and the change in rotation angle of ring D is very small during the subsequent dark transformation to P_r or P_{fr}; (2) the changes in the rotation angles of the other rings (A, B, C) around the single and/or the double bond of their neighboring methyne bridges are small; (3) the conformational changes in the protein moiety of phytochrome around the chromophore is very small.

In the model shown in Fig. 4, only pyrrole ring D is assumed to rotate for the following reasons: (1) dihydropyrrole ring A seems not to rotate easily around the bonds of its neighboring methyne bridge, since ring A is linked to the pro-

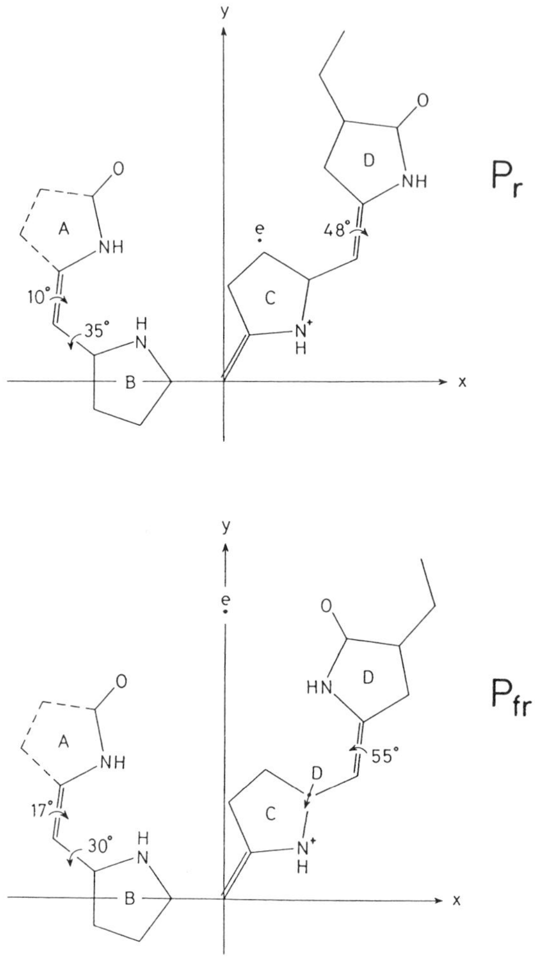

Fig. 4. A model for chromophore structure of phytochrome
which can reproduce the observed light-absorption of phytochrome. This model takes into account the following factors:
the rotation of the pyrrole rings around the bonds shown by
the curved arrows; the interaction of the π-electron system of
the chromophore with a negative point-charge e in the P_r-chro-
mophore model, or with e and a point-dipole D in the P_{fr}-chro-
mophore model. The position of e in the P_r-chromophore model
is (1.0, 3.7, 5.5) Å referring to the coordinate system in the
figure. The position of e and D in the P_{fr}-chromophore model
are (0.0, 7.5, 4.5) Å and (2.3, 2.7, 3.5) Å, respectively.
The magnitude of D is 1.5 debye, and its direction is shown by
the arrow (20 degrees from the negative y-axis and 10 degrees
from the positive z-axis).

tein moiety through a thioether linkage; (2) if more than one
pyrrole rings rotate, it would be inevitable to introduce such
a large conformational change in the protein moiety that contradicts to the observed CD spectrum of phytochrome which
shows no appreciable spectral shift in the wavelength region
of amino acid residues.

The long-wavelength transition moment of P_{fr} chromophore
model was calculated to be almost parallel to that of P_r chromophore model, since the conformational change of the model is
small as a whole.

V. CONCLUDING REMARKS

In the present paper, we mentioned that the change in bond
lengths of phytochrome chromophore in its deformed excited
state plays an important role in its conformational change in
phytochrome phototransformation. And, from calculating the
total energy and optical absorption of the π-electron system
in phytochrome chromophore, we showed the possibility that the
cis-trans isomerization of pyrrole ring D about $C(15)-C(16)$
double bond is the initial reaction of phytochrome phototransformation.

According to our theory of phytochrome phototransformation, its intermediates may be interpreted as dark consequences of a non-adiabatic transition from a certain virtual
ground state to the most stable ground state (i.e. the ground
state of the phototransformed chromophore). And this transition seems to be accompanied by the gradual conformational
changes of both the chromophore and the protein moiety of phytochrome. If these changes in conformation are restricted
within a small region around the chromophore, we will be able
to expect a lifetime of the order of 10^{-6} s for some intermediate. Therefore, the intermediate shows different absorption
spectrum at different time after light absorption corresponding to different point on the process mentioned above.

According to this interpretation of intermediates, it may
be evident that the phototransformation intermediates of phytochrome show different absorption spectra from one another,
and that the intermediates in the process of $P_r \rightarrow P_{fr}$ phototransformation are not identical to those of $P_{fr} \rightarrow P_r$ one.

Finally, it should be emphasized that we need to extend
the method of calculating the deformed excited state to include valence-shell electrons in order to judge whether the
present results are reasonable or not. About eight years ago,
we tried such extension within the framework of INDO-CI method, and we had succeeded in elucidating the optical absorption
of formamide and its related compounds (Sakuranaga et al.,

1979). If we further extend this method to include the change
in bond lengths due to the torsion around bonds, we will be
able not only to judge the present results but also to calculate the vibrational states and vibronic coupling associated
with the deformed excited or virtual ground state. And basing
on these studies and developing non-adiabatic approaches, we
will be able to propose a dynamical theory of the photoisomerization of phytochrome chromophore.

REFERENCES

Inoue, Y., Konomi, K., and Furuya, M. (1982). *Plant Cell Phys-
iol*. 23:731.
Lagarias, J. C., and Rapoport, H. (1980). *J. Am. Chem. Soc*.
102:4821.
Nakachi, K., Ishikawa, K., and Suzuki, H. (1981). *J. Phys.
Soc. Jpn*. 50:617.
Sakuranaga, M., Nakachi, K., and Suzuki, H. (1979). *J. Phys.
Soc. Jpn*. 46:944.
Shimazaki, Y., Inoue, Y., Yamamoto, K. T., and Furuya, M.
(1980). *Plant Cell Physiol*. 21:1619.
Sugimoto, T., Oishi, M., and Suzuki, H. (1977). *J. Phys. Soc.
Jpn*. 43:619.
Sugimoto, T., Inoue, Y., Suzuki, H., and Furuya, M. (1984).
Photochem. Photobiol. 39:697.
Sugimoto, T., Saito, M., and Suzuki, H. (1985). *J. Phys. Soc.
Jpn*. 54:438.
Suzuki, H., Komatsu, T., and Kato, T. (1973). *J. Phys. Soc.
Jpn*. 34:156.
Suzuki, H., Sugimoto, T., and Ishikawa, K. (1975). *J. Phys.
Soc. Jpn*. 38:1110.
Zener, C. (1932). *Proc. R. Soc. London*. A137:696.

SMALL-ANGLE X-RAY SCATTERING,
A USEFUL TOOL
FOR STUDYING THE STRUCTURE OF PHYTOCHROME[1]

Satoru Tokutomi

Division of Biological Regulation
National Institute for Basic Biology
Okazaki, Japan

Mikio Kataoka
Fumio Tokunaga

Department of Physics
Faculty of Science
Tohoku University
Sendai, Japan

I. INTRODUCTION

To elucidate the molecular basis of the primary action of phytochrome-mediated responses, many reserchers have studied the molecular differences between Pr and Pfr of purified phytochrome. Most of the results obtained so far are related to changes induced in a limited area or site within the molecule, such as exposure of amino acid residues, hydrophobic and proton exchangeable sites (see Song, 1983). Few macromolecular changes, however, have been shown until recently, when the changes in the apparent mol. wt (Lagarias and Mercurio, 1985) and Stokes radii (Sarkar et al., 1984) were shown. The former report showed that the apparent mol. wt of Pfr of oat phytochrome is larger than that of Pr by about 20

[1]Partly supported by Grants (86-106) for Cooparative Reserch at the Photon Factory of the National Laboratory of High Energy Physics, Tsukuba, Japan.

167

kDa, suggesting that considerable changes in the molecular shape may be induced.

An understanding of the conformational changes affecting the secondary, tertiary, or quarternary structure is key to understanding the molecular basis of phytochrome action, since such changes may play important roles in the transduction of the light signal by modifying the interaction of phytochrome with other molecules. At the present time, X-ray crystallography is the only method known to obtain structural information regarding proteins at atomic resolution. That method is, however, applicable to only crystallized proteins. The structure of proteins in solution can be studied using small angle X-ray scattering (SAXS), although its resolution is poorer than that of crystallography. SAXS has some advantages over crystallography since the protein need not be crystallized nor does the effect of crystallization in the analysis of the results need consideration.

Recently, the resolution of SAXS has been remarkably improved by the use of synchrotron radiation as a powerful source of X-rays and by the development of a position sensitive detector system. We have begun a structural study of phytochrome by SAXS and have obtained the following preliminary results.

II. STRUCTURAL INFORMATION OBTAINABLE BY SAXS

The general theory of SAXS is treated in several excellent textbooks (Glatter and Kratky, 1982; Kratky, 1983). We briefly consider here the kinds of information which can be obtained by SAXS. By analyzing SAXS data we can obtain information regarding dimension and shape, subunit or domain structure, and secondary structure of proteins in solutions.

Most information related to dimension comes from a parameter called radius of gyration (Rg), which is the square root of the mean square distance of all individual electrons in the molecule from the center of gravity. Rg reflects the three-dimensional expansion of the electrons in the molecule and can be used as a measure of molecular dimension.

Rg is obtained as follows. The distribution of scattering intensity, I(S), at small-angle can be expressed as

$$I(S)=I(0)\exp[-(4/3)\pi^2 R_g{}^2 S^2],$$

where S $(=2\sin\theta/\lambda)$, 2θ, and λ are the magnitude of scattering vector, scattering angle, and wavelength of X-rays, respectively. When $\ln I(S)$ is plotted against S^2, the plot should be approximated by a straight line, which is known as a

Guinier plot (Guinier and Fournet, 1955).

$$\ln I(S) = -(4/3)\pi^2 Rg^2 S^2 + \ln I(0)$$

Rg can be estimated from the slope of the plot. Subunits or domains whithin a molecule which are distinct in their electron density, also give distinct Rg's. When the detailed shape of protein molecules is determined, one can calculate precise parameters which describe the dimensions of an overall molecule, subunits, or domains, using the Rg's.

Maximum distance, which gives the maximum dimension of solutes, can also be obtained from the distance distribution function, which is the Fourier transform of the scattering curve. Molecular weight, volume, degree of hydration, and specific inner surface can be determined directly from the scattering curve or its Fourier transform.

The overall shape of a protein molecule is obtained from a comparison of the observed intensity data with the calculated ones for possible ideal bodies, by a trial and error procedure. In this comparison, a log I(S) vs. log S plot is used, since this comparison depends only on the shape of the molecules. The detailed shape of the protein molecule can be obtained by constructing a more detailed model including information regarding subunit or domain structure, molecular weight, etc., obtained using other techniques.

III. EXPERIMENTAL PROCEDURES

114 kDa phytochrome from etiolated pea tissue was obtained by the procedure for the 121 kDa phytochrome preparation described by Tokutomi et al. (1986). The 62 kDa tryptic peptide of 114 kDa phytochrome was prepared as described by Yamamoto and Furuya (1983). For measurements, sample solutions were concentrated by precipitation with ammonium sulfate. The precipitated samples were resupended in either 100mM potassium phosphate and 1mM Na_2EDTA (pH 7.8), or in 25mM HEPES and 1mM Na_2EDTA (pH 7.8), then dialyzed against the same buffer to remove remaining ammonium sulfate. The dialyzing buffers were used as control samples. Concentration of the dialyzed samples was adjusted to 2.4 - 0.01% (w/v) using the buffers.

Sample solutions were exposed to X-rays for 10-30 min at 7°C. The exposure did not caused any detectable degradation of the sample. It induced, however, some bleaching of the chromophore. We, therefore, avoided measuring SAXS of the sample exposed further to X-ray. Photoreversibility of the changes in SAXS was examined by comparing the data obtained

from three different samples irradiated with far-red, far-red
then red, and far-red, then red, then far-red light, respectively.

   SAXS was measured with the small-angle X-ray scattering
equipment for solutions (SAXES) at BL-10C of the Photon Factory, National Laboratory of High Energy Physics, Tsukuba,
Japan, using synchrotron radiation as light source. The wavelength of X-ray was 1.5 A. The design and the performance of
the SAXES were presented elsewhere (Ueki et al., 1985). A
block diagram of the measurement system is indicated in Fig.
1. The obtained scattering data were analyzed with a microcomputer using the software developed by one of the authors
(M. K.).

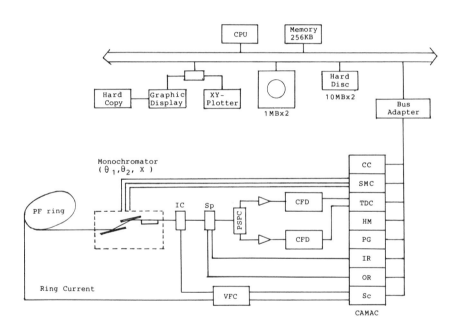

FIGURE 1.   Block diagram of SAXES.

CC	Crate Controller	Sc	Scaler
IC	Ion Chamber	SMC	Stepping Motor Controller
Sp	Specimen	TDC	Time to Digital Convertor
HM	Histogramming Memory	TPG	Timing Pulse Generator
IR	Interrupt Register	OR	Output Register
PSPC	Position Sensitive Proportional Counter		
CFD	Constant Fraction Discriminator		
VFC	Voltage Frequency Convertor		
CAMAC	Computer Aided Measurement And Control system		

IV. PRELIMINARY RESULTS ON PEA PHYTOCHROME

The ultimate goal of this study is to determine what structural changes of intact phytochrome are induced by the phototransformation from Pr to Pfr. A large amount of phyto- chrome is required for the measurements and data analysis. We, therefore, started measurements using 114 kDa phytochrome from etiolated pea tissue, since this phytochrome is easy to prepare in large scale and is more stable than the 121 kDa phytochrome. Here we will describe the preliminary results obained with 114 kDa phytochrome and its tryptic 62 kDa pep- tide.

A. Guinier analysis

The Guinier curve of 114 kDa phytochrome in Pr could not be approximated by a single straight line (Fig. 2), which is due either to heterogeneity of the phytochrome molecules or to subunit or domain structure whithin the phytochrome molecule. The curve can be approximated by a minimum of three straight lines. From each line, Rg can be calculated. For conve- nience, we will designate the Rg's obtained from the lines closest to, second closest to, and farthest from zero angle, as RgI, RgII, and RgIII, respectively (Fig. 2). RgI, RgII,

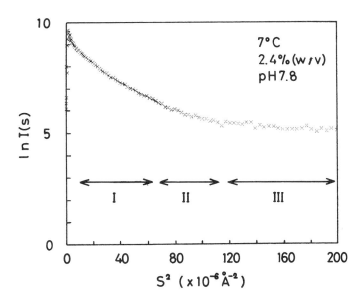

FIGURE 2.   Guinier plot for 114 kDa pea phytochorme in Pr.

and RgIII were calculated as 54, 39 and 15 A, respectively.

RgI is the major component and is independent of the phytochrome concentration between 0.03 to 2.4% (w/v), indicating that RgI is affected neither by interparticle interference nor by aggregation, which is obvious especially at above 0.5% (w/v) judging from the steep increase of I(S) near zero angle (Fig. 2).

The 125 kDa oat phytochrome was shown to exist in dimer form over a concentration range of 0.006–0.066% (w/v) at slightly basic or neutral pH. The amino–terminal region is not required to form the dimer (Jones and Quail, 1986). 118 kDa oat phytochrome, which lacks the 7 kDa amino terminus, is reported also to exist as a dimer (Hunt and Pratt, 1980). 125 kDa oat phytochrome is reported to have the apparent mol. wt of about 300–350 kDa by exclusion column chromatography (Lagarias and Mercurio, 1985; Jones and Quail, 1986). The apparent mol. wt of our 114 kDa phytochrome was measured using a TSK-3000SW column and estimated as 314 kDa, supporting its exsistence in dimer form. RgI, thus, may be correlated to the overall size of the 114 kDa phytochrome dimer. Rg of 47.0 A was obtained for bovine liver gultamate dehydrogenase by SAXS, which had a mol. wt of 312 kDa (Pilz and Sund, 1971). This result supports our interpretation for RgI.

Judging from the SDS gel elecrophoresis pattern and the elution profile on exclusion column chromatography, RgII and RgIII may not come from hetrogeneity of our phytochrome prepa-

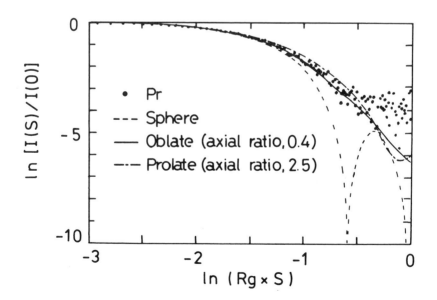

FIGURE 3. Log–log plot of Pr and homogeneous ideal bodies.

rations. RgII seems to be correlated with the dimension of the chromophoric or nonchromophoric domain (52 kDa) in the monomer of 114 kDa pea phytochrome (Yamamoto and Furuya, 1983), since Rg of 41 A was obtained for 62 kDa chromophoric peptide. A Stokes radius of 50 A was reported for small oat phytochrome (Sarkar et al., 1984), which may suppor this interpretation. RgIII is close to the reported Rg of 14.4A for bovine milk α-lactalbumin with mol. wt of 14 kDa (Lee and Timasheff, 1974), suggesting that RgIII may come from the inner structure of the phytochrome monomer.

B. Shape analysis

Shape analysis was performed using log-log plots. The scattering profile of Pr was compared with spheres and ellipsoids of various axial ratios, typical examples of which are shown in Fig. 3. Although the curves are calculated assuming the homogeneous electron density within the bodies, and that the inner structure of phytochrome contributes to the scattering profile , the comparison gives us some idea of the shape of the molecule. The overall shape of the phytochrome molecule (dimer in Pr) can best be approximated by an oblate with axial ratio of 0.4 among the sphere and ellipsoids. The shape and the diameters calculated from RgI are shown in Fig. 4. The maximum diameter estimated from the distance distribution function is about 140 A. That value is reasonable in comparison to the equatorial diamter of 168 A. It has been shown that 118 kDa oat phytochrome in Pr has a Stokes radius of 80 A

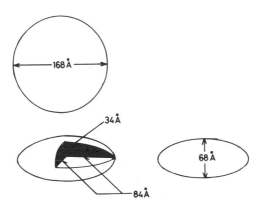

FIGURE 4.   Image of overall molecule of 114 kDa pea
             phytochrome (dimer) in Pr.

(Sarkar et al., 1984). That value is also comparable to the
dimensions assigned in the present model.

C. Photoreversible change

114 kDa pea phytochrome has low Pfr content when in the
red-light-induced photostationary state (Yamamoto and Furuya,
1983) and shows rapid dark processes following illumination
(Tokutomi et al., 1986). The scattering curve of Pfr was
obtained as follows. X-ray scattering was measured in the
dark after red light illumination. The average content of Pr
during exposure to X-ray was determined by monitoring the
absorbance change at 667 nm (absorption maximum of Pr) after
red light illumination using another sample. The contribution
of Pr to the scattering curve was subtracted using the average
content of Pr, which was 67% for 15 min exposure time.
    A pair of Guinier plots for Pr and Pfr is indicated in
Fig. 5. The curves obviously differ from each other. The
sample irradiated with far-red, red and then far-red light,
showed a curve almost superimposable on the curve for the
initial Pr, except for the steep increase near zero angle due
to aggregation of phytochrome molecules. In addition to the
photoirreversible aggregation, photoreversible structural
changes were definitely induced in the phytochrome molecules
which was seen in the Guinier region. Since the plot for Pfr

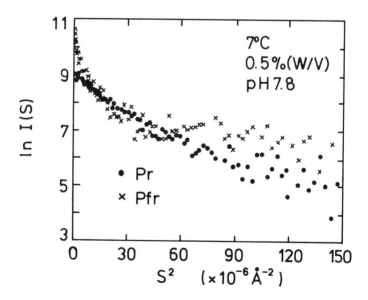

FIGURE 5. Red-light-induced changes in Guinier plot.

is quite noisy due to the subtraction of the contribution of Pr and since the contribution of aggregation is considerable, it is difficult to obtain detailed information for Pfr. RgI and RgIII tend to increase slightly, while RgII tends to decrease slightly. If the interpretation of Rg's are correct, slight expansion or elongation of the overall molecule and of the domain related to RgIII, and slight shrinkage of the chromophoric or nonchromophoric domain are suggested.

V. PROBLEM AND PROSPECTS FOR STRUCTURAL STUDIES OF PHYTOCHROME BY SAXS

A. High value of volume and obscure inner structure

Exclusion chromatography gave a mol. wt of 314 kDa for Pr, suggesting that the dimer is a "nonglobular" protein as is the case for oat 125 kDa phytochrome (Lagarias and Mercurio, 1985; Jones and Quail, 1986). The volume of the oblate obtained for the overall shape of the phytochrome molecule is caluculated as 1,005,000 A^3. Using a partial specific volume of 0.74, the mol. wt of this oblate is calculated as 743 kDa, which is far greater than the mol. wt of 228 kDa for the dimer. A high degree of hydration or dynamic fluctuation of the fragments in the molecule can be a possible cause, but cannot by themselves explain such a high value of the volume. Another explanation for the high value of the volume is that the molecule contains large void space.

The void may come from the crevice between the monomer in the dimeric state, or among the domains or fragments in the monomer. It has been proposed by electron microscopy that phytochrome has an oligomeric structure (Correll et al., 1968). The observed scattering curve, however, showed little periodicity at high angle which is characteristic of the subunit or domain structures, suggesting that monomers and domains contact each other and one another, respectively, over large area and that the overall molecule is tightly built. There is presently no suitable explanation for the high value of the volume. A similar problem will occur also with RgII, 39 A, if RgII reflects the dimension of the chromophoric domain with mol. wt of 62 kDa. For example, Rg of 39.6 A was obserbed with Bakers' yeast malate synthetase of which the mol. wt is 170 kDa (Zipper and Durchschlang, 1978). To understand the origin of these high value of volume and the obscure inner structure, more analysis is required.

B. Low content of Pfr

Another major problem is the low resolution of the scattering profile of Pfr owing to the low content of Pfr after red light illumination. It is shown that 125 kDa oat phytochrome has a high content of Pfr at the photostationary state and that it has no dark reversion (Vierstra and Quail, 1983). Moreover, 125 kDa oat phytochrome seems to be by far more stable than the 121 kDa pea phytochrome (Song, personal communication). The 125 kDa oat phytochrome seems to be better for studies of photoinduced structural changes of phytochrome.

C. Further useful techniques of SAXS

High intensity synchrotron radiation makes it possible to measure SAXS in a very short time. The time resolution of SAXS is now improved to the order of ms. By combining flash photolysis and SAXS, kinetical studies of photoinduced structural changes may be possible. Since phytochrome can photocycle, we can accumulate the changes to improve the signal-to-noise ratio. Although the time resolution is poor compared with other spectroscopic techniques, we can obtain direct information on structural changes in the protein moiety. Flash photolysis SAXS is attractive new technique to be established.

Another useful technique employing SAXS is so-called contrast variation technique. What is observed by X-ray scattering is the difference of electron density between protein and solvent. By changing the electron density of solvent, specific areas can be contrasted against the remaining portion of the protein. Using this technique, we can obtain more precise information regarding the detailed inner structure of phytochrome molecules.

AKNOWLEDGEMENTS

We thank Prof. Furuya for encouraging our study, Mrs. J. Sakai and M. Nakasako for their cooperation in this study, Drs. T. Ueki, Y. Amemiya and Y. Hiragi for giving us useful suggestions and Miss Y. Tsuge for her technical assistance.

REFERENCES

Correll, D. L., Steers, E., Jr., Towe, K. M. and Shropshire, W., Jr. (1968). Biochim. Biophys. Acta 168:46.

Glatter, O. and Kratky, O. (1982). "Small Angle X-ray Scattering." Academic Press, London.

Guinier, A. and Fournet, G. (1955). "Small-Angle Scattering of X-rays." John Wiley & Sons, New York.

Hunt, R. E. and Pratt, L. H. (1980). Biochemistry 19:391.

Jones, A. M. and Quail, P. H. (1986). Biochemistry 25:2987.

Kratky, O. (1983). In "Biophysics." (W. Hoppe, W. Lohmann, H. Markl and H. Zeigler, eds.), p. 65. Springer Verlag, Berlin.

Kumoshinski, T. F. and Pessen, H. (1985). In "Methods in Enzymology vol.117." (C. H. W. Hirs and S. N. Timasheff eds.), p. 154. Academic Press, New York.

Lagarias, J. C. and Mercurio, F. M. (1985). J. Biol. Chem. 25:2415.

Lee, J. C. and Timasheff, S. N. (1974). Biochemistry 13:257.

Pilz, I. and Sund, S. (1971). Eur. J. Biochem. 20:561.

Sarkar H. K., Moon D. K., Song, P. S., Chang, T. and Yu, H. (1984). Biochemistry 23:1882.

Song, P. S. (1983). Annu. Rev. Biophys. Bioengin. 12:35.

Tokutomi, S., Inoue, Y., Sato N., Yamamoto, K. T., and Furuya, M. (1986). Plant Cell Physiol. 27:765.

Ueki, T., Hiragi, Y., Kataoka, M., Inoko, Y., Amemiya, Y., Izumi, Y., Tagawa, H. and Muroga Y. (1985). Biophys. Chem. 23:115.

Vierstra, R. D. and Quail, P. H. (1983). Biochemistry 22:2489.

Yamamoto, K. T. and Furuya, M. (1983). Plant Cell Physiol. 24:713.

Zipper, D. and Durchschlang, H. (1978). Eur. J. Biochem. 87:85.

DICHROIC ORIENTATION OF PHYTOCHROME
- THEORIES AND EXPERIMENTS

Christer Sundqvist[1]

Department of Plant Physiology
University of Göteborg
Göteborg, Sweden

Hiro-o Hamaguchi

Department of Chemistry
Faculty of Science
University of Tokyo
Tokyo, Japan

I. INTRODUCTION

Plants are influenced by light in different ways. Besides photosynthesis the morphological features are altered by light as well as the cell metabolism. All different light-governed processes are dependent on the wavelength and the intensity of the light. The direction of the light which impinges on the plant can be determining for some responses, and further more in some cases the plane of polarization of the incoming light can be of vital importance for the response. Phenomena thus dependent on the direction of the electrical vector of the light are found in fungi (Jesaitis, 1974), algae (Haupt, 1983), mosses (Nebel, 1969), ferns (Furuya, 1978) and higher plants (Zurzycki and Lelatko, 1969). In fungi only blue light photoreceptors are responding to the direction of the electrical vector of the light. In other plant groups both blue and red light photoreceptors can act. For some responses phytochrome is the most prominent photoreceptor. The orientation of chlo-

―――――――――――――――――――――――
[1]Supported by NFR grant B-BU 4259-114.

roplasts within the cell of the alga Mougeotia and the growth
response of the filamentous protonema of ferns are examples
of plant responses oriented by the electrical vector where
phytochrome is clearly involved (Haupt, 1959; Bünning and
Etzold, 1958; Kadota et al., 1982). This chapter deals with
the functional role of phytochrome in processes dependent on
the electrical vector of the light. The presentation is concentrated on the reactions directly connected to the phytochrome molecule. The information gathered from the biological
systems is of fundamental importance in this connection. A
thorough penetration of the reaction chain leading to e.g.
chloroplast reorientation in Mougeotia is made by W. Haupt in
chapter 18 of this volume.

In the case of phytochrome it is well established that we
deal with a photoreceptor which consists of a comparatively
small chromophore and a large apoprotein. During phototransformation the transition moment of the chromophore will reorient. The question can be raised if the reorientation is a
result of a change in the chromophore itself, a conformational
change in the protein or a reorientation of the protein in a
membrane. Many of the presentations and comments given at the
round table discussion titled "Dichroic orientation of phytochrome - theories and experiments" at the XVI Yamada conference can be regarded as a contribution to solve this question.
The present chapter is an account of these discussions. They
comprised all from the physiological effect in whole plant
material to analysis of phytochrome in vitro with laser-Raman
spectroscopy. It is based, besides on the references cited at
the end, on contributions at this round table discussion by
P. Eilfeld, H. Hamaguchi, W. Haupt, Y. Inoue, A. Kadota, M.
Kraml, R. Scheuerlein, P.-S. Song, T. Sugimoto and C. Sundqvist.

II. THE WHOLE CELL SYSTEMS

A. Mougeotia chloroplast reorientation

The Mougeotia cell contains a large single chloroplast
which can rotate to expose its face or its edge towards the incoming light (for a review see Haupt, 1983). The exposure of
the face to the light is a low-intensity response mediated by
phytochrome. Experiments with a microbeam show the photoreceptor to be situated in the cytoplasmic layer close to the cell
wall rather than at the chloroplast. When linearly polarized
light was used for inducing the reaction, chloroplasts reorien-
ted in cells which had the long axis oriented perpendicular to
the electrical vector of the light. It can be deduced from

those and similar experiments that the transition moments of
the red-absorbing form of phytochrome (Pr) are oriented parallel to the cell surface and the molecule alinged along a helical pattern.

The irradiation of the cells with red light induces a
conversion of Pr to the far-red absorbing form of phytochrome
(Pfr). At low fluence rates and short irradiation times a gradient in phototransformation within the cell can easily be understood. A Pfr gradient is necessary to explain movement of
the chloroplast which seems to orient with its edges wherever
a depletion of Pfr can be found. The conversion of Pr to Pfr
leads to the reorientation of the chromophore and the transition moments. This reorientation results in an increased light
absorption of phytochrome at the flanks of the cell where the
transition moment now is parallel with the electrical vector
of the light. It is now mainly Pfr that absorbs light and thus
the equilibrium between Pr and Pfr is adjusted towards Pr. This
leads to the orientation of the chloroplast in a face position
even under continuous irradiation.

Determination of the time course for the reorientation of
the phytochrome chromophore can reveal the phytochrome intermediate which changes the direction of the transition moment and
can also give information about the chemical nature of the
process. The results would have been more easily interpreted
if the physical and chemical background to the existence of
intermediates were known. The time course of the reorientation
of phytochrome from Pr orientation to Pfr orientation in Mougeotia has been studied using a double flash technique (Kraml
and Schäfer, 1983; Kraml et al., 1984). The Mougeotia cells
were irradiated sequentially with unpolarized flashes of 1 ms
red and 1 ms far-red light. The light pulses were separated by
dark periods of different length. Full far-red reversibility
was not attained until all dark relaxation to Pfr was completed after 50 ms.

Using this technique with polarized light, phytochrome
photoconversion was performed in Mougeotia by a red flash vibrating parallel to the cell axis (Kraml et al., 1984). This
established a high level of phytochrome intermediates and
later on of Pfr. After a certain dark interval (during which
the conversion of intermediates to Pfr commences) a non saturating far-red light vibrating perpendicular to the cell axis
was applied. Under the experimental conditions chloroplast movement could only occur if either Pfr or the intermediates
differed in orientation from Pr. Chloroplast reorientation was
not observed until the irradiations were more than 5 ms apart
and was fully developed after 30 ms. This indicates that the
dichroic change of phytochrome orientation in the Mougeotia
cells occurs between 5 and 30 ms and is thus a rather slow

process compared to the light absorption in itself. With laser
flashes it is possible to irradiate with high fluence and with
as short periods as 15 ns. Germination of lettuce seeds can be
induced by 15 ns pulses (Scheuerlein and Braslavsky, 1985).
The wavelengths of the laser flashes used so far have been 620,
660 and 690 nm. A response in germination was obtained with
flashes of 620 and 660 nm but not with 690 nm. This was interpreted as an induction of a photochromic system Pr $\rightleftharpoons$ lumi-R
with 620 and 660 nm irradiations. Nevertheless it was still
possible to produce lumi-R in amounts large enough to permit
dark relaxation to Pfr. At 690 nm the photochromic system was
shifted to a high degree to Pr and the building up of lumi-R
was strongly reduced. This experimental method was applied to
the alga Mougeotia to obtain information about the dichroic
orientation of the phytochrome (Scheuerlein and Braslavsky,
1987). More specifically the first approach was a test to establish the equilibrium with polarized laser pulses to reveal
if the chromophore reorients already at the intermediate
lumi-R.

It turned out that single flash irradiations with the
wavelengths 620, 660 and 690 nm could not induce a change from
profile to face position even if the fluence was high (100
J/m^2). In contrast a full response was obtained with 30 s irradiations with both 660 and 690 nm even if the dose of light
was not higher. When 2-5 flashes with 2 s dark interval were
used 620 and 660 nm gave a full response while 690 nm light had
no effect. The results could best be explained if a nearly similar transition moment for Pr and lumi-R was assumed. With
single flashes, in consequence of the photochromic system, only a weak (not sufficient) gradient was established, as small
amounts of Pfr was formed by dark relaxation of lumi-R to Pfr.
Poly-flash induction, 620 and 660 nm produced gradually an
effective gradient and a movement of the chloroplasts took
place. Because lumi-R has a strong absorption at 690 nm, no
active gradient can be formed by irradiation with 690 nm light
even after poly-flash induction; the system is always adjusted
towards Pr.

If, in contrast, a reorientation was assumed for lumi-R
similar to Pfr, an active gradient would have been expected in
all experiments as the photochromic system would not have been
effective due to the polarization of the laser flashes. This
is, however, not in accordance with the experimental results.

From results obtained with whole cell material it can
thus be concluded that the chromophore in the transient lumi-R
has an orientation similar to that of Pr and that reorientation occurs in a later step during the transformation cycle.
This is in agreement with other experimental data (Kadota et
al., 1986; Kraml et al., 1984).

B. Adiantum protonemata

The protonemata of the fern Adiantum show a polarotropic response that is regulated by phytochrome. Single cell protonemata irradiated with a microbeam at the subapical region, turned its growth direction towards the irradiated side (Kadota et al., 1982). The active phytochrome is located in the ectoplasm or in the plasmamembrane. This was shown by irradiation of the protonemata after a series of 3 repeated centrifugations, which removed all cell components except the ectoplasm, plasmamembrane and microtubuli from the apical region of the protonemata (Wada et al., 1983). The growth reaction was most prominent when light with an electrical vector parallel to the cell surface was used. For reversion of the red light effect the best result was obtained if the direction of the electrical vector of far-red light was perpendicular (Kadota et al., 1982). In detail, however, the arrangement of the phytochrome molecules within the cell is not the same in Adiantum and Mougeotia. The tropic response in Adiantum is controlled by Pfr in both the extreme tip and the subapical region (Kadota et al., 1985). These differences in phytochrome organisation may indicate that also other differences are plausible. The dichroic orientation of phototransformation intermediates were analysed with a polarization plane-rotatable double laser flash irradiator (Kadota et al., 1986). The flash duration was 0.1 µs and two flashes were given 2 µs apart as the shortest delay time. An irradiation with 640 nm light with the electrical vector parallel to the protonemal cell axis caused a tropic growth response. Polarized 710 nm light vibrating normal to the cell axis reversed the polarotropic effect. Such a reversion was also found if the flashes were given 30 s apart but not if they were given 2 µs or 2 ms apart. When the short irradiations were used a reversion was found if the far--red light was polarized parallel to the cell surface. The reorientation of the phytochrome seems to take place between 2 ms and 30 s after the first irradiation in agreement with the results in Mougeotia. As a summary from the results obtained with whole cell material it is possible to state that the reorientation of the phytochrome occurs in the millisecond range, i.e. within a period of time of such a length that it probably reflects not only a reorientation of the chromophore but also a conformational change of the protein.

An interesting prospective for the future would be the application of the double flash technique to plant materials exposed to different temperatures. The temperature should be varied around the membrane transition point as to reveal the interaction between the phytochrome molecule and a possible host membrane.

C. Dichroic effect on seed and spore germination

Several attempts have been made to show light induced
linear dichroism in well-known phytochrome regulated processes
in higher plants. One such experiment has been to reveal a dichroic effect in lettuce seed germination (Björn personal communication). This has mostly not been successful. This may be
due to the complexity of the plant tissue. A less complex system is the fern spore whose germination is also governed by
red light irradiations. Initial experiments with this system
were reported by Haupt (Haupt and Björn, 1987). The fern spore
system is assumed not to have an oriented response in contrast
to the earlier studied systems. From the fluence response curves it could be calculated that the best action dichroism
should be found when 40% Pfr is present. The fern spores are,
however, very sensitive to Pfr and in standard experiments the
induction of germination is saturated at this level. However,
even if spores are made less sensitive, as to be saturated at
much higher Pfr levels, no action dichroism has been possible
to detect. A slightly more promising approach seems to be to
study the reversal of red induced germination. Experiments are
started and the results are awaited with excitement.

II. IMMOBILIZED PHYTOCHROME

The whole cell system has of course the highest biological
relevance although the complexity of the system makes the results difficult to interpret. A more uniform system is found
in immobilized purified phytochrome. The advantage of simplicity is of course correlated with a loss of relevance for the
biological system. For example no information is obtained about
a possible rotation of the protein in relation to the cell
membranes as the protein molecule is attached to agarose beads
during these experiments. Furthermore the immobilization itself can influence the properties of the phytochrome molecule.
The original method (Björn, 1981; Sundqvist and Björn, 1983a)
made use of phytochrome immobilized on polyclonal antibodies
attached to CNBr-activated agarose. The absorption of the
sample was measured in a dual-wavelength spectrophotometer with
the two beams polarized in right angles to each other. The
sample was irradiated with polarized red light and the absorption measured again. Molecules, during the immobilization,
accidentally oriented with the transition moment parallel to
the electrical vector of the irradiation light were preferentially photoconverted and were no longer registered by the
spectrophotometer. This causes a decreased signal during the

absorption measurements in one direction and hence an absorption dichroism. The change in absorption could be used to calculate the rotation angle for the phytochrome chromophore.

The ratio R between the change in linear dichroism at 730 nm and 660 nm was used to calculate the rotation angle of the transition moment according to the formula:

$$R = \frac{8.33(1 - 1.5 \sin^2 \theta) - 0.27}{4.63(1 - 1.5 \sin^2 \theta) - 14.0}$$

R is determined experimentally and θ is the angle of rotation. The value obtained was $32°$ (or $180° - 32°$) for large phytochrome (Sundqvist and Björn, 1983a) and $31°$ (or $180° - 31°$) for native phytochrome (Ekelund et al., 1985).

The attachment of the phytochrome molecule to antibodies might influence the photoconversion process, and the bonds established might not stop the rotation of the whole molecule (protein + chromophore). An indirect method to evaluate the interaction between the phytochrome and the antibody on the rotation angle is to immobilize phytochrome directly to the sepharose beads. For this purpose CNBr-activated Sepharose, phenyl-Sepharose and octyl-Sepharose were used. The spectral properties of the samples were not identical to those of phytochrome in solution. To be able to make the calculation of the angle of rotation, relative absorption values had to be calculated. The results obtained gave values around $31.5°$ (or $180° - 31.5°$) for the change in the angle of the transition moment of phytochrome (Sundqvist and Ekelund, 1986). For the immobilized phytochrome a change in linear dichroism induced with red polarized light could be reversed with polarized far-red light. The most effective far-red light was obtained when the angle between the planes of polarization of red and far-red light was approximately $23°$. This corresponds again to a rotation angle of approximately $32°$ (or $180° - 32°$) (Sundqvist and Björn, 1983b).

A promising approach which in the future might give valuable information about the conformational changes coupled to the Pr $\rightleftharpoons$ Pfr transformation was presented by Y. Inoue. Isolated phytochrome was immobilized on tresyl-activated Sepharose beads with the help of monoclonal antibodies. The advantages of the use of monoclonal antibodies are the knowledge about the binding site and the consequence a binding to a specific part of the phytochrome protein might have on the angle of chromophore rotation. This gives the possibility to determine the part of the protein moiety which is involved in conformational changes during photoconversion. To be able to reach this conclusion, different values for the angle of the

chromophore rotation must be obtained. The preliminary results
were partly discouraging on this point. With the method used
all tested antibodies (mAP-1, mAP-5, mAP-6, and mAP-10) gave
similar results, which approached an angle for the reorientation of 50°. The experimental values obtained could also be
interpreted as if some free rotation of the phytochrome molecule existed. This would be possible if free phytochrome is
present in the sample or if the bound phytochrome still has a
possibility of free rotation. At present it is not possible to
reach any conclusive decision on this point as too little is
known about the interaction between the antibodies and the
phytochrome molecule. A free rotation is possible if the phytochrome is attached at only one place. Experimental confirmation of the immobility of the phytochrome molecules is needed
irrespective of which types of antibodies (monoclonal or polyclonal) are used.

III. THE CHROMOPHORE AND THE APOPROTEIN

The change in the angle of rotation of the transition moments of Pr and Pfr found in the immobilized phytochrome might
be refered to a change in the chromophore or a conformational
change in the protein induced by the photochemical reaction in
the chromophore. The primary photoreaction may be a Z-E photoisomerization of the 15-16 double bond (Rüdiger, 1983; Rüdiger
et al., 1985; Sugimoto et al., 1984). There is probably a free
rotation around the 14-15 single bond. The ring D could, however, not rotate as this would bring peripheral methyl groups
too close. Such a rotation could possibly occur during a
simultaneous Z-E photoisomerization at the 15-16 double bond.
A rotation of the single bond by 45° could change the transition moment by 14°. During the transformation of Pr to Pfr no
additional change in the transition moment takes place. Even
if it is not possible to exclude the chromophore isomerization
as the only reaction it is more probable that this reaction is
followed by an interaction between the N-terminus and the
chromophore. This is supported by a change in the CD-signal,
as the first reaction in the phototransformation also gives
rise to a change in the CD-signal. This would indicate a strong
conformational change which could be interpreted as a result
of the Z-E isomerization at the 15-16 double bond (Eilfeld and
Rüdiger, 1985; Eilfeld et al., 1986 a,b). This change in CD-
-signal is photoreversible and represents the step from Pr to
lumi-R. For the next intermediates, meta-Ra and meta-Rc, the
CD-signals are weak. Also Pfr shows a weak CD-signal but of
opposite sign. The Pr form and the intermediates lumi-R and

meta-Ra seem to represent the same area of interaction between the chromophore and the protein part. The conversion to meta-Rc might involve a small conformational change, but up to this stage the transition moment has kept its orientation. During the transition from meta-Rc to Pfr the chromophore changes its orientation and this results in the interaction with new protein domains probably the N-terminus. This will also include a rotation which brings the chromophore into a more exposed position as phytochrome in the Pfr form is more easily degraded by bilirubin oxidase (Song et al., 1986). The chromophore may turn out of the hydrophobic environment, where the Pr form was located. In the Pr form the long wavelength transition moments are parallel with each other and they are almost parallel to the long axis of the molecule. The change in the transition moment during phototransformation seems to be small or close to 14° (Song unpublished results, Sugimoto, 1986). Conformational changes within the protein might lead to a larger change of the transition moment. According to the results of Sundqvist and Björn (1983a) it can be approximately 32°. The conformational change might then lead to a rotation of the whole phytochrome molecule to give a total rotation that approaches 90°.

There are several indications for a conformational change leading to the reorientation of the long wavelength transition moment of phytochrome (Eilfeld et al., 1986a; Chai et al., 1986). The quantitative documentation of the change is, however, still limited to the results obtained with dichroic measurements on immobilized phytochrome. This is an artificial system, however, and much has to be clarified about the physical and chemical properties of this system. It is encouraging, that the results obtained in vivo can be interpreted in accordance with an in vitro system (Björn, 1984; Sugimoto, 1986). Even if the light used for reversal of a red light induced effect is most effective when the plane of polarization is rotated by 90° this can still mean that the long wavelength transition moment has rotated approximately 35° from Pr to Pfr.

Further detailed discussion on the dichroic properties of phytochrome needs more precise knowledge on the nature of the long wavelength absorption of the chromophore. Although the two absorption peaks are conventionally called Qx and Qy from the analogy with the porphyrins, there is a possibility that they are due to only one electronic transition and its vibrational structure. The potentiality of Raman spectroscopy in elucidating this matter was pointed out by Hamaguchi at the discussion. The application of ordinary Raman spectroscopy to phytochrome has been hindered by strong fluorescence from the chromophore itself. More efforts should be directed to the use

of non-linear Raman spectroscopy including IRS (inverse Raman
spectroscopy) and CARS (coherent anti-Stokes Raman spectroscopy).

IV. CONCLUDING REMARKS

On the first view the response of plants towards polarized
light is difficult to explain in terms of usefulness for the
plant. However, in Mougeotia for example the action dichroism
is the basis for detecting light direction. Further explora-
tion of the phenomenons regulated by action dichroism will
give a better understanding for their importance.

The phenomenon of action dichroism has, however, been a
tool to compare the chemical relation between phycochrome b
and phytochrome (Björn et al., 1984). The rotation angle is,
however, 0° for phycochrome b but approximately 32° for phytochrome which indicates that the molecules probably have
different origin. A similar and even more decisive conclusion
has been reached recently by comparing the amino acid sequences of the protein part of phytochrome and phycobilins
(Hershey et al., 1985).

The action dichroism in itself might furthermore be an
important tool in the research efforts to elucidate the mechanism of the phytochrome action. To obtain a response to polarized light the photoreceptor (phytochrome) must be immobilized
in the cell or it must be localized in an area where a con-
straint in free rotation can be imposed on the molecule. A
freely rotating molecule would easily ·lose the information
about light polarization. This gives a possibility to distinguish between bound and free phytochrome and reactions
governed by the different types of phytochrome. This is a
possibility that probably can be more explored in the future.

ACKNOWLEDGMENTS

The authors are much indebted to Professor W. Haupt for
valuable comments on the manuscript.

REFERENCES

Björn, L.O. (1981). Photochem. Photobiol. 33:707.
Björn, L.O. (1984). Physiol. Plant. 60:369.

Björn, G., Ekelund, N.G.A. and Björn, L.O. (1983). Physiol.
    Plant. 60:253.
Bünning, E. and Etzold, H. (1958). Ber. Deutsch. Bot. Ges.
    71:304.
Chai, Y.G., Song, P.-S., Cordonnier, M.-M. and Pratt, L.H.
    (1986). In "Proceedings of the XVI Yamada conference
    Phytochrome and Plant Photomorphogenesis" (M. Furuya,
    ed.), p. 66. Okazaki, Japan.
Eilfeld, P. and Rüdiger, W. (1985). Z. Naturforsch. 40c:109.
Eilfeld, P., Eilfeld, P. and Rüdiger, W. (1986a). In "Pro-
    ceedings of the XVI Yamada conference Phytochrome and
    Plant Photomorphogenesis" (M. Furuya, ed.), p. 62.
    Okazaki, Japan.
Eilfeld, P., Eilfeld, P. and Rüdiger, W. (1986b). Photochem.
    Photobiol. 44:761
Ekelund, N.G.A., Sundqvist, C., Quail, P.H. and Vierstra, R.D.
    (1985). Photochem. Photobiol. 41:221.
Furuya, M. (1978). Bot. Mag., Tokyo, 1:219.
Haupt, W. (1959). Planta 53:484.
Haupt, W. (1983). In "The Biology of Photoreception" (D.J.
    Cosens and D. Vince-Prue, eds), p.423. Society for Ex-
    perimental Biology, Great Britain.
Haupt, W. and Björn, L.O. (1987). J. Plant Physiol. in press.
Hershey, H.P., Barker, R.F., Idler, K.B., Lissemore, J.L., and
    Quail, P.H. (1985). Nucleic Acids Research 13:8543.
Jesaitis, A.J. (1974). J. Gen. Physiol. 63:1.
Kadota, A., Wada, M. and Furuya, M. (1982). Photochem. Photo-
    biol. 35:533.
Kadota, A., Wada, M. and Furuya, M. (1985). Planta 165:30.
Kadota, A., Inoue, Y. and Furuya, M. (1986). Plant Cell Phy-
    siol. 27:867.
Kraml, M. and Schäfer, E. (1983). Photochem. Photobiol. 38:461.
Kraml, M., Enders, M. and Burkel, N. (1984). Planta 161:216.
Nebel, B.J. (1969). Planta 87:170.
Rüdiger, W. (1983). Phil. Trans. R. Soc. Lond. B 303:377.
Rüdiger, W., Eilfeld, P., Thümmler, F. (1985). In "Optical
    Properties and Structure of Tetrapyrroles (G. Blauer and
    H. Sund, eds.), p. 349. Walter der Gruyter, Berlin, New
    York.
Song, P.-S., Kwon, T.-I., Singh, B.R., Choi, J. and Chai, Y.G.
    (1986). In "Proceedings of the XVI Yamada conference
    Phytochrome and Plant Photomorphogenesis" (M. Furuya, ed.),
    p.35. Okazaki, Japan.
Scheuerlein, R. and Braslavsky, S.E. (1985). Photochem. Photo-
    biol. 42:173.
Scheuerlein, R. and Braslavsky, S.E. (1987). In press.
Sugimoto, T., Ito, E. and Suzuki, H. (1986). In "Proceedings
    of the XVI Yamada conference Phytochrome and Plant Photo-

morphogenesis" (M. Furuya, ed.), p. 58. Okazaki, Japan.

Sugimoto, T., Inoue, Y., Suzuki, H. and Furuya, M. (1984). Photochem. Photobiol. 39:697.

Sundqvist, C. and Björn, L.O. (1983a). Photochem. Photobiol. 37:69.

Sundqvist, C. and Björn, L.O. (1983b). Physiol. Plant. 59:263.

Sundqvist, C. and Ekelund, N.G.A. (1986). Physiol. Plant. 66:185.

Wada, M., Kadota, A. and Furuya, M. (1983). Plant Cell Physiol. 24:1441.

Zurzycki, J. and Lelatko, Z. (1969). Acta Soc. Bot. Pol. 38:493.

IV. PROBLEMS AND PROSPECTS OF PHYSIOLOGICAL APPROACHES

PHYTOCHROME INTERACTIONS WITH PURIFIED ORGANELLES

Stanley J. Roux[1]

Department of Botany
The University of Texas at Austin
Austin, Texas 78713 USA

INTRODUCTION

What are the critical early steps in the transduction
chain leading from the photoactivation of phytochrome to the
altered growth and development of the irridiated plants?
Several discussions and symposium talks addressed this
question during the XVI Yamada Conference. In this context,
the long-standing debate on whether Pfr-induced membrane
alterations were important for photomorphogenesis was renewed.
This chapter will review recent experiments designed to test
questions raised in this debate, focusing on experiments that
examined phytochrome-induced responses in purified organelles.

To understand the context of these experiments it will be
useful to comment briefly on the debate about the membrane
hypothesis of phytochrome action. In its simplest form this
hypothesis proposes that Pfr-induced changes in the function
of certain cellular membranes are important early events (time
scale of seconds) in photomorphogenesis. The debate about the
validity of this hypothesis has a long history, and the reader
is referred to any one of many previous reviews (e.g.,
Hendricks and Borthwick, 1967; Marmé, 1977; Quail, 1983; Roux,
1986) for a detailed exposition of the main arguments pro and
con. The key data often quoted in the debate may be
summarized in an oversimplified form as follows. Favoring the
notion that membrane changes may be important early events in

[1]Supported by NSF grant PCM 8402526 and by USDA grant
86-CRCR-1-2001.

photomorphogenesis are the observations of rapid membrane or
surface potential changes induced by Pfr (Racusen, 1976;
Weisenseel and Ruppert, 1977; Newman, 1981), or rapid ion
fluxes induced by Pfr (Hale and Roux, 1980; Das and Sopory,
1985), and the evidence that one specific ion flux, that of
Ca^{2+}, may act as a second messenger for certain specific
phytochrome repsonses (Wayne and Hepler, 1984; Serlin and
Roux, 1984; Roux et al., 1986). Those skeptical of the
membrane hypothesis emphasize correctly that there is no
definitive in vivo evidence that phytochrome interacts
directly with membranes and that for many phytochrome
responses, such as altered gene expression, there is no
compelling evidence that membranes are involved at all.

For many phytochrome responses, the lack of compelling
evidence one way or another on the question of membrane
involvement reflects mainly the lack of critical
experimentation testing the question. Among the relatively
few laboratories designing experiments to test the involvement
of membranes in phytochrome responses, some are using intact
cellular systems; others are using cell-free systems. The
advantages and disadvantages of both approaches have been
evaluated (Roux, 1986). In this chapter we will restrict
our emphasis to the results and implications of the most
recent experiments that have examined phytochrome-organelle
interactions. These studies utilized as test systems mainly
purified mitochondria and nuclei, so most of our discussion
will concern these organelles.

EVIDENCE OF PHYTOCHROME ASSOCIATION WITH ORGANELLES IN CELLS

Ideally, a key rationale for studying phytochrome-
organelle interactions in vitro would be founded on in situ
immunocytochemical evidence that phytochrome is associated
with one or more of the cellular organelles. Unfortunately,
the results of immunocytochemical surveys are equivocal
(Pratt, 1986), neither confirming nor disproving phytochrome-
organelle associations. The first ultrastructural-level study
of the subcellular localization of phytochrome indicated that
significant levels of anti-phytochrome immunolabel could be
detected in association with mitochondria, nuclear envelopes,
endoplasmic reticulum and plasma membranes (Coleman and Pratt,
1974; Fig. 1a). Subsequent ultrastructural surveys (e.g.,
Speth et al., 1986; McCurdy and Pratt, 1986; Fig. 1b) have
not found definitive evidence of the association of immunolabel with any cellular membrane. From the more recent work,
it seems apparent that if phytochrome is present at all on
membranes, the level of it detectable with the antibodies and

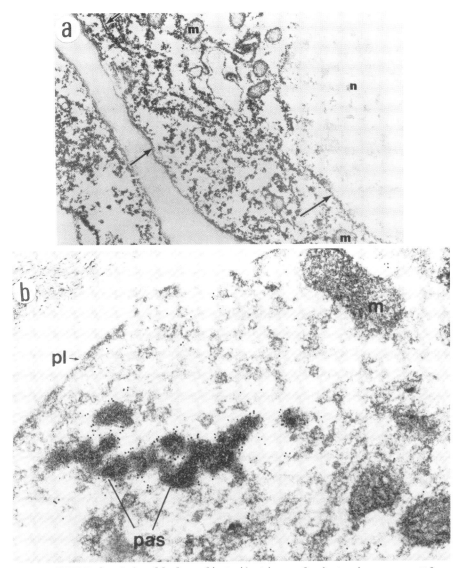

FIGURE 1. Subcellular distribution of phytochrome as Pfr
in dark-grown <u>Avena</u> coleoptile tissue.
(a) From Coleman and Pratt (1974), using peroxidase
immunolabeling. As the section is otherwise unstained,
electron density is associated only with phytochrome.
Plasma membrane and mitochondria are stained.
(b) From McCurdy and Pratt (1986). Phytochrome-specific
immunolabel (20 nm colloidal gold) is associated with
undefined structures (PASs); no immunolabel is found in
association with mitochondria (m) or the plasmalemma (pl).

fixation techniques now being used is near or below the limit
of sensitivity of the method.

On the other hand there is no question that Pfr induces
functional alterations in a number of organelles in intact
cells in vivo, including mitochondria (Bajracharya et at.,
1976), chloroplasts / etioplasts (Schopfer and Apel, 1983),
and nuclei (Tobin and Silverthorne, 1985). Based on these
results it is certainly valid to ask what is the minimal
cellular or subcellular unit that will support red-light-
induced changes in organelle function. During the last ten
years, experiments designed to answer this question have been
carried out using a variety of purified organelles as test
systems, mostly mitochondria, plastids, and nuclei. As
reviewed in Roux (1986), photoreversible responses have been
reported in most of the purified organelles tested. Though
these results certainly allow for alternative explanations,
they are consistent with the interpretation that phytochrome
can have physiologically significant effects, directly or
indirectly, on a variety of plant cell membranes.

DETECTION OF PHYTOCHROME IN ISOLATED ORGANELLE PREPARATIONS

Mitochondria

Though the presence of phytochrome in preparations of
mitochondria, plastids and nuclei could be inferred from the
observation of photoreversible spectral of physiological/
biochemical changes in the preparation, only in the case of
purified mitochondria was phytochrome actually detected
immunocytochemically in assocation with the organelle (Roux
et al., 1981). All authors who have reviewed this result
have discussed the possibility that this association could
be artifactual, and there is general agreement that at least
some of the phytochrome found in purified organelle
preparations probably ended up there as a result of non-
specific associations that occured during the isolation
procedure. However, the assumption that all of the
phytochrome in organelle preparations is there only
incidentally should not be taken as proven or even probable.

In the case of phytochrome associations with
mitochondria, there are at least two reasons to believe they
may have biological significance. First, one of the responses
induced by Pfr in purified mitochondria--the release of Ca^{2+}
from them (Roux et al., 1981)--would have a biological
consequence in vivo consistent with observed phytochrome
actions: viz., Pfr induces an increased cytosolic $[Ca^{2+}]$
(Miyoshi, 1986), and some Pfr responses appear to utilize
Ca^{2+}-mediated transduction pathways (Wayne and Hepler, 1985;

Roux et al., 1986).

A second reason is that the association of phytochrome with mitochondria which is induced by an in vivo light treatment is almost certainly more integral than the low-affinity surface binding that would be predicted of a non-specific interaction. Most of the phytochrome that associated with mitochondria randomly during the purification would be expected to have a low affinity attachment to the mitochondrial surface and thus should be readily removable by an extensive protease treatment. The results of Serlin and Roux (1986) indicate that this is not the case. Extensive proteolysis of mitochondria isolated from red-light irradiated oat seedlings removes only about 25% of the immunoprecipitable phytochrome from the preparation, while the remaining 75% all retains the full native monomer molecular weight of 124 kDa (Fig. 2:Table I). In these experiments, the association of phytochrome with mitochondria required that it first be photoactivated to Pfr in vivo, for without the in vivo actinic irradiation there was no phytochrome detectable in the mitochondrial preparation (Fig. 2). These findings are in agreement with most of the in situ immunocytochemical studies to date, which indicate that Pr has a primarily cytosolic locale, with little or no detectable associations with any cellular membranes (Pratt, 1986). Control experiments incubating exogenous Pr with mitochrondria isolated from unirradiated plants showed that any phytochrome that bound to this preparation (see Cedel and Roux, 1980a) was rapidly degraded by proteases (Fig. 2). Thus, apparently only Pfr can be "sequestered" or otherwise protected from proteases in purified oat mitochondrial preparations. The protease treatments used by Serlin and Roux were essentially the same as those used to study the incorporation of cytoplasmically translated proteins into their target organelles. These results are consistent with the interpretation that Pfr may be incorporated into mitochondria in vivo, but such a conculsion would be premature without substantiating evidence. However, these data do render improbably the postulate that most of the phytochrome bound to purified mitochondria is there only as a consequence of the renowned propensity of Pfr to stick to charged surfaces.

Are mitochondria, then, appropriate model systems for studying phytochrome action in cells? The crucial missing data needed to answer this question is a resolution of whether Pfr ever associates with mitochondria in vivo. Although the early results of Coleman and Pratt appeared to answer this question in the affirmative, the more recent results are not supportive of this conclusion (Speth et al. 1986; McCurdy and Pratt, 1986). Approaches to resolving this issue will be discussed further in the last section of this chapter.

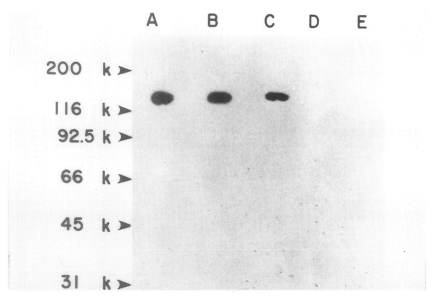

FIGURE 2. Fluorogram of ^{125}I-labeled phytochrome immunoprecipitated from mitochondria that were extracted from irradiated (lanes A and C) or unirradiated (lanes D and E) tissue. Lane A, phytochrome recovered from proteinase-treated mitochondria; Lane B standard of purified phytochrome (200 ng of protein loaded); Lane C, phytochrome recovered from proteinase-treated mitochondria; Lane D, phytochrome recovered from untreated mitochondria (tissue not irradiated); Lane E, phytochrome recovered from proteinase-treated "dark" mitochondria which had been incubated with exogenously added phytochrome in its inactive Pr form. (from Serlin and Roux, 1986).

Table I. Immunoprecipitated Phytochrome from Untreated and Proteinase-Treated Mitochondria[a]

Sample	Immunoprecipitated Phytochrome (cpm)		Average B/A (%)
	Untreated (A)	Proteinase Treated (B)	
I	102660	77850	73.5
II	101840	72570	

[a]Samples I and II were from the same preparation of mitochondria, but were iodinated (^{125}I) and subjected to immunoprecipitation as separate samples. All cpm values were normalized for equivalent amounts of mitochondrial protein.

More biochemical studies could also help to determine
whether phytochrome has specific and/or biologically relevant
interactions with mitochondria. For example, it should be
possible to identify the receptor(s) in purified mitochondrial
preparations to which phytochrome binds with a high affinity
in vitro (Cedel and Roux, 1980a), and to determine whether
this receptor is in or on mitochondria in situ. Also possible
would be to establish whether actinic irradiations given in
vivo induce changes in the same mitochondrial enzyme
activities that have been reported to be photoreversibly regulated in vitro (Cedel and Roux, 1980b; Serlin et al., 1984).

Nuclei

Just as in the case of mitochondria, nuclei were
identified as a site of phytochrome localization by the early
immunocytochemical results of Coleman and Pratt(1974) but not
by more recent surveys (Speth et al., 1986; McCurdy and Pratt,
1986). Also paralleling certain mitochondrial results (Serlin
and Roux, 1986, highly purified preparations of pea nuclei
have associated with them a significant level of phytochrome
which does not appear to be merely surface bound, for it
remains associated with the nuclei even after they are
extracted with the detergent Triton X-100 (Nagatani et al.,
1986). Additionally, there are at least two Pfr-induced
nuclear responses that can be observed in the isolated
organelles as a consequence of an in vivo irradiation.
One of these is definitively related to subsequent
photomorphogenic changes in plants. This is the enhanced
run-off transcriptions of chlorophyll a/b protein and of the
small subunit of ribulose bisphosphate carboxylase that are
measured in isolated nuclei following a red-light treatment
given in vivo (see review, Tobin and Silverthorne, 1985).
These transcriptional responses are clearly involved in the
regulation of chloroplast morphogenesis by light. Of
themselves, they do not imply any direct interaction of
phytochrome with nuclei, but the related report of Ernst and
Oesterhelt (1984) does. They found that purified phytochrome
added to a preparation of isolated rye nuclei can influence
the rate of transcription occuring in these nuclei. These
important results need to be substantiated with more detailed
studies.
A second nuclear response that can be observed in vitro
following an in vivo phytochrome photoactivation is the
stimulation of a nuclear ATPase. This was first reported by
Wagle (1980) (Fig. 3a) and was subsequently confirmed by Chen
(unpublished data). Chen and Roux (1986) have found that
this same enzyme activity (more accurately termed an NTPase,
since it can also utilize GTP, UTP, and CTP as substrates) can

A

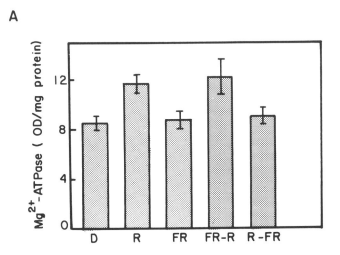

B

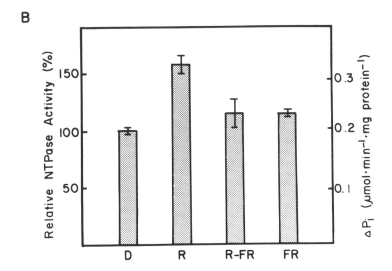

FIGURE 3.   Phytochrome-regulated ATPase activity in pea
nuclei.
    (a) Irradiations given in vivo, activity measured in
nuclei isolated from etiolated pea seedlings. (from Wagle,
1980).
    (b) Irradiations given to purified nuclei isolated from
etiolated pea seedlings.   (from Chen and Roux, 1986).

be photoreversibly stimulated also by actinic irradiations given in vitro to purified pea nuclei (Fig. 3b). These results imply that phytochrome is at least present in the preparation of nuclei, if not directly bound to them. As noted above, Nagatani et al. (1986) find that the phytochrome in their preparation of pea nuclei remains bound with nuclear matrix material even after a detergent and salt wash of the nuclei. An even more convincing demonstration of phytochrome-nuclear association would be direct immunocytochemical localization of phytochrome in either purified pea nuclei, or, preferably, in the nuclei of intact cells of fixed pea tissue. As yet this has not been done (see Saunders et al., 1983 and discussion below), and so the question remains unresolved.

In an attempt to further characterize how phytochrome regulates nuclear ATPase activity, Chen and Roux (1986) tested how the enzyme was affected by various cations, inhibitors and chelators when pea nuclei were incubated in them. They found that 20-30 micromolar Ca^{2+} could stimulate the Mg^{2+}-dependent nuclear NTPase several fold, and that the stimulation of this enzyme by light was blocked by EGTA and by calmodulin antagonists. These results indicated that the biochemical mechanism by which Pfr stimulated the nuclear NTPase required Ca^{2+}, quite possibly for the activation of Ca^{2+}-binding regulatory proteins such as calmodulin.

Recent fractionation studies of Chen and Roux (manuscript submitted) show that as much as half the NTPase activity in purified pea nuclei is due to a single chromatin-associated protein with an apparent monomer molecular weight of 47 kDal. The activity of this enzyme is stimulated over 3-fold by purified oat calmodulin in the presence, but not in the absence, of Ca^{2+}. According to these characteristics, this enzyme is probably the same as the Ca^{2+} and calmodulin regulated ATPase assayed in a chromatin-enriched fraction of pea seedlings by Matsumoto et al. (1984) (Fig.4). It is also very likely to be the same nuclear NTPase as the one that is stimulated by Pfr through a mechanism that is Ca^{2+}-dependent and sensitive to calmodulin antagonists (Chen and Roux, 1986). Pea nuclei in intact cells contain calmodulin (Dauwalder et al., 1986), and so the elements needed to have a Ca^{2+}/calmodulin-based mechanism for controlling NTPase and other nuclear activities are in place.

The relevance of the light-regulated NTPase activity to light-regulated transcription rates in nuclei is unknown. Grossman et al. (1981) have postulated that nuclear NTPases could play an important role in the regulation of transcription by controlling the pool size of nucleotides needed for RNA polymerase II activity. However by this mechanism, the stimulation of nuclear NTPases by light could only have a generalized impact on transcription. Some additional control factor, such as the selective binding or association of

the NTPase with only certain regions of the chromatin, would
have to be invoked to account for the specific effects of
Pfr on the synthesis of selected messages. A library
of monoclonal antibodies has been raised against the
calmodulin-regulated nuclear NTPase, and immunogold
localization studies are now underway to examine the
ultrastructural distribution of this enzyme in pea nuclei.

In the nuclei of SV40-3T3 cells, one of the major NTPases
is regulated by a phosphorylation-dephosphorylation mechanism
(Purrello et al., 1983). Since there are precedents for
enzymes being under the dual control of calmodulin and
phosphorylation (Malencik and Fischer, 1982), Datta et al.
(1985) examined whether the actinic red light that stimulated
the pea nuclear NTPase affected its phosphorylation status.
Their preliminary results indicate that, indeed, light does
stimulate the phosphorylation of a 47 kDa protein in pea
nuclei in a Ca^{2+} dependent fashion (Datta et al., 1985).
However, immunological analyses will be necessary to determine
for certain whether this protein is the major nuclear NTPase.
Interest in the general question of the regulation of protein
phosphorylation by phytochrome, both in nuclei (Datta et al.,
(1985) and throughout the cell in general (Otto and Schäfer,
1986), has been heightened by the discovery that protein
kinase activity is associated with highly purified
preparations of phytochrome (Wong et al., 1986).

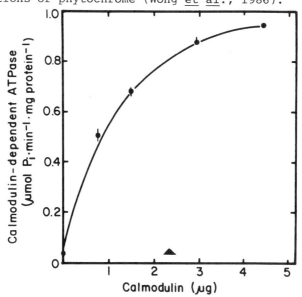

Figure 4. Activation of ATPase activity in the chromatin
fraction of pea nuclei by calcium and pea calmodulin. (▲),
Activity with 2.2 ug pea calmodulin + 5 mM EGTA (no Ca^{2+}).
(from Matsumoto et al., 1984)

As reviewed above, Ca^{2+} appears to mediate the Pfr-induced stimulation of both NTPase activity and nuclear protein phosphorylation. This implies that Pfr somehow modulates the availability of Ca^{2+} in the nuclear preparation. Experimentally determining if Pfr does this, and if so, how?, would clarify whether phytochrome-membrane interactions were involved in the light-induced enzyme responses in nuclei. Although none of the Ca^{2+}-dependent light responses in pea nuclei have been linked to Pfr-stimulated changes in transcription rates, there is currently much interest in the question of whether Ca^{2+} can serve as a second messenger for regulating mRNA synthesis. Resendez et al. (1985, 1986) have shown that a chemically-induced uptake of Ca^{2+} into Chinese hamster fibroblast cells stimulates the enhanced transcription of glucose-regulated genes. Matsumoto and Taku (1986) recently found that Ca^{2+} could induce changes in the absorption spectrum, sedimentation, and melting profile of pea chromatin. They interpreted their results to mean that Ca^{2+} could promote changes in the structure of chromatin that could alter its template activity. A crucial technical advance that would permit researchers to critically test the role of Ca^{2+} in transcription, and, indeed, to understand transcriptional regulation in general, would be the development of techniques for reliably initiating transcription in isolated nuclei.

To summarize this section, isolated nuclei exhibit two photoreversible responses that are potentially relevant to understanding photomorphogenesis. Whether nuclei are relevant model systems for testing the membrane hypothesis cannot yet be determined, largely, again, because of the lack of definitive data on whether phytochrome is associated with nuclei in situ. Since this problem seems to be the major stumbling block to reaching a consensus on the utility of studying phytochrome responses in isolated organelles, we will discuss it more in detail in the following section.

IN VITRO RESULTS VS. IN SITU IMMUNOCYTOCHEMISTRY: RECONCILIATION?

In vitro results favor the conclusion that phytochrome (or, at least, Pfr) associates with organelles; recent in situ immunocytochemistry surveys do not. The simplest reconciliation of these seemingly conflicting results would be to show that one or other of these results is an artifact of the techniques used.

The easiest artifact to imagine is that the association of photoactivated phytochrome with organelles is a low-affinity surface phenomenon arising out of the relatively greater propensity of Pfr to stick to charged surfaces. As reviewed above, this explanation does not appear to hold for Pfr-mitochondria associations.

A somewhat less obvious artifact is the encapsulation of soluble cytoplasmic components into membrane vesicles generated during tissue homogenization and the co-sedimentation of some of these vesicles with organelles. Some percentage of the bulk soluble phytochrome would be expected to be randomly encapsulated during the vesiculation of membranes that are torn in the initial homogenization step, and, given the heterogeneity of these vesicles, some fraction of them would probably contaminate every preparation of organelles, even those that were extensively purified. Such encapsulated phytochrome would be inaccessible to the proteases used by Serlin and Roux to remove surface-bound proteins. However, it seems likely that encapsulated phytochrome made up only a small fraction of the total phytochrome present in the mitochondrial preparation. The encapsulation process should be "blind" to any soluble protein, encapsulating, for example, Pr as readily as any other soluble protein. Yet very little Pr co-purifies with mitochondria isolated from unirradiated plants (Serlin and Roux, 1986). The option of a selective encapsulation of Pfr into vesicles by some artifactual mechanism has not been experimentally ruled out and should be tested, however unlikely it may seem.

The failure of immunocytochemical techniques to detect phytochrome associated with organelles in situ could also be an artifact of the techniques used. This possibility has been mentioned in earlier reviews (e.g., Pratt, 1986), though usually it has not been discussed in any detail. The most obvious potential artifact is damage to phytochrome antigenicity during tissue preparation. It would be surprising if the epitopes of phytochrome remained unmodified after chemical fixation, dehydration, incubation above $50^{\circ}C$ and other treatments that are standard in many immunocytochemical protocols. Alternative procedures that minimize denaturing steps, such as freeze-substitution fixation (McCurdy and Pratt, 1986) still have prolonged periods of tissue exposure to acetone and the disadvantage of ice-crystal damage that somewhat limits the field of unaltered tissue that can be examined. As yet one can only speculate on the relative loss of phytochrome antigenicity, if any, that would result from each of the several potentially denaturing treatments commonly used in immunocytochemistry. It should be possible with quantitative immunotechniques to estimate what this loss might be relative to each specific antibody preparation used, and such information could provide valuable insights on the relative merits of various methodologies.

Another potential problem is that current immunocytochemical techniques are thought to detect all cellular phytochrome indiscriminately. Thus, if membrane-bound phytochrome comprised a sufficiently small percentage of the total cellular phytochrome, it would probably appear as "background noise" in the immunostain compared to the bulk phytochrome stained in the cytoplasm. Many phytochrome responses are saturated at relatively low ratios of Pfr/Ptot. If the relatively rare "active" phytochrome is localized differently than the bulk "inactive" phytochrome, then current in situ localizations of phytochrome, by virtue of highlighting the sites of the bulk phytochrome, could very well be obscuring the sites of "active" phytochrome. Ironically, if "active" phytochrome is membrane bound, an effective way to detect it there without the obscuring competition of bulk phytochrome would be to immunocytochemically localize it in purified organelles (Roux et al., 1981)

A related matter that deserves some discussion is that all the antibodies thus far used to localize phytochrome in etiolated tissues have been raised against soluble phytochrome. Presumably, then, most of these antibodies detect epitopes that are accessible in soluble phytochrome. It is certainly possible, if not highly probable, that the epitopes most accessible in soluble phytochrome would not be accessible, or would be far less accessible, in membrane-bound phytochrome. Independent of whether phytochrome binds to membranes in cells, it definitely has the physical potential to bind tightly to and alter the permeability of lipid bilayers (Furuya et al., 1981; Kim and Song, 1981; Georgevich and Roux, 1982). Given this perspective, it would be reasonable to select from libraries of monoclonal antibodies those with an ability to recognize lipsome-bound phytochrome, and then test whether these antibodies detect a distribution of phytochrome in cells distinct from those depicted in past surveys. On the chance that the antibodies currently available do not detect certain epitopes that become accessible only when phytochrome binds to membranes, it would also be useful to raise antibodies against liposome-bound phytochrome, and to test what distribution of immunoreactive sites they detect in cells. Such antibodies would at least yield useful biochemical information on the epitopic structure of phytochrome in a hydrophobic environment. They would also be a novel approach to minimizing the interference of soluble phytochrome in localization studies, if indeed this bulk phytochrome masks membrane-bound phytochrome from detection by the antibodies currently being employed.

The use of purified organelles has an illustrious history of providing valuable insights to plant biologists, especially

those studying the mechanisms of photosynthesis and
respiration. Although it is not yet clear whether phytochrome
has functional associations with organelles in vivo, it does
appear to have functional associations with them in vitro. A
more detailed characterization of these associations,
especially in purified mitochondria and nuclei, should help
determine the value of isolated organelles as models for
studying the cellular mechanisms of photomorphogenesis.

REFERENCES

Bajracharya, D., Falk, H., and Schopfer, P. (1976). Planta
    131:253.
Cedel, T., and Roux, S. J. (1980a). Plant Physiol. 66:696.
Cedel, T., and Roux, S. J. (1980b). Plant Physiol. 66:703.
Chen, Y.-R., and Roux, S. J. (1986). Plant Physiol. 81:609.
Coleman, R. A., and Pratt, L. H. (1974). J. Histochem.
    Cytochem. 11:1039.
Das, R. and Sopory, S. K. (1985). Biochem. Biophys. Res.
    Commun. 128:1455.
Datta, N., Chen, Y.-R. and S. J. Roux (1985). Biochem.
    Biophys. Res. Commun. 128:1403.
Dauwalder, M., Roux, S. J., and Hardison, L. (1986). Planta
    168:461.
Ernst, D. and Oesterhelt, D. (1984). EMBO J. 3:3075.
Furuya, M., Freer, J. H., Ellis, A., and Yamamoto, K. T.
    (1981). Plant Cell Physiol. 22:135.
Georgevich, G., and Roux, S. J. (1982). Photochem. Photobiol.
    36:663.
Grossman, K., Haschke, H. P., and Seitz, H. U. (1981). Planta
    152:457.
Hale, C. C. II, and Roux, S. J. (1980). Plant Physiol. 65:658
    Hendricks, S. B., and Borthwick, H. A. (1967). Proc. Natl.
        Acad. Sci. U.S.A. 58:2125.
Kim I.-S., and Song, P.-S. (1981). Biochemistry 20:5482
Malencik, D. A., and Fischer, E. H. (1982). In "Calcium and
    Cell Function", Vol. III (W.-Y. Cheung, ed.), p. 161.
    Academic Press, New York.
Marmé, D. (1977). Ann. Rev. Plant Physiol. 28:173.
Matsumoto, H., Yamaya, T. and Tanigawa, M. (1984). Plant Cell
    Physiol. 25:191.
Matsumoto, H., and Takyu, T. (1986) Plant Cell Physiol.
    27:293.
McCurdy, D. W., and Pratt, L. H. (1986). J. Cell Biol.
    103:2541.
Miyoshi, Y. (1986). In: "Proceedings of the XVI Yamada

Conference. Phytochrome and Photomorphogenesis. (M.Furuya, ed.). Abstracts, p. 101.

Nagatani, A., Jenkins, G. I., and Furuya, M. (1986). In: "Proceedings of the XVI Yamada Conference Phytochrome and Photomorphogenesis. (M. Furuya, ed.). Abstracts, p. 98.

Newman, I. A. (1981). Plant Physiol. 68:1494.

Otto, V., and Schäfer, E. (1986). In: "Proceedings of the XVI Yamada Conference. Phytochrome and Photomorphogenesis. (M. Furuya, ed.). Abstracts, p. 106.

Pratt, L. H. (1986). In: "Photomorphogenesis in Plants." (R. E. Kendrick and G. H. M. Kronenberg, eds.), p. 61. Martinus Nijhoff, Dordrecht.

Purrello, F., Burnham, D. B. and Goldfine, I. D. (1983). Proc. Natl. Acad. Sci. U.S.A. 80:1189.

Quail, P. (1983). Encycl. Plant Physiol., New Series, Vol. 16A (W. Shropshire and H. Mohr, eds.), p. 178. Springer-Verlag, Berlin.

Racusen, R. H. (1976). Planta 132:25.

Resendez, Jr., E., Attenello, J. W., Grafsky, A., Chang, C. S., and Lee, A. S. (1985). Mol. Cell. Biol. 5:1212

Resendez, Jr., E., Ting, J., Kim, K. S., Wooden, S. K., and Lee, A. S. (1986). J. Cell Biol. 103:2145.

Roux, S. J. (1986) In: "Photomorphogenesis in Plants" (R. Kendrick and G. H. M. Kronenberg, eds.), p. 115, Martinus Nijhoff, Dordrecht.

Roux, S. J., McEntire, K., Slocum, R. D., Cedel, T. E., and Hale, C. C. II. (1981). Proc. Natl. Acad. Sci., U. S. A. 78:283.

Roux, S. J., Wayne, R. O., and N. Datta. (1986). Physiol. Plant. 66:344.

Saunders, M. J., Cordonnier, M.-M., Palevitz, B. A., and Pratt, L. H. (1983). Planta 159:545.

Schopfer, P., and Apel, K. (1983). Encyclo. Plant Physiol., New Series, Vol. 16A, (W. Shropshire and H. Mohr, eds.), p. 258. Springer-Verlag, Berlin.

Serlin, B. S., and Roux, S. J. (1984). Proc. Natl. Acad. Sci. U. S. A. 81:6368.

Serlin, B. S., and Roux, S. J. (1986). Biochem. Biophys. Acta 848:372.

Serlin, B. S., Sopory, S. K., and Roux, S. J. (1984). Plant Physiol. 74:827.

Speth, V., Otto, V., and Schäfer, E. (1986). Planta 168:299.

Tobin, E., and Silverthorne, J. (1985). Ann. Rev. Plant Physiol. 36:569.

Wagle, J. C. (1980). Ph. D. Thesis, Ohio U., Athens.

Wayne, R. O. and Hepler, P. K. (1984). Planta 160:12.

Weisenseel, M. H., and Ruppert, H. K. (1977). Planta 137:225.

Wong, Y.-S., Cheng, H.-C., Walsh, D. A, and Lagarias, J. C. (1986). J. Biolog. Chem. 261:12089.

PROPERTIES OF MEMBRANE-ASSOCIATED
PHYTOCHROME IN PEAS

Katsushi Manabe

Division of Biological Regulation
National Institute for Basic Biology
Okazaki Japan

Department of Biology
Yokohama City University
Yokohama, Japan

I. INTRODUCTION

Physiological analysis of many phytochrome dependent reactions led to the conclusion that membrane-associated phytochrome has physiological roles especially in lower plants such
as algae (Haupt, 1972), ferns (Furuya, 1983) and mosses (
Hartmann and Jenkins, 1984). Also in higher plants, some phenomena which appeared shortly after Pfr formation were thought
to be consequence of phytochrome induced changes of membrane
properties.

Direct detection of phytochrome in membrane fractions extracted from etiolated plant was first reported by Rubinstein
et al., 1969). Light induced re-distribution of phytochrome in
cells has been studied extensively using cell fractionation
techniques since Quail's work (Quail et al., 1973). Red light
irradiation of etiolated plants caused an increase of phytochrome recovered in pelletable fraction. This phenomenon
(light enhanced phytochrome pelletability) itself is induced
by phytochrome system. Physiological significance and biochemical and biophysical features of the phenomenon are yet
unknown. In this article author describes first the biochemical aspects of membrane-associated phytochrome, then the putative binding partner proteins of phytochrome in the membrane
fraction which were detected by chemical cross-linking technique.

209

II. BIOCHEMICAL AND SPECTROPHOTOMETRICAL PROPERTIES
   OF MEMBRANE-ASSOCIATED PHYTOCHROME

   Purified mitochondria (Manabe and Furuya, 1975b) and mi-
crosomal fractions (Manabe and Furuya, 1975b) extracted from
totally etiolated pea stems contain small amount of phytochrome which can not be removed by repeated washings. Red
light irradiation of pea stems enhanced the phytochrome contents in the membrane fractions (Manabe and Furuya, 1975a).
Kinetic features of the above reaction which was shown in pea
stems resemble "light enhanced phytochrome pelletability"
first reported by Quail et al. (1973) in corn. Whereas they
used high concentration of Mg++ in extraction and the resus-
pension buffers (Pratt and Marmé, 1976), we used a buffer
without or with very low concentration of the divalent cation.
In Quail's system, up to half phytochrome in cell was pelletable (most of which was pelletable only when Mg++ concentration
was more than 1 mM, Quail, 1978), while about 10 % of phytochrome was pelletable in our system (Manabe and Furuya,
1975a). In our system, as Watson and Smith (1983) pointed out
in similar cell fractionation method using oat as the plant
material, the pelletable phytochrome seems to bind firmly with
the membrane. Treatment of the pelletable phytochrome with
high concentration of salt (150 mM KCl) or high pH (9.0) did
not cause dissociation of the phytochrome. The pelletable
phytochrome can be solubilized by detergent treatments (SDS or
CHAPS). These data suggests that the pelletable phytochrome
associates with membrane by relatively strong interaction
rather than non-specific ionic or electrostatic binding.
   Difference spectra of pelletable pea phytochrome were measured in 1,000 - 20,000 g (Shimazaki and Furuya, 1980) and
1,000 - 7,000 g pelletable fractions (Manabe, 1983). Spectral shape and the both Pr and Pfr peak positions in the difference spectra of phytochrome were very similar to those of
purified 121 kDa (native) soluble phytochrome (Lumsden et al.,
1985). The phototransformation pathways analysed by low temperature spectrophotometry both from Pr to Pfr (Manabe, 1983)
and from Pfr to Pr (Manabe, 1987) were essentially the same
as that of phytochrome measured in vivo (Kendrick and Spruit,
1977). Molecular size of pelletable phytochrome obtained from
red and far-red light irradiated pea stems was measured by SDS
gel electrophoresis, Western blotting, followed by immunochemical detection (Manabe et al., 1984) using monoclonal
anti-pea phytochrome IgGs (Nagatani et al., 1984). The mobility of the phytochrome was identical to that of soluble one
(121 kDa). Peptide mapping pattern of the pelletable phytochrome cleaved by papain was also identical to that of soluble
one. Therefore, biochemical properties of pea phytochrome so

far studied show no difference between soluble and pelletable
one except the solubility and stability for incubation
(Shimazaki and Furuya, 1980).

III. CHEMICAL CROSS-LINKING OF
PELLETABLE PEA PHYTOCHROME

As mentioned above, studies of solubility of pelletable
pea phytochrome may suggest that phytochrome binds specifically to a protein receptor which is integral to the membrane
protein. If such a receptor protein exists, it can be a
mediator protein of phytochrome function. To examine this
possibility, chemical cross-linking techniques were applied
to the pelletable phytochrome. A bifunctional reagent, DTBP
(Dimethyl-3,3'-dithiobispropionimidate), which has a cleavable
disulfide bridge in the middle of the molecule (Wang and
Richards, 1974) was used for the cross-linking. When pelletable phytochrome was incubated with DTBP in darkness, at least
two phytochrome-conjugated bands, with molecular weights of
235 and 280 kDa, appeared. While control experiment without
DTBP yielded no or very little conjugated bands. The procedures are briefly shown in Fig. 1. In the cross-linking

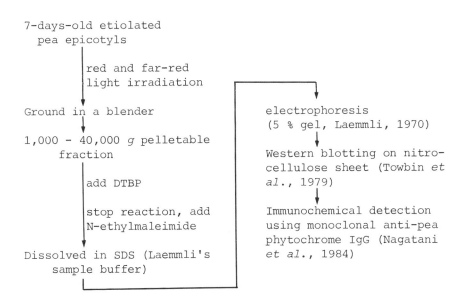

7-days-old etiolated
   pea epicotyls

   | red and far-red
   | light irradiation

Ground in a blender   electrophoresis
                                       (5 % gel, Laemmli, 1970)

1,000 - 40,000 *g* pelletable
     fraction   Western blotting on nitro-
                                       cellulose sheet (Towbin *et*
   add DTBP   *al.*, 1979)

   stop reaction, add   Immunochemical detection
   N-ethylmaleimide   using monoclonal anti-pea
                                       phytochrome IgG (Nagatani
Dissolved in SDS (Laemmli's   *et al.*, 1984)
   sample buffer)

Fig. 1.   Detection methods of cross-linked pelletable
pea phytochrome.

experiments, we mainly used 1,000 - 40,000 g pelletable phytochrome rather than 1,000 - 7,000 g fraction, because both
pelletable fractions showed qualitatively same properties in
these cross-linked experiments.

For further analysis, the cross-linked pelletable phytochrome was then purified and analysed by two dimensional gel
electrophoresis in order to determine what protein(s) was
cross-linked to phytochrome. The SDS dissolved sample of the
cross-linked phytochrome complex was purified by affinity
chromatography using monoclonal anti-pea phytochrome antibody-
conjugated agarose. The pufified sample was electrophoresed in
Laemmli's system using 5 % gel (Laemmli, 1970). The gel was
then sliced and incubated with dithiothreitol to cleave the
disulfide bridge of the cross-linker. The gel slice was put
on the stacking gel of the 2-nd dimensional gel (10 % gel).
After the electrophoresis, the gel was either stained by
silver staining (Oakley et al, 1980) or immunoblotted with
anti-pea phytochrome antibody. The silver staining profile
(Fig. 2) of the 2-nd dimensional gel showed that the cleavage
of the disulfide bridge yielded at least two off-diagonal
spots (in addition to the monomer phytochrome bands), each of

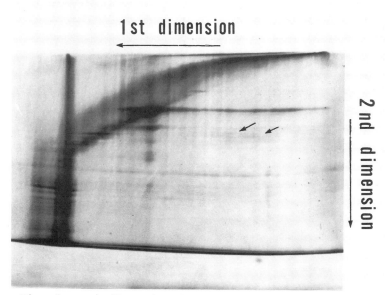

1st dimension

2nd dimension

Fig. 2.   Two-dimensional gel electrophoresis of the purified cross-linked pelletable pea phytochrome. Gel was stained
by silver according to Oakley et al. (1980). Arrows indicate
ca. 90 kDa off-diagonal spots (K. Manabe, K. T. Yamamoto and
M. Furuya, unpublished).

which appeared to be derived from the cross-linked phytochrome bands. The molecular weights of the spots were similar and estimated to be about 90 kDa. There are other faint spots that corresponded to also around 90 kDa molecular weight. These spots were not detected by immunochemical staining using anti-pea phytochrome IgGs (data not shown).

If the dithiothreitol-reduced gel slice of the first dimensional gel was put on the second dimensional stacking gel in the presence of <u>Staphyllococcus</u> V-8 protease, the silver-stained gel no longer yielded the 90 kDa off diagonal spots.

These data suggest that phytochrome molecule exist near the 90 kDa protein(s) in the pelletable fraction. These protein(s) seem to be membrane protein(s) since they were not removed out by washing with high salt concentration buffer (200 mM KCl). Whether they have specific binding capacity for phytochrome is not known yet.

IV. CONCLUSION

Biochemical and spectrophotometrical properties of pelletable phytochrome are essentially same as that of soluble phytochrome. Pelletable phytochrome seems to associate firmly with the membrane. Chemical crosslinking studies of pelletable phytochrome revealed that membrane protein(s) of ca. 90 kDa exist near the phytochrome molecule.

REFERENCES

Furuya, M. (1983). In "Photomorphogenesis (Encyclopedia of plant Physiology. New series Vol. 16B)" (W. Shropshire, Jr. and H. Mohr. ed.), p. 569. Springer-Verlag, Berlin.

Hartmann, E. and Jenkins, G. I. (1984). In "The experimental Biology of Bryophytes" (A. F. Dyer ed.), p. 203. Academic Press, London.

Haupt, W. (1972). In "Phytochrome" (K. Mitrakos and W. Shropshire, Jr. ed.), p. 349. Academic Press, London.

Kendrick, R. E. and Spruit, C. J. P. (1977). Photochem. Photobiol. 69:413.

Laemmli, U. K. (1970). Nature 227:680.

Lumsden, P. J., Yamamoto, K. T. and Furuya, M. (1985). Plant Cell Physiol. 26:1313.

Manabe, K. (1983). Plant Cell Physiol. 24:1225.

Manabe, K. (1987). Physiol. Plant. 69:413.

Manabe, K. and Furuya, M. (1974). Plant Physiol. 53:343.

Manabe, K. and Furuya, M. (1975a). Plant Physiol. 56:772.

Manabe, K. and Furuya, M. (1975b). Planta 123:207.

Manabe, K., Nagatani, A., Yamamoto, K. T. and Furuya, M. (1984). Photochem. Photobiol. Supple. 39:86s.

Nagatani, A., Yamamoto, K. T., Furuya, M., Fukumoto, T. and Yamashita, A. (1984). Plant Cell Physiol. 25:1059.

Oakley, B. R., Kirsch, D. R. and Morris, N. R. (1980). Anal. Biochem. 105:1059.

Pratt, L. H. and Marmé, D. and Schäfer, E. (1973). Plant Physiol. 58:686.

Quail, P. H. (1978). Photochem. Photobiol. 27:759.

Quail, P. H., Marmé, D. and Schäfer, E. (1973). Nature New Biol. 245:189.

Rubinstein, B., Drury, K. S. and Park, R. B. (1969). Plant Physiol. 44:105.

Shimazaki, Y. and Furuya, M. (1980). Planta 154:121.

Towbin, H., Staehelin, T. and Gordon, J. (1979). Proc. Natl. Acad. Sci. USA 76:4350.

Wang, K. and Richards, F. M. (1974). J. Biol. Chem. 249:8005.

Watson, P. J. and Smith, H. (1983). Planta 154:121.

PHOTOCONTROL OF ION FLUXES AND MEMBRANE
PROPERTIES IN PLANTS

Richard E. Kendrick
Margreet E. Bossen[1]

Laboratory of Plant Physiological Research
Agricultural University
Wageningen
The Netherlands

I. INTRODUCTION

A casual glance at this book will reveal that phytochrome
research has gone full circle in the last 20 years and gene
expression is once more the centre of discussion. The
introduction of molecular biology techniques to higher plants
has now enabled the original hypothesis of Mohr (1966) to be
tested in vitro. There is now no doubt that phytochrome
regulates the transcription of genes, although how this
precisely takes place remains a mystery. Nevertheless, the
rapid effects of phytochrome photoconversion on membrane
processes, could still form a functional part of the
transduction chain leading to differential gene expression.
This chapter reviews the literature concerned with phytochrome
regulated membrane phenomena, briefly outlines the new results
presented at the XVI Yamada Conference and summarizes the
discussions of the workshop under the same title.

---

[1]Supported by the Foundation for Fundamental Biological
Research (BION), which is subsidized by the Netherlands
Organization for the Advancement of Pure Research (ZWO).

II. LITERATURE REVIEW

It is not yet clear whether phytochrome in its far-red (FR) absorbing form (Pfr) is bound to, or associated with, membranes. In fact, the evidence to date, supports the idea that most of the phytochrome in dark-grown plants is concentrated at non-membrane sites after red light (R) irradiation (Speth et al., 1986; McCurdy and Pratt, 1986). Despite these observations, a number of fast membrane related responses to R (lag time of seconds) have been observed: e.g. changes in membrane potential, Ca^{2+} and water permeability. These changes may be important intermediate steps in the transduction chain(s) between light absorption by the phytochrome molecule and the ultimate physiological responses. Extrusion of H^+ and modification of K^+ transport have been reported with longer lag times (minutes) and could be, themselves, directly associated with physiological processes such as elongation growth of coleoptiles and hypocotyls, opening of stomata and control of hook-opening.

Rapid changes in membrane potential have been measured with micro-electrodes. Racusen (1976) and Weisenseel and Ruppert (1977) measured depolarization of the plasma membrane in response to R in dark-grown oat coleoptiles and internodes of Nitella respectively. For Nitella internodes the lag phase was 1 s, whereas in the case of oat coleoptiles it was less than 10 s. The magnitude of the plasma membrane depolarization in Nitella was correlated with the $[Ca^{2+}]$ of the medium; a maximum being obtained with 5 mM Ca^{2+}. The depolarization could be accounted for by Ca^{2+} influx or Cl^- efflux. Newman (1981) measured a hyperpolarization of the potential between the surface of intact dark-grown oat coleoptiles and the bathing medium of the roots, with a lag of 4.5 s, after a 1 s flash of R. However, this hyperpolarization was preceded by a small depolarization during the R flash. Ten minutes after R, a FR flash induced a depolarization. The hyperpolarization could be due to an efflux of Ca^{2+} and the small depolarization to a small influx of Ca^{2+}. Fluxes of other ions (e.g. Cl^-) cannot be excluded. The zeta-potential of light-grown algal cells of Mesotaenium was reduced by R (Stenz and Weisenseel, 1986) and the effect was FR reversible. In this case, Ca^{2+} also seemed to be involved. If 1 mM Ethylene glycol-bis (β-aminoethyl ether) N,N',N'-tetraacetic acid (EGTA) was added to the medium, no change after R was observed. The same was true if the Ca^{2+}-channel-blocker Verapamil (10 μM) was applied.

Fluxes of Ca^{2+} over the plasma membrane after R and/or FR have been measured in different ways and in different plant systems. In protoplasts from dark-grown maize leaves Das and Sopory (1985), measured a transient enhancement of $^{45}Ca^{2+}$

uptake within 30 s after R. This could be prevented by FR given directly after R. In protoplasts treated with 5-hydroxytryptamine a $^{45}Ca^{2+}$ uptake equal to that after R was observed, indicating an involvement of inositol metabolism. No change in long-term uptake of $^{45}Ca^{2+}$ after R in comparison to darkness was found in protoplasts from dark-grown wheat leaves (Åkerman et al., 1983). For oat coleoptile protoplasts a Ca^{2+} efflux was measured after R and subsequent FR reversed the effect (Hale and Roux, 1980). $[Ca^{2+}]$ in the surrounding medium was measured using the Ca^{2+}-sensitive dye murexide, so that only net fluxes could be observed. A Ca^{2+} efflux after R was only found when Ca^{2+} was present in the medium during irradiation or when the protoplasts were pre-incubated in Ca^{2+} prior to irradiation.

For spores of Onoclea sensibilis a transient increase of Ca^{2+} uptake after R was observed by Wayne and Hepler (1985). The total amount of Ca^{2+} within the spores after R, as measured by atomic absorption spectroscopy, was also observed to increase. External Ca^{2+} is essential for germination of the spores. In Mougeotia, R-induced uptake of $^{45}Ca^{2+}$, which is prevented by subsequent FR, has been observed (Dreyer and Weisenseel, 1979).

Ca^{2+} fluxes associated with organelles have also been reported. In a mitochondrial fraction from etiolated oat shoots, Ca^{2+} fluxes, measured with murexide, are under photocontrol, being modified within 40 s of irradiation. R diminished net uptake and subsequent FR restored the dark control rate. When ruthenium red was included in the medium to block the active uptake of Ca^{2+}, an efflux of Ca^{2+} was measured after R irradiation. These results suggest that after R the $[Ca^{2+}]$ in the cytoplasm increases (Roux et al., 1981). According to Serlin and Roux (1986), phytochrome, as Pfr, is localized interior to the surface of the outer mitochondrial membrane. Phosphorylation of specific nuclear proteins was promoted by R in isolated nuclei if Ca^{2+} (7 μM) was present in the medium (Datta et al., 1985). Transport of Ca^{2+} over the nuclear membrane might be involved since ethylene diaminotetraacetic acid (EDTA) blocked the R promotion of phosphorylation.

Two reports concerning investigations of possible phytochrome-mediated H^+ fluxes have recently been published. During spore germination of Onoclea sensibilis after R, no change in intracellular pH was observed. There was also no measurable efflux of H^+ (Wayne et al. 1986). In the case of cucumber hypocotyl sections which included the hook, an enhanced acidification of the medium was observed after R, when Mg^{2+} and K^+ were present in the medium. Ca^{2+} could not replace Mg^{2+}. A Mg^{2+} dependent K^+ stimulated ATPase appears to be involved (Roth-Bejerano and Hall, 1986).

In the R-dependent elongation of oat coleoptiles, protons seem to play a role. Pike and Richardson (1977) observed a R-induced, FR reversible stimulation, of H^+ efflux from apical segments of etiolated oat coleoptiles, 2 h after irradiation. Lürssen (1976) however, measured an alkalization of the medium after R with sub-apical segments. Shinkle and Briggs (1985) have indicated a relationship between R irradiation, external pH and $[K^+]$ in the case of elongation of sub-apical oat coleoptile segments.

Mesophyll protoplasts from the primary leaves of dark-grown wheat shrank less after R than in darkness when they were transferred to a medium of higher osmotic potential (Blakeley et al., 1983). K^+ appeared to be necessary for maximal effect, but in all experiments Ca^{2+} was present in the medium. Sometimes a rapid transient influx of K^+ ($^{86}Rb^+$) was observed after R, with a lag of less than 1 min. However, the final concentration was lower after R than in the dark control.

Changes in water permeability after R have also been reported. In Mougeotia, an acceleration of plasmolysis of the cytoplast was observed after R when cells were transferred to a medium of higher osmotic potential (Weisenseel and Smeibidl, 1973). They attributed these observations to changes in water permeability. In oat coleoptiles and bean hypocotyls R changed water permeability, measured as efflux of preloaded ^3H_2O (Lecock and Buffel, 1986). Dependent on fluence rate, the change could be an increase or a decrease. However, Pike (1976) observed no influence of phytochrome on efflux of ^3H_2O preloaded etiolated bean buds, pea epicotyl segments and oat coleoptile segments.

There also appear to be phytochrome mediated changes in permeability of the plasma membrane to gibberellins (Kepler and Mertz, 1986). When GA_3 was applied immediately after R a 50% reduction in uptake of this gibberellin by tobacco protoplasts was observed. After 15-30 min no difference was found in $[GA_3]$ between the R treatment and the dark control.

III. SUMMARY OF WORKSHOP DISCUSSIONS

A. Regulation of Ion Fluxes in vivo

1. Tissues and Cells. The germination of fern spores of Dryopteris sp. was demonstrated to be phytochrome controlled and relies on the presence of inorganic nitrogen in the form of NH_4^+ and NO_3^- ions for germination at subsaturating fluences of R (Haupt et al., 1986). Very low NH_4NO_3

concentrations are effective (10 μM), suggesting that inorganic nitrogen functions as a signal, not as a nutrient source. Both NH_4^+ and NO_3^- are effective. The effects of light and nitrogen were temporally separated. Addition of EDTA prevented the effect of R stimulation of germination, as did the application of the calmodulin inhibitor trifluorperazine, indicating Ca^{2+} also plays a role. Experiments in which Ca^{2+} is supplied to cells at different times after R would be very interesting. The response is characterized by a rather long escape time (8-16 h) for FR reversibility. Such a response is difficult to reconcile with a rapid mechanism of phytochrome action at the level of membranes.

    2. Protoplasts. Phytochrome controls the rate of rotational cytoplasmic streaming in Vallisneria mesophyll protoplasts(Takagi and Nagai, 1986), the rate being fast after R and slow after FR. The rate of streaming is also increased in the presence of the chelator EDTA. The capacity of FR to reverse the effect of R was lost in the presence of La^{3+}, suggesting fluxes of Ca^{2+} across the plasma membrane are involved. Measurements of the $[Ca^{2+}]$ in the medium showed that an efflux occurred after R and an uptake after FR(response time 10 min). Murexide was used as assay method and the effect of R was sensitive to the ATPase inhibitor vanadate, as well as proton pump inhibitors diethylstilbestrol (DES) and N,N'-dicyclohexylcarbodiimide (DCCD). The maximum change in $[Ca^{2+}]$ measured was 5 μM. The results suggest that phytochrome regulates an energy-dependent Ca^{2+} transport system which involves H^+ movement across the plasma membrane.

    The swelling of etiolated wheat leaf mesophyll protoplasts was shown by Bossen et al.(1986) to take place if they were exposed to R while being maintained in 0.5M sorbitol and at the physiological temperature of $25^\circ C$. This contrasts with all the previously published work on the swelling of protoplasts in which a temperature of $4^\circ C$ was utilized making physiological relevance of the response questionable (Blakeley et al., 1983; Kim et al., 1986). The final volume of protoplasts was also observed to be greater after R upon transfer to a range of higher and lower osmotic potentials. The response to both light and osmotic shock are complete within 10 min at $25^\circ C$. The standard incubation medium contained both 1 mM Ca^{2+} and K^+. Experiments in which the protoplasts were pre-washed with EDTA demonstrated that the response had a prerequisite for Ca^{2+} and that this could not be replaced by Mg^{2+}. Contrary to the suggestion in the literature that K^+ was required for the swelling response of protoplasts in response to R (Blakely et al., 1983) for this response it was not required. The response was measured in two ways: (i)by photographing populations of protoplasts and measuring their diameter to enable calculation of the mean protoplast volume; (ii)by measuring the diameter of individual protoplasts held on a suction pipette utilizing a microscope and television camera. Both systems yielded essentially similar results, but the latter method enables the determination of kinetics of the response with some accuracy, which will assist our understanding of the nature of the response. Clearly an investigation of the uptake process using an assay for Ca^{2+} in the medium, as well as measurement of the intracellular $[Ca^{2+}]$ are required. Problems associated with the possibility of utilizing fluorescent probes for the latter measurement are discussed below.

Miyoshi(1986) measured the cytoplasmic $[Ca^{2+}]$ after R and FR in protoplasts of Vigna mungo and Pisum sativum using the fluorescent probe Fura 2-AM. This probe was introduced by electroporation(one pulse of 90 V) and followed by 30 min incubation in the same solution at $25^{\circ}C$. Slow changes in the fluorescence were observed that were R/FR reversible and which depended on the $[Ca^{2+}]$ and pH of the surrounding medium. He concluded that the effects were due to transport of Ca^{2+} across the plasma membrane, not to their release and uptake from internal stores such as organelles. This is probably the first report of a direct measurement of a phytochrome regulated change in cytosolic $[Ca^{2+}]$ in a higher plant.

Fluorescent probes have been used successfully in animal systems to measure the cytosolic $[Ca^{2+}]$. However in the case of plant cells there are serious problems. When the probe is successfully introduced into the cell there is a chance that a large proportion enters the endoplasmic reticulum and vacuole where the $[Ca^{2+}]$ is perhaps an order of magnitude higher than in the cytoplasm. This means that the change in fluorescence of interest is only small compared to the large background fluorescence. Fura 2-AM and Quin 2 may both be suitable and a possible method to check subcellular localization is to utilize microspectrofluorimetry.

B. What Is the Role Played by Calcium in Photomorphogenesis?

Of the systems under discussion in the workshop nearly all had either a requirement for Ca^{2+} or were associated with Ca^{2+} fluxes. While the evidence for calmodulin activation as a common feature of the transduction chain is good, nobody would commit themselves to say that the primary ion moved as a result of phytochrome phototransformation was Ca^{2+}. The vast knowledge of calcium regulation from animal systems sets a framework for understanding the parallel role in plant cells. Evidence is now accumulating that there are other calcium-binding proteins present in plant cells which have different molecular masses and are immunologically distinct from calmodulin.

The involvement of ion fluxes as a prerequisite of gene expression is not entirely circumstantial. In animal systems it has been shown the Ca^{2+}-ionophore A23187 can result in specific induction of gene expression. This is now a priority area of research in plants.

C. Ion Fluxes in Model Systems

Little work is at present taking place in this area of research, at least partially, due to the difficulty of obtaining research funds for such speculative projects. However, the earlier work of Roux and Yguerabide (1973) with proteolytically degraded phytochrome revealed the possibilities that this technique presents. It remains a feasible method of studying the possible conformational interaction between phytochrome and membranes despite the body of data which indicates that their only possible interaction is subtle (Vierstra and Quail, 1986). However, a reconstituted functional membrane will almost certainly require the inclusion of yet unknown proteins.

D. Regulation of Other Membrane Properties

There were no new reports at the workshop on electrical effects in relationship to phytochrome. However, the possibilities that protoplasts provide for patch-clamp studies were pointed out. Katoaka (1986) presented data on blue light-enhanced current influx at the apex of the coenocytic alga Vaucheria terrestris which precedes the light-growth response and phototropism. The current changes were measured using the vibrating probe and occurred within 10 s, whereas the first growth changes were only detectable after 2 min. The fact that Verapamil interacts with the response suggests that Ca^{2+} of extracellular origin regulates the magnitude of this response.

Two groups, Hartmann and Pfaffmann (1986) and Sopory et al. (1986) independently implicated phytochrome action with phosphatidyl inositol metabolism. By analogy to animal systems this provides a possible link between photoreceptor and membrane resulting in a rapid breakdown of membrane lipid: where phosphatidylinositol-4,5-bisphosphate is broken down into diacylglycerol and inositol triphosphate, leading to stimulation of Ca^{2+} uptake (Nishizuka, 1984).

E. Sensitization of Phytochrome Responses to Very Low Fluences: The Involvement of Membranes

At the end of the workshop VanDerWoude presented evidence to support that his dichromophoric model (VanDerWoude, 1985) of phytochrome action necessitated the involvement of membranes. This model was proposed on the basis of experiments which show biphasic fluence-response relationships, with the fluence ranges of response being approximately four orders of magnitude apart (referred to as very low fluence and low fluence responses). In addition, his model is based on the observation that the phytochrome molecule is present in solution as a dimer. The dimeric model itself does not require the involvement of membranes, but the process by which sensitization occurs, enabling response in the very low fluence range, where response can occur as a result of photoconversion of one. in a million phytochrome molecules to Pfr, appears to do so. This process of sensitization can be achieved by temperature shock, as well as by treatment with antibiotics and ethanol, all factors known to influence membrane properties. The sharp optimum for sensitization of many systems at 28-30°C is characteristic of a membrane transition. The sensitization of <u>Kalanchoë</u> seeds by treatment with gibberellins, so that they are induced to germinate in response to light in the very low fluence range, may also function through their action on membranes (Rethy et al., 1986).

III. PRIORITIES FOR FUTURE RESEARCH

To date, the study of phytochrome effects at the membrane level has been hampered by the lack of a reproducible system, or by the fact that the systems studied were so complex that they defied critical analysis. It is therefore a necessary prerequisite for understanding any possible membrane function of phytochrome, to develop a system which is both reproducible and simple. This could be achieved in two ways. By rebuilding a simple biological system i.e. reconstituting a membrane system with functional phytochrome, but this may be too great a step at this stage with our limited knowledge. Perhaps the most promising approach at the present time is the utilization of simple cell systems such as protoplasts and organelles. It is vital not only to measure one ion in isolation, but to build up a detailed picture of the functional membrane system. Attention should be paid to acquiring kinetic parameters which have been too often lacking in the research published to date.

REFERENCES

Åkerman, K.E.O., Proudlove, M.O., and Moore, A.L. (1983).
    Biochem. Biophys. Res. Commun. 113:171.
Blakely, S.D., Thomas, B., Hall, J.L., and Vince-Prue, D.
    (1983). Planta 158:416.
Bossen, M.E., Dassen, H.H.A., Latour, J.B., Kendrick, R.E.,
    and Kronenberg, G.H.M. (1986). Proceedings of the XVI
    Yamada Conference Photomorphogenesis in Plants, p. 126.
    Okazaki.
Das, R., and Sopory, S.K. (1985). Biochem. Biophys. Res.
    Commun. 128:1455.

Datta, N., Chen, Y.-R., and Roux, S.J. (1985). Biochem.
    Biophys. Res. Commun. 128:1403.
Dreyer, E.M., and Weisenseel, M.H. (1979). Planta 146:31.
Hale, C.C., and Roux, S.J. (1980). Plant Physiol. 65:658.
Hartmann, E., and Pfaffmann, H. (1986). Proceedings of the XVI
    Yamada Conference Photomorphogenesis in Plants, p. 107.
    Okazaki.
Haupt, W., Scheuerlein, R., Mische, S., and Mader, U. (1986).
    Proceedings of the XVI Yamada Conference
    Photomorphogenesis in Plants, p. 119. Okazaki.
Katoaka, H. (1986). Proceedings of the XVI Yamada Conference
    Photomorphogenesis in Plants, p. 99. Okazaki.
Kepler, L.D., and Mertz, O. (1986). Plant Cell Physiol.
    27:861.
Kim, Y.-S., Moon, D.-K., Goodin, J.R., and Song, P.-S. (1986).
    Plant Cell Physiol. 27:193.
Lecock, F.E., and Buffel, K.A. (1986). Biochem. Physiol.
    Pflanzen 181:459.
Lürssen, K. (1976). Plant Sci. Lett. 6:389.
McCurdy, D.W., and Pratt, L.H. (1986). Planta 167:330.
Miyoshi, Y. (1986). Proceedings of the XVI Yamada Conference
    Photomorphogenesis in Plants, p. 101. Okazaki.
Mohr, H. (1966). Z. Pflanzenphysiol. 67:63.
Newman, I.A. (1981). Plant Physiol. 68:1494.
Nishizuka, Y. (1984). Trends Biochem. Sci. 9:163.
Pike, C.S. (1976). Plant Physiol. 57:185.
Pike, C.S., and Richardson, A.E. (1977). Plant Physiol.
    59:615.
Racusen, R.H. (1976). Planta 132:25.
Rethy, R., Dedonder, A., Van Wiemeerisch, L., De Petter, E.,
    Fredericq, H., and De Greef, J. (1986). Proceedings of the
    XVI Yamada Conference Photomorphogenesis in Plants, p.
    120. Okazaki.
Roth-Bejerano, N., and Hall, J.L. (1986). J. Plant Physiol.
    122:329.

Roux, S.J., and Yguerabide, J. (1973). Proc. Natl. Acad. Sci.,
    U.S.A. 70:762.
Roux, S.J., McEntire, K., Slocum, R.D., Cedel, T.E., and Hale,
    C.C. (1981). Proc. Natl. Acad. Sci. U.S.A. 78:283.
Serlin, B.S., and Roux, S.J. (1986). Biochim. Biophys. Acta
    848:372.
Shinkle, J.R., and Briggs, W.R. (1985). Plant Physiol. 79:349.
Sopory, S.K., Das, R., Sharma, A.K., Sreedhara, S., and
    Mukarram, M. (1986). Proceedings of the XVI Yamada
    Conference Photomorphogenesis in Plants, p. 102. Okazaki.
Speth, V., Otto, V., and Schäfer, E. (1986). Planta 168:299.
Stenz, H.-G., and Weisenseel, M.H. (1986). J. Plant Physiol.
    122:159.
Takagi, S., and Nagai, R. (1986). Proceedings of the XVI
    Yamada Conference Photomorphogenesis in Plants, p. 100.
    Okazaki.
VanDerWoude, W.J. (1985). Photochem. Photobiol. 42:655.
Vierstra, R.D., and Quail, P.H. (1986). In "Photomorphogenesis
    in Plants" (R.E. Kendrick and G.H.M. Kronenberg, eds.)
    p. 35. Martinus Nijhoff/Dr. W. Junk Publishers, Dordrecht.
Wayne, R., and Hepler, P.K. (1985). Plant Physiol. 77:8.
Wayne, R., Rice, D., and Hepler, P.K. (1986). Dev. Biol.
    113:97.
Weisenseel, M.H., and Ruppert, H.K. (1977). Planta 137:225.
Weisenseel, M.H., and Smeibidl, E. (1973). Z. Pflanzenphysiol.
    70:420.

PHYTOCHROME CONTROL OF INTRACELLULAR MOVEMENT

Wolfgang Haupt

Institut für Botanik und Pharmazeutische Biologie
Universität Erlangen-Nürnberg
Erlangen, Federal Republic of Germany

I. INTRODUCTION

Light control of intracellular movement in plants is usually mediated by blue light-absorbing pigments. However, there are a few examples where phytochrome is known as photosensory pigment, viz. cytoplasmic streaming (rotation, cyclosis) in Vallisneria gigantea (Takagi and Nagai, 1985), chloroplast redistribution in fern gametophytes (fig. 1c, d; Yatsuhashi et al., 1985), and chloroplast reorientation in the algae Mougeotia and Mesotaenium (fig. 1a and b; cf. reviews by Haupt 1983; Haupt and Wagner, 1984; Wagner and Grolig, 1985). In the latter cases, especially in Mougeotia, considerable progress has been made in elucidating the whole transduction chain from light perception to the final response. Yet, in detail many questions are still open, and the possibility for generalizations is limited.

In this chapter, a short survey will be given about the presumed transduction chain in Mougeotia according to our present knowledge. Some open problems in this model will be pointed out, and interesting deviations as found in the other organisms will be referred to.

II. CHLOROPLAST REORIENTATION IN MOUGEOTIA: THE SIGNAL TRANSDUCTION MODEL

The cylindrical cell of the filamentous alga Mougeotia contains one large ribbon-shaped chloroplast only, which nearly fills the cell area when viewed from the face. This chloro-

225

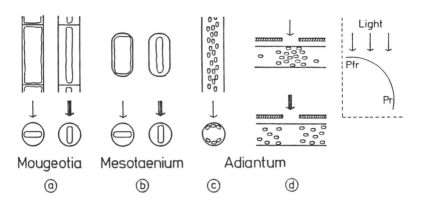

FIGURE 1. Chloroplast orientation in unilateral light of
low and high fluence rate (single and triple arrows, respectively); surface view and cross section. In d, longitudinal
movement induced by partial irradiation is shown. Inset:
Cross section of a cell with the Pfr gradient controlling
chloroplast movement.

plast can rotate so as to expose its face to low-intensity
light or its profile to high-intensity light, as far as we
are dealing with white light (fig. 1a). Most of our knowledge
refers to the low-intensity movement, which will be the main
topic of this chapter (cf. Haupt, 1983; Haupt and Wagner,
1984; Wagner and Grolig, 1985).

For the low-intensity orientation to be induced, a short
light pulse is sufficient, which can be shortened, in proper
conditions, into the nanosecond range (Scheuerlein and
Braslavsky, 1987). The response, then, proceeds in darkness
during the next 15 to 45 min. Thus, we can clearly separate
the signal input, i.e. light perception, from the output,
i.e. the movement proper.

A. Signal Perception

Chloroplast orientation to face position has maximal sensitivity in the long-wavelength range of light, i.e. in the
red. The inductive effect of a red pulse is fully reversible
by a subsequent far-red pulse, and thus phytochrome is the
photosensory pigment. By irradiation with microbeams it has
been shown that light perception occurs in the cytoplasm.
Moreover, a pronounced action dichroism can be found, with
cells responding to polarized light only, if its electrical
vector contains a component perpendicular to the long axis of

the cell. This action dichroism is interpreted to be based on a dichroic orientation of phytochrome, with the transition moments of Pr being oriented parallel to the surface and aligned in a helical structure. Pfr, on the other hand, has its transition moment normal to the cell surface. The strict orientation of both Pr and Pfr requires that phytochrome be associated with relatively stable cell structures, most probably with the cell membrane (plasmalemma).

Perception of light direction is based on the dichroic orientation of Pr: The component of unpolarized light that vibrates parallel to the cell axis is absorbed all around the cell, but the component that vibrates perpendicularly to it is preferentially absorbed at the proximal and distal faces of the cell as compared to its flanks. Thus, a gradient of Pfr is established, in which the chloroplast orientates (cf. fig. 1, inset; for detailed explanation see Haupt, 1983). Levelling of the Pfr gradient in saturation is avoided by the different orientation of Pr and Pfr, by which the photoequilibrium Pr $\rightleftharpoons$ Pfr is shifted towards Pfr at the proximal and distal faces, but towards Pr at the flanks, thus ensuring a Pfr gradient even in saturation.

In summary: phytochrome is the photosensory pigment for perception of light; its dichroic orientation is the basis for perception of light direction, and a levelling effect in saturation is avoided by the change in orientation upon photoconversion Pr $\rightleftharpoons$ Pfr.

B. The Response System

Good evidence has been accumulated for assuming actin-myosin interaction as underlying the movement (cf. Wagner and Grolig, 1985). Actin has been detected in the cell by fluorescent phalloidine and by decoration with heavy meromyosin; movement can reversibly be inhibited by cytochalasin B, which is a relatively specific inhibitor of actin activity, and likewise the myosin inhibitor N-ethyle-maleimide can block the response. Finally, microfilaments of the expected dimensions can be detected close to the chloroplast's edges in the peripheral cytoplasm. Thus, actin-myosin interaction most probably provides the mechanical force for the movement.

C. The Black Box

The question remains to be answered how the activity of the actin-myosin system is regulated by the Pfr gradient and whether an internal signal can be found as connecting input and output. As an extrapolation from muscle actomyosin,

calcium could be a good candidate for a regulatory function.
Several approaches support this assumption (cf. Wagner and
Grolig, 1985).

The response appears to depend on calcium. If Mougeotia
is grown in calcium-free medium, its content of a water-
insoluble fraction of Ca^{2+} decreases within a few days. Synchronously the response to a light pulse is reduced. Upon
providing the cells with calcium, both calcium content and
response increase again.

Investigations with the electron microprobe reveal that
most of the calcium is concentrated in microbodies (sometimes
called tannin vesicles), and these calcium vesicles are most
abundant near to the chloroplast's edges. Thus, they are
located close to the microfilaments, which they are assumed
to regulate in their activity.

The regulatory function of calcium is supported furthermore by the observation that calmodulin, which has been found
in Mougeotia, appears to be involved in the response: several
calmodulin inhibitors, viz. trifluoperazine and a chemically
completely different group of substances (e.g., "W 7" and
"W 13") can block the response (Serlin and Roux, 1984; Wagner
et al., 1984). It is proposed that calcium-calmodulin acts as
a myosin phosphokinase (Wagner and Grolig, 1985).

Thus, there is good evidence for calcium being an internal signal for regulating the motor apparatus. It then remains to demonstrate a Pfr control on calcium redistribution,
and indeed this has been shown: uptake of ^{45}Ca is enhanced
after red irradiation, and this effect is far-red reversible
(Dreyer and Weisenseel, 1979). Moreover, release of Ca^{2+} from
the calcium vesicles is found as an effect of Pfr (Wagner et
al., 1987). Thus, Ca^{2+} fulfills all requirements to be an
internal signal that links the Pfr gradient to a localized
regulation of actin-myosin activity.

In summary, the following signal transduction is proposed:
DIRECTIONAL LIGHT SIGNAL → gradient of Pfr → local change in
calcium fluxes → calcium redistribution → calcium-calmodulin
interaction → activation of actin or myosin → actin-myosin
interaction → ORIENTED CHLOROPLAST MOVEMENT.

III. COMPLICATIONS AND UNSOLVED PROBLEMS
IN THE SIGNAL-TRANSDUCTION MODEL

A. Ageing of Pfr

After a pulse of red light, the chloroplast movement in
Mougeotia proceeds and continues in darkness, and this can
take as much as about 45 min. Thus, once the Pfr gradient has

been established, it must be fairly stable. This becomes still more evident, if the response is temporarily inhibited by blocking the energy-delivering metabolism or by inhibiting the actin-myosin system (see above); in those experiments, response is resumed and still under Pfr control, if the inhibitors are withdrawn after several hours (Fetzer, 1963; Wagner and Klein, 1981). Finally, Kraml et al. (1985) found Pfr to be functional for an amazingly long time: On the basis of evenly distributed Pfr, a non-saturating far-red pulse locally reverts Pfr to Pr, thus establishing a Pfr gradient; the chloroplast is repelled by the remaining Pfr even if this had been formed 40 h before.

In a strong contrast, Mesotaenium usually cannot respond to a single red pulse, but requires red light being continuously applied during the whole response, or at least as repetitive pulses. In the latter case, with dark intervals as short as 10 min the response level is already reduced. It has been shown that under these conditions Pfr is still present after the dark interval, but steadily loses its physiological activity, which is restored upon cycling Pfr → Pr → Pfr (Haupt and Reif, 1979). Recently, Kraml (1986) found that a Pfr gradient, originating from one red pulse, can be rendered active for a full response by blue light. It is not yet known whether this blue light acts on the "ageing" process or elsewhere in the transduction chain.

Nevertheless, the difference between Mougeotia and Mesotaenium is not as fundamental as it appears: Even in Mougeotia some ageing of Pfr has been reported with the result that the chloroplast orientates in a gradient of "aged" versus "newly formed" Pfr (Fischer, 1963; see below). This ageing can be observed already within minutes, but it obviously remains a quantitative rather than a qualitative effect: even after 40 hours Pfr can still control movement (see above).

B. Additional Photoreceptor Systems

The action spectrum in Mougeotia shows also a smaller peak in the blue region, which originally was assumed to be due to the short-wavelength absorption band of phytochrome (Haupt, 1959). However, this blue-light effect is not influenced by a simultaneous background irradiation of strong far-red, which completely abolishes the effect of strong red light (Gabryś et al., 1984). Thus, the blue-light effect is independent of phytochrome, a preliminary action spectrum peaks around 450 nm (Walczak et al., 1984). Except for this difference in signal perception, the transduction chain appears to be the same, making use of calcium, calmodulin, and actin-myosin interaction (Wagner et al., 1987).

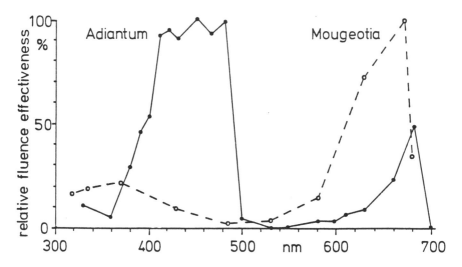

FIGURE 2. Action spectra of low-intensity chloroplast reorientation in Mougeotia and Adiantum. The maxima are normalized to 100%. After Haupt (1959) and Yatsuhashi et al. (1985).

Likewise, a phytochrome-independent blue-light effect has
been found also in Adiantum (Yatsuhashi et al., 1985). Interestingly, however, blue light is double as effective as red
light in Adiantum, in strong contrast to its low effect in
Mougeotia (fig. 2).

Besides this orientating vectorial effect, blue light can
also exhibit a scalar effect: If blue light of a sufficiently
high fluence rate is given simultaneously with continuous red
light, or as repetitive pulses alternating with red pulses,
the response is reversed, the chloroplast now moves with its
edges towards the highest Pfr levels, i.e. the "high-intensity
movement" ensues, the chloroplast turns its profile to the
red-light source, irrespective of the direction of the blue
light (cf. Schönbohm, 1980). For the same response obtained by
strong blue light alone, it has to be assumed that the scalar
blue-light effect reverts the vectorial blue-light signal, be
it absorbed by phytochrome (Schönbohm, 1980) or by the alternative blue-light system mentioned above. The action spectrum
of the scalar blue-light effect is similar to that of the vectorial effect, but the wavelength resolution of both of them
does not yet allow a precise comparison.

At least qualitatively the same situation has been found
in Mesotaenium, but not yet in Adiantum.

C. Tetrapolar Versus Bipolar Pfr Gradients

According to the model, the chloroplast of Mougeotia orients in a tetrapolar phytochrome gradient: two regions of lowest Pfr concentration are approached by the edges, and two
regions of highest Pfr concentrations are avoided by them.
Due to the cell anatomy, difficulties should arise in a bipolar gradient, i.e. if there is only one region of highest
Pfr concentration, being opposite to that of lowest concentration. In this case the chloroplast has to find a compro-
mise. There is good evidence that one edge leaves the region
of highest Pfr level, but the opposite edge has to move upstream in the Pfr gradient (fig. 3a; cf. Schönbohm, 1980). In
contrast, Frank (1977) assumes that in a bipolar gradient one
edge could approach the lowest Pfr level, thus forcing its
counterpart to tolerate approaching the highest Pfr level
(fig. 3b). To resolve these discrepancies, it has to be known,
for each of the experimental approaches, at what Pfr level
the physiological effects are saturated.
   The bipolar gradient of Schönbohm, as discussed above, has
been obtained by irradiating the cell with parallel vibrating
light, but with an incident angle of 45° to the normal (fig.
3a). If the interpretation is correct, the result confirms the
helical-like orientation of the Pr transition moments. It is
tempting to apply a corresponding irradiation to Adiantum,

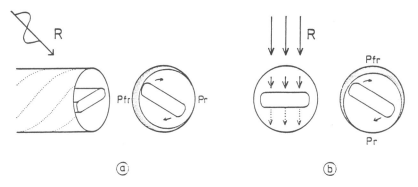

$$(a) \qquad\qquad\qquad (b)$$

   FIGURE 3. Generation of a bipolar Pfr gradient in Mougeotia by polarized red light, vibrating parallel to the cell
axis. a. The helical orientation of Pr transition moments is
indicated; red light (R) is given with an incident angle of
45°. The predicted Pfr gradient (dark crescent) is shown in
the cross section, together with the observed chloroplast reorientation. b. The two cross sections show attenuation of
light by the chloroplast (incident angle of 0°), and the predicted Pfr gradient with the observed chloroplast reorientation.

where the independent small chloroplasts should arrange in a
bipolar pattern. Yet, Yatsuhashi et al. (personal communication) could not detect any tendency to such a pattern.

D. Effects of Polarized Light Vibrating Parallel
to the Cell Axis

It has been pointed out that perception of light direction depends on the dichroic orientation of phytochrome, and
no gradient of Pfr should be established by polarized light
with the electrical vector parallel to the cell axis. Nevertheless, occasionally chloroplast orientations as induced by
parallel polarized light have been reported in Mougeotia and
Adiantum, and not all these cases have been explained yet.

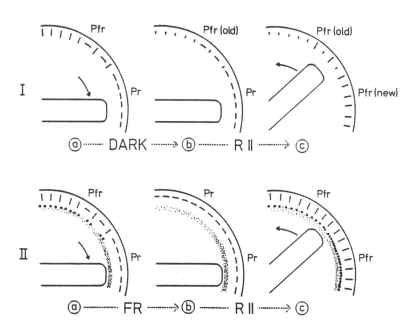

FIGURE 4. Chloroplast movement in Mougeotia as oriented
by polarized light, vibrating parallel to the cell axis (R ‖).
In the cross sections, tangential dashes denote Pr, radial
dashes Pfr, and short radial dashes "old" Pfr. I. Orientation
in a gradient of new versus old Pfr. II. Due to differential
adaptation, the response system has different sensitivity
(density of small dots) to the uniform Pfr level; its resulting activity is shown by the large dots. a. After preorien-
tation of the chloroplast. b. Before the test irradiation.
c. Movement after the test irradiation.

1. RED-LIGHT RESPONSE IN MOUGEOTIA. If a chloroplast is preoriented through a regular low-intensity response, subsequent irradiation with parallel vibrating R from any direction (test irradiation) results in reorientation by 90° (fig. 4I). These results led to the concept of ageing of Pfr (Fischer, 1963): That fraction of Pfr that has already been formed by the preirradiation, is less active than the newly-formed Pfr from the test irradiation; the chloroplast orients in a gradient of newly-formed versus aged Pfr. Accordingly, the test irradiation becomes completely ineffective, if saturating FR is given between preirradiation and test irradiation, or if the test irradiation is given in saturation, thus transforming all "old Pfr" to "new Pfr" by cycling (Heller, 1976).

However, under certain conditions (which still have to be specified), similar reorientations are induced by parallel vibrating red even if the preexisting Pfr is abolished by far-red (Kraml et al., 1984; Scheuerlein and Braslavsky, 1987). It is assumed that the Pfr gradient, established by and kept constant during the preirradiation, results in a differential "adaptation", i.e. differential sensitivity of the system to Pfr (fig. 4II). A subsequent parallel vibrating test irradiation establishes a uniform Pfr level all around the cell, but this Pfr is more effective in the sensitive regions, thus pushing away the chloroplast's edge from there (Kraml et al., 1984). It is highly desirable to know the nature of the "adaptation" and to locate it in the transduction chain. It might be that even diffusion of Pr has to be considered, by which a gradient of Pr would be levelled. Thus, an unequal distribution of Ptot would result (Grum-Heller, 1977).

There are additional orientating effects of parallel vibrating red light, which cannot be explained yet on the basis of the above hypotheses. This holds especially for responses requiring unusual long irradiation times (Frank, 1977; Schön-bohm and Brühl, 1979) or being induced by flashes in the nanosecond range (Haupt and Polacco, 1979).

2. RED-LIGHT RESPONSE IN ADIANTUM. As has been found by Yatsuhashi et al. (personal communication), parallel vibrating red light directs the chloroplasts in the protonema towards the flanks. Obviously, this irradiation establishes a gradient of Pfr activity with the highest levels at the flanks. Since the cells were preirradiated from above, it is tempting to compare this response with the ageing or adaptation model of Mougeotia, although in Adiantum the time course of pretreatment (6 h light + 48 h darkness) is much longer than in Mougeotia. Crucial experiments should include additional far-red preirradiations and variation of the direction of test light in relation to that of preirradiation.

3. BLUE-LIGHT RESPONSES IN MOUGEOTIA. The effectivity of
parallel vibrating light is still more pronounced in the blue
region. For the low-intensity movement, it is even slightly
more effective than normal vibrating light. Neither can an explanation be given for this unusual action dichroism, nor for
the reversion of it, if the blue test irradiation is applied
at low temperature (Schönbohm, 1980).

Similarly unexplained, also for high-intensity movement
(as induced by blue light without additional red), parallel
vibrating light is more effective in a certain fluence-rate
range (Schönbohm, 1980).

E. Coupling of Pfr to the Internal Signal

Whether or not the above presented model (section IIC) is
correct in detail, the internal signal finally regulates chloroplast movement so that its edges appear to be repelled by
the highest concentrations of Pfr (see above). This is true
for Mougeotia and Mesotaenium. In Adiantum, however, the small
chloroplasts accumulate at the regions of highest Pfr level
(Yatsuhashi et al., 1985). Thus, one of the internal signals in
the transduction chain must depend on Pfr in an inverse way in
the algae and in the fern (cf. also fig. 1).

Moreover, the effect is reversed also in Mougeotia and
Mesotaenium, if strong blue light is applied simultaneously
with R: the edges now approach the high Pfr levels (cf. Schönbohm 1980). And vice versa, in Adiantum the chloroplasts can
be induced to disappear from the high-level Pfr regions, if
the fluence rate of red is increased (Yatsuhashi et al., 1985).
Thus, the "switch" signal, transforming the low-intensity
orientation to high-intensity orientation, is operated differently in both kinds of systems. Much comparative work appears
to be needed to elucidate and explain the differences.

F. The Calcium Problem

The regulation of calcium fluxes by Pfr is still an open
question, at least in detail. In Mougeotia, calcium influx is
activated by Pfr (Dreyer and Weisenseel, 1979), but in Vallisneria, Pfr enhances its efflux (Takagi and Nagai, 1985). Moreover, calcium influx through the cell membrane in Mougeotia
cannot be the essential link between Pfr and movement: The
responsivity of the chloroplast depends on the level of cellular calcium rather than directly on the actual calcium concentration of the environment (Wagner and Klein, 1978). Thus,
more probably Pfr-facilitated release of calcium from the vesicles could be critical for transduction, and this phyto-

chrome effect has recently been demonstrated (Wagner et al., 1987). This is consistent with the location of these calcium stores close to the actin microfilaments. Unfortunately, till now differential effects on calcium redistribution upon partial irradiation of a cell have not yet been demonstrated.

With these uncertainties in mind, it can be questioned whether indeed Pfr control of calcium fluxes is as strictly localized as required for calcium being the directional internal signal. Alternatively, an overall calcium-calmodulin activation may be a basic prerequisite for the response, whereas an additional, still unknown internal signal determines the directionality, connecting the Pfr gradient with the local activity of the motor apparatus (Grolig and Wagner, 1988).

Finally, it is questioned by Schönbohm and Schönbohm (1984), whether calcium is an indispensable link in the transduction chain at all. In their experiments the low-intensity response is not impaired by depleting Mougeotia of calcium for several days. Authors, instead, suggest that phenolic compounds, which are found in the tannin vesicles, too, might be taken into consideration as possible internal signals. However, as long as no concomitant measurements of intracellular calcium content are available, the calcium hypothesis need not be abandoned.

G. Regulation of the Motor Apparatus

In Mougeotia, calcium is assumed to activate the actin-myosin system. In contrast, cytoplasmic streaming in Vallisneria is inhibited by calcium; lowering of intracellular calcium, which is required for streaming, can be caused by Pfr (Takagi and Nagai, 1985). But even in Mougeotia, there is an apparent contradiction. Whatever the final decision about calcium as a link, the activity of the actin-myosin system is under phytochrome control. However, Schönbohm (1973) concluded, from centrifugation experiments, that Pfr fastens the chloroplast at its place, whereas in a gradient the chloroplast is pushed away by Pfr.

These inconsistencies become less serious, if we consider the complexity of actin-myosin interaction as underlying the response. Intracellular movement requires that actin and myosin are anchored to respective structures (cf. cytoplasmic streaming in Nitella; Nagai and Hayama, 1979), e.g. actin to the outer layers of cytoplasm. Pfr, then, may either activate directly actin-myosin interaction, or it may anchor actin to its supporting structures, thus only establishing the basic requirement for movement. Besides, it might be the presence of actin microfilaments at all, which is critical for move-

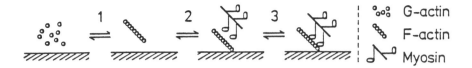

FIGURE 5. Hypothetical possibilities for calcium action
on the actin-myosin system: 1. Polymerization/depolymerization
of actin. 2. Anchoring/releasing of F-actin to and from cell
structures. 3. Activation/inactivation of actin-myosin interaction.

ment, and this depends on the polymerization of G-actin to F-
actin. The Pfr-controlled internal signal, then, could regu-
late the G $\rightleftharpoons$ F-actin equilibrium as a still more fundamental
prerequisite for movement (Wohlfahrt-Bottermann, 1977).

Thus, three levels of regulation of the motor apparatus
by Pfr can be assumed (fig. 5), and it would depend on the
experimental conditions, which of the three processes is the
limiting factor for movement.

IV. GENERAL CONCLUSIONS

Although a reasonable transduction chain can be presented
for the light-induced chloroplast orientation in <u>Mougeotia</u>, in
detail many problems remain unsolved. They concern all levels
of the transduction chain: photoperception, internal signal,
mechanism of movement. These problems become still more complicated if we realize that the model cannot simply be transferred to other systems, but that even in apparently very
similar systems there are basic differences. Detailed analysis
of those differences may be an important step in an approach
to finally elucidate the transduction chains in a few systems,
and thus to improve knowledge about possible mechanisms of
phytochrome action.

REFERENCES

Dreyer, E.M., and Weisenseel, M.H. (1979). Planta 146:31.
Fetzer, J. (1963). Zeitschr. Bot. 51:468.
Fischer, W. (1963). Zeitschr. Bot. 51:348.
Frank, M. (1977). Zeitschr. Pflanzenphysiol. 82:210.
Gabryś, H., Walczak, T. and Haupt, W. (1984). Planta 160:21.
Grolig, F., and Wagner, G. (1988). Botanica Acta 101: in press.
Grum-Heller, S. (1977). Zeitschr. Pflanzenphysiol. 81:212.

Haupt, W. (1959). Planta 53:484.

Haupt, W. (1983). Progr. Phycol. Res. 2:227.

Haupt, W., and Polacco, E. (1979). Plant Science Letters 17:67.

Haupt, W., and Reif, G. (1979). Zeitschr. Pflanzenphysiol. 92:153.

Haupt, W., and Wagner, G. (1984). In "Membranes and Sensory Transduction" (G. Colombetti and F. Lenci, eds.), p. 331. Plenum Press, New York - London.

Heller, S. (1976). Zeitschr. Pflanzenphysiol. 79:8.

Kraml, M. (1986). Proc. XVI Yamada Conf., Okazaki, Japan, 94.

Kraml, M., Enders, M., and Bürkel, N. (1984). Planta 161:216.

Kraml, M., Leopold, K., and Winkler, B. (1985). Book of Abstracts, Europ. Symp. Photomorphogenesis in Plants, Wageningen, The Netherlands, 143.

Nagai, R., and Hayama, T. (1979). J. Cell. Sci. 36:121.

Scheuerlein, R., and Braslavsky, S. (1987). Photochem. Photobiol., submitted.

Schönbohm, E. (1973). Ber. Deutsch. Bot. Ges. 86:423.

Schönbohm, E. (1980). In "The Blue Light Syndrome" (H. Senger, ed.), p. 69. Springer-Verlag, Berlin - Heidelberg.

Schönbohm, E., and Brühl, K.-L. (1979). Ber. Deutsch. Bot. Ges. 92:305.

Schönbohm, E., and Schönbohm, E. (1984). Biochem. Physiol. Pflanzen 179:489.

Serlin, B.S., and Roux, S.J. (1984). Proceed. Natl. Acad. Sci. USA, 81:941.

Takagi, S., and Nagai, R. (1985). Plant Cell Physiol. 26:941.

Wagner, G., and Grolig, F. (1985). In "Sensory Perception and Transduction in Aneural Organisms" (G. Colombetti and F. Lenci, eds.), p. 281. Plenum Press, New York - London.

Wagner, G., and Klein, K. (1978). Photochem. Photobiol. 27:137.

Wagner, G., and Klein, K. (1981). Protoplasma 109:169.

Wagner, G., Valentin, P., Dieter, P., and Marmé, D. (1984). Planta 162:62.

Wagner, G., Grolig, F., and Altmüller, D. (1987). Photobiochem. Photobiophys., in press.

Walczak, T., Gabryś, H., and Haupt, W. (1984). In "Blue Light Effects in Biological Systems" (H. Senger, ed.), p. 454. Springer-Verlag, Berlin - Heidelberg.

Wohlfarth-Bottermann, K.E. (1977). In "Cell Differentiation in Microorganisms, Higher Plants and Animals" (L. Nover and K. Mothes, eds.), p. 564. Fischer-Verlag, Jena.

Yatsuhashi, H., Kadota, A., and Wada, M. (1985). Planta 165:43.

PHOTO- AND POLAROTROPISM IN FERN PROTONEMATA

Masamitsu Wada
Akeo Kadota

Department of Biology
Tokyo Metropolitan University
Tokyo, Japan

I. INTRODUCTION

The processes of light perception and the following photosensory transduction in photomorphogenesis are still unclear despite the recent great advances in molecular studies of phytochrome (see Chapter II.-1-7). Since at this moment there is no reproducible in vitro phenomenon of photomorphogenesis at the organelle level, systems at the cell level such as green algae, fern and moss protonema, fungus hypha and culture cells of higher plants are the most promising materials for kinetic studies of photomorphogenesis, provided such cells are simple in their organization and sensitive and prompt in their photoresponses. Fern protonema provide one of the best materials since, in addition to the above characteristics, the elementary processes of the development of fern haplophase are regulated mainly by light, as has been shown by many studies since the last century (Miller 1968, Furuya 1983). Each step of the development is independently and experimentally inducible by light.

Fern protonemata display phototropism (Mohr, 1956) and polarotropism (Bünning and Etzold, 1958). Polarotropism has been studied because of its curious nature in which protonemata grow not toward the light source but perpendicularly to the incident light and also to the electrical vector (E-vector) of linearly polarized light (Etzold, 1965; Steiner, 1969). The dichroic effects of polarized light have provided an insight into the orientation of photoreceptor pigments in plant cells. Although relatively long periods of light were needed to achieve the response in most of the earlier works, we have established an experimental system with protonemata of the maiden-hair fern, Adiantum capillus-veneris L., in which phototropic and polarotropic responses can be induced by a pulse irradiation of stimulus light (Wada et al., 1981). This system has made it possible to examine the photoresponses in detail photomorphogenetically. We have been analyzing the tropic responses induced by microbeam irradiation in an effort to determine the intracellular photoreceptive site, the dichroic orientation of photoreceptors, and the regulation of the growing direction by the level of Pfr.

II. DEFINITION OF PHOTO- AND POLAROTROPISM

Polarotropism has been studied with protonemata cultured between the surface of agar media and a glass plate (i.e. bottom of Petri dish or cover slip) placed perpendicular to the incident light and immediately next to the protonemata. The glass plate prevents protonemal growth toward the light source, resulting in typical polarotropic growth. When protonemata are grown in agar media and are briefly irradiated with polarized red light, the protonemata show neither true polarotropism nor true phototropism, but grow obliquely toward the incident light, showing a mixture of polarotropic and phototropic responses. We measured the response in such cases, if necessary, as the angle between appropriate vectors deduced from cell axes before and after light treatment.

Phototropism and polarotropism are somewhat difficult to define in the case of partial irradiation of a cell with a microbeam. Irradiation of one side of a protonema induces cell growth toward the irradiated side, but not simply toward the light source, even with polarized or unpolarized light. Here we call the light-induced protonemal growth toward the side irradiated with a higher fluence as "phototropism" or "phototropic response" (Fig. 1).

III. PHOTORECEPTORS

Action spectra for polarotropism have been measured in Dryopteris (Etzold, 1965; Steiner,1969) and in Adiantum (Kadota et al, 1984). In Dryopteris protonemata, both red light and blue-near UV light induce the response, the latter being much more effective. While phytochrome and a blue light-absorbing pigment were assigned as the receptor pigments, the involvement of phytochrome was unclear because

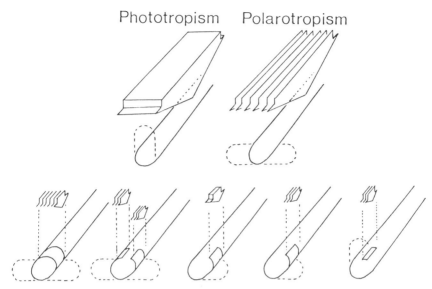

Phototropic responses induced by local irradiation

FIGURE 1.   Scheme showing the methods of irradiation, and the responses induced by whole cell irradiation (top) and local irradiation (bottom). ⬦:unpolarized light, ⬦:polarized light.

long term irradiation was used to induce a response and therefore only a partial reversal of the red light effect was obtained with far-red light. The action spectrum in Adiantum, on the other hand, showed the effectiveness of red light, but no effectiveness in the blue/near-UV region. The action maximum was around 680 nm and thus shifted to a longer wavelength as compared with the usual absorption peak of Pr. The difference spectrum of phytochrome in vivo in fern gametophytes was found to have a peak at around 660 nm (Grill and Schraudolf, 1981; Tomizawa et al, 1982), consistent with that of phytochrome difference spectra from higher plants. Nevertheless, the response is clearly mediated by phytochrome since the 680 nm light effect was fully reversed by subsequent far-red light. Typical red/far-red reversibility of phototropism of Adiantum, which was induced by a microbeam given to a flank of the subapical region, was also achieved.

Furthermore, the peak of the action spectrum for this phototropic response was shown to be around 660 nm (Kadota et al, 1984) which is consistent with the usual Pr absorption spectrum.

The reason for the difference in the peak wavelengths of the action spectra between polarotropism and phototropism is not clear at present. However, it is interesting to note that 680 nm was also the action maximum for the phytochrome-mediated chloroplast photoorientation under polarized light in Adiantum protonemata (Yatsuhashi et al, 1985). Moreover, the polarized red light-induced chloroplast rotation in Mougeotia also has an action peak at 680 nm, while the action peak of Mougeotia phytochrome was estimated to be about 660 nm from the action spectrum for another type of phytochrome response (Haupt, 1968). Haupt explained the result by stating that 680 nm light would be more effective than 660 nm light in creating a steep gradient of Pfr in the cell because of the different dichroic orientation of Pr and Pfr at the cell periphery. In Adiantum protonemata, phytochrome molecules have been shown to have the same dichroic orientation as in Mougeotia and the intracellular gradient of Pfr has been found to be responsible for the tropic response (see V, VI of this chapter). Thus, the same explanation may be applicable to Adiantum.

While blue light had no effect on the polarotropism in Adiantum, it was shown recently that phototropic response was effectively induced by blue light when applied to the half side of a protonema (Hayami et al, 1986). The blue light-induced tropic response was partly reversed by subsequent far-red light, and blue/far-red photoreversibility was repeatedly observed. However, full reversal of the response was not observed even with a higher fluence of far-red light, indicating that both phytochrome and a blue light-absorbing pigment were involved in this blue light effect. Detailed studies of the far-red light effect on the fluence-response curve for this blue light effect revealed that the simultaneous involvement of the two receptors was limited in the relatively high fluence range and that the blue light-absorbing pigment alone was responsible in the lower fluence range. In the higher fluence range, the response mediated by the blue light receptor became saturated and the phytochrome response became evident, indicating a difference in the sensitivities of the two receptor pigments to blue light.

The reason why the involvement of blue light receptors was not detected in the action spectrum for Adiantum polarotropism is not clear at present, but it is likely that the intracellular arrangement of blue light receptors is not appropriate to give a polarotropic response under short term irradiation.

IV. LOCALIZATION OF PHOTORECEPTORS

The intracellular photoreceptive site for the red light-induced polarotropic response was studied with a microbeam irradiator. Both flanks of the subapical region (apical 5-15 μm region) of the protonema were shown to be responsible for the polarotropism (Wada et al, 1981). Further studies of the blue light-induced phototropism revealed that phytochrome and a blue light-absorbing pigment were located on the same part (apical 5-25 μm region) of a cell (Hayami et al, 1986), suggesting that the transduction chains after light perception might be largely the same for both receptor systems.

Phytochrome molecules regulating the tropic response in Adiantum may be located in the ectoplasm and/or plasma membrane because a pulse microbeam irradiation at the subapical photoreceptive site effectively induced a response even in the centrifuged protonema in which oil droplets, chloroplasts and other cytoplasm except ectoplasm were removed from apex (Wada et al, 1983) (Fig. 2).

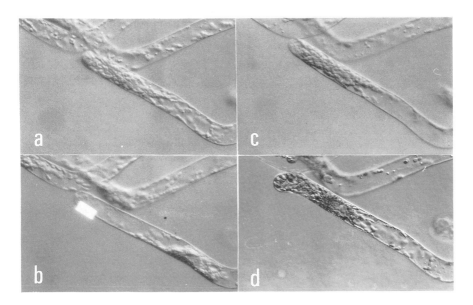

FIGURE 2. Induction of phototropic response in a centrifuged protonema. a. before centrifugation. b. after centrifugation and immediately before microbeam irradiation. Microbeam (20 x 10 μm) of infrared light was shown. c. after spinning back the cytoplasm. d. phototropic response at 24 h after the microbeam irradiation. (from Wada et al, 1983.)

V. DICHROIC ORIENTATION

When a flank of the subapical region of the protonema was irradiated with a microbeam of linearly polarized red and far-red light, red light with an E-vector parallel to the cell surface was most effective in the induction of the phototropic response. However, the far-red light effect was most prominent when its E-vector was normal to the cell surface (Kadota et al, 1982), indicating a different dichroic orientation of Pr and Pfr in the cell. This had been assumed by Etzold (1965) in Dryopteris protonemata and was experimentally proved for the first time by Haupt et al (1969) in Mougeotia cells for the phytochrome-mediated rotation of chloroplast.

Dichroic orientation of phytochrome at the protonemal apex was also studied using microbeam irradiation (Kadota et al 1985). The response inducible by irradiation of the subapical part was nullified by subsequent red light given at the extreme tip (see, VI of this chapter). On the other hand, far-red light irradiation of the extreme tip after irradiation of the whole cell with depolarized red light effectively induced a tropic response. Each of these red and far-red light effects on the protonemal tip showed an action dichroism, suggesting the same orientation of transition moment of Pr and Pfr in the tip of protonema as described above.

The orientation of the transition moment of Pr in a surface view of protonema was further studied using the phototropic response induced by a polarized red light microbeam irradiation focused on the central area of the subapical part (Wada et al, unpublished data). The results showed no dependency of the response on the polarization plane of red light applied, indicating that Pr orientation in the plane of the membrane is not single-handed spiral but random or double- handed spiral, different from the case of Mougeotia (Haupt and Bock, 1962).

A non-single-handed spiral arrangement of Pr was also suggested by experiments with chloroplast photoorientation (Yatsuhashi et al, unpublished data). When protonemata are irradiated from an oblique direction, keeping the E-vector parallel to the protonemal axis, and if we assume that the transition moment of the photoreceptor is arranged in a single-handed spiral like Mougeotia, light absorption should be null at one flank where the transition moment of the photoreceptor and the light incidence coincide with each other, and the maximum at the opposite flank, causing one-sided accumulation of chloroplasts. However, such one-sided accumulation was not observed at any incident angle tested.

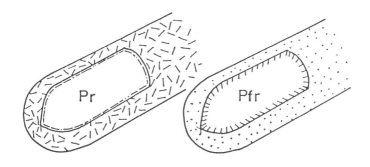

FIGURE 3. Schematic drawing of three dimensional
orientation of phytochrome molecules at the apical part of a
protonema of Adiantum capillus-veneris. The dashes show the
absorption vector.

The three dimensional orientations of Pr and Pfr are
schematically drawn in Fig. 3.

The timing of the change in the orientation of phytochrome
molecules during photoconversion from Pr to Pfr was recently
analyzed by double laser flash experiments (Kadota et al,
1986). The change in the orientation of phytochrome was found
to occur between 2 ms and 30 s after the red flash. Assuming
the phototransformation pathway of Adiantum phytochrome is the
same as that reported for purified phytochrome from higher
plant (Furuya, 1983), the result indicates that the
reorientation of phytochrome occurs between I_{692} and Pfr, that
is, in the final step of the pathway. The time scale of this
change in the orientation of phytochrome is so slow that the
change cannot be ascribed to the conformational change of the
chromophore. Rather, the change may be attributed to the
conformation change of the protein moiety, or if there is a
membrane receptor for phytochrome, to a change in the
interaction between phytochrome and the receptor.

Etzold (1965) assumed that the blue light-absorbing
pigment also had a dichroic orientation in Dryopteris. This
assumption was recently confirmed in Adiantum protonema for
the blue light-induced phototropic response (Hayami et al,
unpublished data). As both phytochrome and a blue light-
absorbing pigment are involved in this blue light effect, the
orientation of the two photoreceptors was investigated
separately. The results indicated that the transition moments
of both the blue light-absorbing pigment and the blue
absorption band of Pr were almost parallel to the cell
surface.

VI. REGULATION OF GROWTH DIRECTION

The tropic responses of tip-growing cells are classified as "bowing", in which the growth rates of both flanks differ, and "bulging", in which a new growth zone arises on the bending side (Green et al, 1970). The tropic responses in fern protonemata has been ascribed to the latter (Etzold, 1965; Davis, 1975). Etzold (1965) explained the photo- and polarotropism in Dryopteris protonemata as being a consequence of the movement of a growing point from the center of the apical dome to another region where the light absorption was highest. According to this model, photoreceptive sites for polarotropism should be different depending upon the direction of the electrical vector of polarized light. However, our results in Adiantum showed that the receptive site of phytochrome-mediated polarotropism in Adiantum is mainly in the protonemal flank of the apical 5-15 μm region and this localization of the photoreceptive site was not affected by the direction of the polarization plane (Wada et al, 1981). Since this apparently contradicts Etzold's model, we must consider an alternate hypothesis to explain the result. When each half of the receptive site was irradiated with polarized red light of different fluences, the tropic response became dependent not on the polarization plane of the light but on the difference in fluence between the two sides (Wada et al, 1981). This indicates that the direction of apical growth may be controlled by a difference in Pfr level between the two flanks of the subapical region. This assumption was tested kinetically with a custom-built microbeam irradiator that allows simultaneous irradiation of two adjacent small portions of the protonema with different fluences of red light. The results suggest that the phototropic response would be a function of the difference in Pfr concentration generated across the two opposite sides of the protonemata (Wada et al, 1986).

Furthermore, irradiation of the apical 0-2.5 μm region with red light was found to inhibit the tropic response which was induced by irradiation of the subapical 10-30 μm part with parallel-vibrating polarized red light. However, far-red light irradiation of the extreme tip after depolarized red light irradiation of the whole cell effectively induced a tropic response (Kadota et al, 1985). These results indicate that: 1) When the level of Pfr is high in the subapical region and there is none or a small amount present in the tip, the tropic response can be induced even if there is no difference in the Pfr level between the right and left sides of the subapical region (in this case, the response occurs randomly in either direction because there is no difference between the two sides). 2) The tropic response is suppressed when the Pfr level is high in both the tip and the subapical region. Thus, the difference in Pfr level between the extreme tip and the subapical region also appears to be crucial in regulating the direction of apical growth.

In conclusion, the direction of apical growth in <u>Adiantum</u> is regulated by the differences in the Pfr level between the extreme tip and the subapical part and between opposite sides of the protonema.   Under polarized light irradiation these differences are caused by the different dichroic orientation of Pr and Pfr at the cell periphery.   However, more detailed studies are needed to clarify how these two factors control the position of the growing point in the apex.

VII   PROSPECTS

Like <u>Mougeotia</u>, fern protonemata have brought us an insight into physiologically active photoreceptors, in terms of their intracellular localization, dichroic orientation and their function based on Pfr concentration.   However, no part of the photosensory transduction chains has yet been clarified, as is the case in all other photomorphogenic responses in plants. The phenomena controlled by phytochrome might be classified into two categories; those which are and those which are not mediated by direct gene expression.   In the latter case, such as fern phototropic responses, active molecules of phytochrome, while not being transmembrane protein (Hershey et al, 1985), are likely to be localized on or very close to the membrane systems such as plasma membrane (Wada et al, 1984). To understand the black box of the photosensory transduction chain, therefore, the real state of phytochrome molecules in relation to biomembranes and/or the cytoskeleton in the cell must be one of the most important and urgent problems to be solved, in addition to the identification of second messengers of these photoresponses. Until we have good experimental <u>in vitro</u> systems of membrane-mediated photomorphogenesis at the organelle level, fern protonemata would seem to be good material for carrying out such experiments.

ACKNOWLEDGMENTS

We would like to thank Dr. P. J. Lumsden for careful reading of the manuscript.

REFERENCES

Bünning, E., and Etzold, H. (1958). Ber. Dtsch. Bot. Ges. 71:304.
Davis, B. D. (1975). Plant Cell Physiol. 16:537.
Etzold, H. (1965). Planta 64:254.
Furuya, M. (1983). In "Encyclopedia of Plant Physiology, New Series, Vol. 16, Photomorphogenesis" (W. Shropshire, Jr. and H. Mohr, ed.), p. 569. Springer-Verlag, Berlin, Heidelberg.
Furuya, M. (1983). Philos. Trans. R. Soc. Lond. Ser. B303:361.
Green, P. B., Erickson, R. O., and Richmond, R. A. (1970). Ann. New York Acad. Sci. 175:712.
Grill, R., and Schraudolf, H. (1981). Plant Physiol. 68:1.
Hayami, J., Kadota, A., and Wada, M. (1986). Plant Cell Physiol. 27: 1571.
Haupt, W. (1968). Z. Pflanzenphysiol. 58:331.
Haupt, W., and Bock, G. (1962). Planta 59:38.
Haupt, W., Mörtel, G., and Winkelnkemper, I. (1969). Planta 88:183.
Hershey, H. P., Baker, R. F., Idler, K. B., Lissemore, J. L., and Quail, P. H. (1985). Nucl. Acids Res. 13:8543.
Kadota, A., Wada, M., and Furuya, M. (1982). Photochem. Photobiol. 35:533.
Kadota, A., Koyama, K., Wada, M., and Furuya, M. (1984). Physiol. Plant. 61:327.
Kadota, A., Wada, M., and Furuya, M. (1985). Planta 165:30.
Kadota, A., Inoue, Y., and Furuya, M. (1986). Plant Cell Physiol. 27:867.
Miller, J. H. (1968). Bot. Rev. 34:361.
Mohr, H. (1956). Planta 17:127.
Steiner, A. M. (1969). Photochem. Photobiol. 9:507.
Tomizawa, K., Manabe, K., and Sugai, M. (1982). Plant Cell Physiol. 23:1305.
Wada, M., Kadota, A., and Furuya, M. (1981). Plant Cell Physiol. 22:1481.
Wada, M., Kadota, A., and Furuya, M. (1983). Plant Cell Physiol. 24:1441.
Wada, M., Shitanishi, K., Kadota, A., and Iino, M. (1986). Proceed. XIV Yamada Conference, p. 145.
Yatsuhashi, H., Kadota, A., and Wada, M. (1985). Planta 165:43.

APPLICATION OF THE DIMERIC MODEL OF PHYTOCHROME ACTION TO HIGH IRRADIANCE RESPONSES

William J. VanDerWoude

Plant Photobiology Laboratory
Plant Physiology Institute
Agricultural Research Service, USDA
Beltsville, Maryland

I. INTRODUCTION

Explanations of phytochrome-mediated behavior often encounter the difficulty of relating responses to [Pfr] or [Pfr]/[Ptot] (Hillman, 1972; Smith, 1983). The initially reasonable assumption that phytochrome exists and acts as a monomer having a single chromophore is now known to be incorrect. Recent biochemical evidence (Jones and Quail, 1986) clearly demonstrates phytochrome to be a dimer composed of identical subunits. Support for the dimeric nature of phytochrome in vivo is provided by fluence-response studies of red light-induced pelletability (Pratt and Marmé, 1976). The establishment of low Pfr levels leads to the pelletability of nearly equal amounts of Pr and Pfr. This behavior is consistent with the existence of two chromophores per dimer and indicates that only one need be in the Pfr form to elicit pelletability of the entire molecule.

Knowledge of the dimeric nature of phytochrome provides new opportunities to examine relationships between phytochrome photoconversion and photomorphogenic responses. Such examination of the germination responses of photodormant lettuce seeds, induced by brief irradiations of very low fluence (VLF) or low fluence (LF), supported a dimeric model for the action of phytochrome (VanDerWoude, 1985). Simulations of the proposed mechanism closely fit the biphasic fluence-response germination behavior of seeds that had been sensitized by prechilling or other pretreatments.

The natural environment of continuous daily irradiation
promotes phytochrome-mediated photomorphogenic behavior
characterized as "high irradiance responses" (HIR)
(Mancinelli and Rabino, 1978). It has proven difficult to
fully explain the HIR on the basis of monomeric phytochrome
(Fukshansky and Schäfer, 1983). This paper presents an
initial evaluation of the dimeric model in relation to the
HIR. Although subject to refinement, the analysis supports
an explanation of the HIR based on the dimeric model.

II. THE DIMERIC MODEL OF PHYTOCHROME ACTION

The dimeric model (Fig. 1) recognizes the interphoto-
transformations of the three possible species of the dimer,
PrPr, PrPfr, and PfrPfr, and assumes: (1) Interaction of the
dimeric species with the next component of the transduction
chain, a specific receptor, X; (2) Very low abundance of X
relative to phytochrome; (3) High stability of the receptor
complexes, PrPfr-X and PfrPfr-X. Simulations of the model
have tentatively assumed the independent phototransformation
of each chromophore, uninfluenced by the state of its
partner or by association with X.
An iterative approach for calculating the behavior of
the model as it relates to responses to brief irradiations
has been described (VanDerWoude, 1985). Simulations of the
model fit fluence requirements for observed responses very
well and supported the following: (1) PrPfr-X is established
by fluences in the VLF range; (2) Saturation of VLF
responses in biphasic fluence-response profiles is related
to saturation of X by PrPfr; (3) [X] is very low, about
0.001 [Ptot] in dark-grown/incubated systems; (4) Since VLF
lead to the occupation of X by PrPfr, and [PrPfr-X] is also
very low, fluences in the LF range, 3 to 4 orders of magnitude greater than VLF, are required to phototransform
PrPfr-X to PfrPfr-X; (5) Responses to LF are related
linearly to [PfrPfr-X] whereas VLF responses correlate well
with log [PrPfr-X]; (6) The activity of PrPfr-X, but not
PfrPfr-X is increased by sensitization.
Kinetics for the promotion and decay of thermally-
induced sensitization in seeds are similar to those of
membrane thermal adaptation (VanDerWoude and Toole, 1980).
Such kinetics, as well as sensitization by ethanol and other
anesthetics (Taylorson and Hendricks, 1979), support the
hypothesis that the activity of PrPfr-X is a function of its
lateral mobility as a membrane-associated complex. Sensitization induced by decreases in extracellular pH (Shinkle and
Briggs, 1985) may also involve mobilization of PrPfr-X.

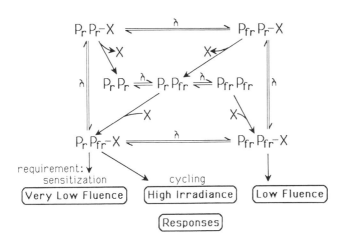

FIGURE 1. Primary elements of the dimeric model of
phytochrome action. The activity of PrPfr-X is increased by
sensitization and by photochemical cycling.

III. THE DIMERIC MODEL AND HIGH IRRADIANCE RESPONSES

An initial evaluation of the applicability of the
dimeric model to the HIR is presented here in brief. Calculations for the simulations were based on extensions of the
approach used in earlier studies of VLF and LF responses.
The full analysis and details of the mathematical approach
will be presented in a future report.

Calculations of the occupation of free X by PrPfr and
PfrPfr considered the individual and relative concentrations
of dimer species and assumed competition by the formation of
unstable PrPr-X and PfrPr-X complexes. They assumed that,
relative to their concentration, PfrPfr has a two-fold
advantage over PrPfr for association with X. At high
fluence rates and/or low [PrPfr] or [PfrPfr], X is expected
to be unsaturated. A rate factor was therefore introduced
that set PrPfr-X near 50% of total X after a 12 h, 704 nm
irradiation at 1 μmol m^{-2}s^{-1}, in the presence of
phytochrome synthesis and destruction.

The parameters and assumptions of the simulations were:
(1) 1,000,000 dimers per phytochrome-containing cell in
dark-grown tissues; (2) 1,000 X per cell; (3) Phytochrome
extinction coefficients and quantum efficiencies for
photoconversion were those reported for dark-grown oat
(Largarias et al., 1987); (4) Half-life of inactive PrPr-X

and PfrPr-X complexes = 1 s; (5) Hourly rate for synthesis of PrPr = 1.5% of the dark level of Ptot; (6) Half-lives for destruction: PrPr = 46 h; PrPfr = 60 min; PfrPfr = 30 min; (7) Nondestruction of dimers at X; (8) No dark reversion.

As information increases, future simulations may beneficially consider also the influence of sequestering, the dependence of synthesis and destruction on fluence rate and wavelength, and the presence and properties of multiple phytochrome species.

The model was examined in relationship to several characteristics of the HIR: (1) Spectral action maxima are displayed in the blue and far red that are strongly dependent on fluence rate; (2) Responses to red are relatively independent on fluence rate; (3) Reduction of initial Ptot by preirradiation reduces HIR responses to blue and far red, but does not diminish responses to red (Beggs et al., 1980; Holmes and Schäfer, 1981).

The spectral dependence of model components at photoequilibria established by 12 h at 3 $\mu mol\ m^{-2}s^{-1}$ is shown in Fig. 2. This simulation assumes the absence of phytochrome synthesis and destruction and 100,000 dimers per cell. The equilibria spectral dependence of both PfrPfr and PfrPfr-X are qualitatively similar to spectrophotometrically detectable "Pfr". PrPfr and PrPfr-X display a broad plateau in the blue and green, and a sharp maxima at 700 nm. Above 750 nm, the competition for X between PrPfr and very low levels of PfrPfr produces increased PrPfr-X (Fig. 2B).

The fluence rate dependence of the HIR has in part been attributed to an HIR requirement for the photochemical cycling of phytochrome (Johnson and Wall, 1983; Fukshansky and Schäfer, 1983; Gaba and Black, 1985). The correspondence of PrPfr-X and HIR maxima in the far red suggest examination of the cycling behavior of PrPfr-X (Fig. 2C). The cycling considered here is the turnover, due to photoconversion, and replacement of PrPfr-X. It is a very small component of the photochemical cycling of the entire phytochrome system. Maxima in the rate of PrPfr-X cycling occur near 404 nm and 696 nm. The multiplicative interaction of [PrPfr-X] and its rate of cycling provides maxima near 412 nm and 700 nm. Although they occur at somewhat lower wavelengths, the relative position and magnitudes of the blue and far red maxima correspond reasonably well with observed HIR behavior.

Inclusion of the process of phytochrome synthesis and destruction in simulations of the model changes the spectral dependencies of "Pfr" and PrPfr most greatly (Fig. 3A). After a 12 h irradiation at 3 $\mu mol\ M^{-2}s^{-1}$ they each display a maximum near 716 nm. Similar behavior of "Pfr" has been discussed in the analyses of other HIR models (Fukshansky and Schäfer, 1983; Johnson and Wall, 1983). This behavior

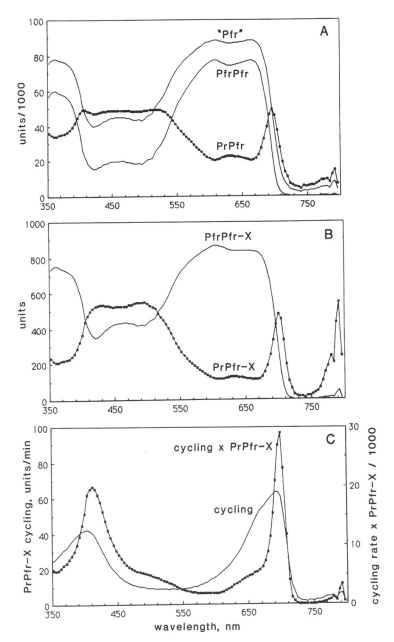

FIGURE 2. Spectral dependence of equilibrium values for components of the dimer model in the absence of phytochrome synthesis and destruction. The simulation is for a 12 h irradiation at 3 μmol m^{-2}s^{-1}. Total phytochrome dimer was set at 10% of the level in darkness.

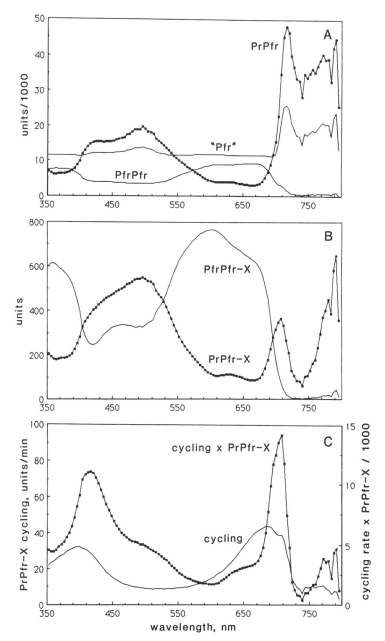

FIGURE 3. Spectral dependence of components of the
dimeric model in the presence of phytochrome synthesis and
destruction after irradiation for 12 h at 3 μmol m⁻²s⁻¹. At
12 h the model is at or near equilibrium over the spectral
region studied.

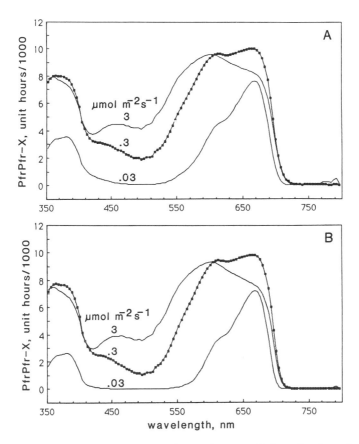

FIGURE 4. The spectral dependence of hourly levels of
PfrPfr-X summed over 12 h of irradiation at the indicated
fluence rates. Initial level of total dimer was (A),
1,000,000 per cell, or (B), reduced to 50,000 per cell, as
would be expected by appropriate red preirradiation.

leads to blue and far red maxima nearer to those characteristic of the HIR. The far red maximum for PrPfr-X occurs
near 708 nm (Fig. 3B). The product of PrPfr-X and its
cycling rate displays maxima at 416 nm and 708 nm (Fig. 3C).
    In response systems that have achieved competence
(Mohr, 1983), HIR responses are likely related to photomorphogenic activity that occurs throughout the irradiation.
To examine the equivalent of HIR activity, subsequent simulations present the 12 h sum of hourly values. Such activity
for PfrPfr-X (Fig. 4A) displays little dependence on red
fluence rate. A reduction of total initial dimer to 5% of
dark levels has little influence on PfrPfr-X (Fig. 4B).

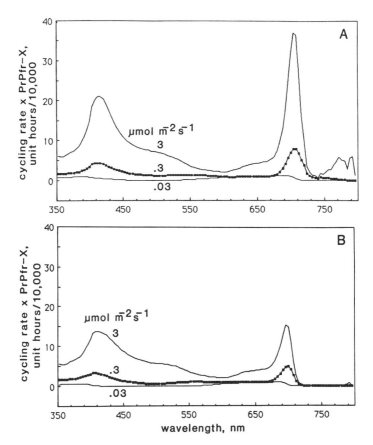

FIGURE 5.   Spectral dependence of hourly levels of the
product of PrPfr-X and its cycling rate, summed over 12 h of
irradiation at the indicated fluence rates. The initial
level of total dimer was (A) 1,000,000 or (B) 50,000/cell.

In contrast to the behavior of PfrPfr-X, the product of
PrPfr-X and its cycling rate is highly dependent on fluence
rate at both blue and far red maxima (Fig. 5A). A reduction
of the initial level of total dimer (by preirradiation)
reduces the magnitude of both maxima (Fig. 5B). Explanation:
Preirradiation reduces the opportunity for far red and, to a
lesser extent, blue, to limit destruction. The availability
of PrPfr to form PrPfr-X is therefore reduced. Such saving
of phytochrome is not a requirement of PfrPfr-X formation,
given the rate constants and assumptions employed.
    Action spectra for the model (Fig. 6A) were developed
from 12 h simulations at from .01 to 10 μmol m^{-2}s^{-1}. Relative photon effectiveness was calculated as the normalized

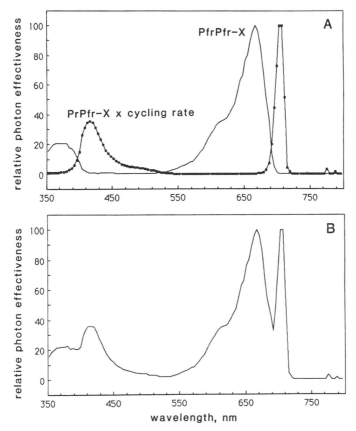

FIGURE 6. Action spectra for the behavior of the
dimeric model under continuous irradiation. (A) Action
spectra for PfrPfr-X and for PrPfr-X x its cycling rate,
each normalized to its maximum. (B) Normalized sum of the
two action spectra in A.

reciprocal of the fluence rate required for 55% of maximum
levels, based on maximum slopes of logarithmic fluence rate-
response plots of action summed over 12 h. The PfrPfr-X
action maxima are 668 nm and a low plateau centered on 372
nm. Maxima for PrPfr-X x cycling rate are positioned at 416
and 706 nm. Figure 6B presents a simulated action spectrum
for responses to the combined action of PfrPfr-X and PrPfr-X
x cycling, if it is assumed they share the same effective-
ness-response relationship. The action spectrum for the
behavior of the model corresponds well with the generalized
characteristics of the HIR.

The simulations demonstrate the suitability of the dimeric model to explain the complex behavior of the HIR. The assumptions employed in evaluating the model remain to be examined and tested. However, the ability of the dimeric model to account for VLF, LF, and HIR response behavior indicate its potential to provide a unified mechanism for phytochrome-mediated responses.

Analysis of the model supports involvement of the formation and action of PrPfr-X in HIR promoted by blue or far red light, as well in VLF responses. It suggests that cycling of PrPfr-X increases its action in the HIR, whereas its activity in VLF responses is increased by physical, chemical, or hormonal sensitizations that may increase its mobility. Increased cycling and mobility of PrPfr-X may serve similarly to accelerate interactions of PrPfr-X with the next component of the transduction chain.

REFERENCES

Beggs, C.J., Holmes, M.G., Jabben, M., and Schäfer, E. (1980). Plant Physiol. 66:615.

Fukshansky, L., and Schäfer, E. (1983). In Encyclopedia of Plant Physiology, New Series, Vol. 16A, "Photomorpho-genesis" (W. Shropshire Jr. and H. Mohr, ed.), p. 69. Springer-Verlag, Berlin, Heidelberg, New York.

Gaba, V., and Black, M. (1985). Plant Physiol. 75:1011.

Hillman, W.S. (1972). In "Phytochrome" (K. Mitrakos and W. Shropshire Jr., ed.), p. 573. Academic Press, New York.

Holmes, M.G., and Schäfer, E. (1981). Planta 153:267.

Wall, J.K., and Johnson, C.B. (1983). Planta 159:387.

Jones, A.M., and Quail, P.H. (1986). Biochemistry 25:2987

Lagarias, J.C., Kelly, J.M., Cyr, K.L., and Smith, W.O., Jr. (1987). Photochem. Photobiol. 45: In Press.

Mancinelli, A.L., and Rabino, I. (1978). Bot. Rev. 44:129.

Mohr, H. (1983). In Encyclopedia of Plant Physiology, New Series, Vol. 16A, "Photomorphogenesis" (W. Shropshire Jr. and H. Mohr, ed.), p. 336. Springer-Verlag, Berlin, Heidelberg, New York.

Pratt, L.H., and Marmé, D. (1976). Plant Physiol. 58:686.

Shinkle, J.R., and Briggs, W.R. (1985). Plant Physiol. 79:349.

Smith, H. (1983). Phil. Trans. R. Soc. Lond.B 303:443.

Taylorson, R.B., and Hendricks, S.B. (1979). Planta 145:507.

VanDerWoude, W.J. (1985). Photochem. Photobiol. 42:655.

VanDerWoude, W.J., and Toole, V.K. (1980). Plant Physiol. 58:686.

ROLES OF PHYTOCHROME IN PHOTOPERIODIC FLORAL INDUCTION

Daphne Vince-Prue

Department of Botany
University of Reading
Reading, UK

Atsushi Takimoto

Laboratory of Applied Botany
Faculty of Agriculture
Kyoto University
Kyoto, Japan

Photoperiodism is a special case of photoregulation in which the perception of light is coupled to a time-measuring mechanism in such a way that the direction of the response is determined by the duration of light received in every 24-h cycle. From the evidence of R/FR reversibility, involvement in photoperiodism was one of the earliest demonstrations of phytochrome action. Nevertheless a complete understanding of the roles of phytochrome in bringing about photoperiodic responses has yet to be achieved. It is the intention of the present chapter to focus attention on some of the unresolved problems.

PHOTOPERCEPTION IN SHORT-DAY PLANTS

A. Paradoxes and Problems

Under natural conditions, the durations of light and darkness are absolutely related and the critical daylength could, therefore, be detected by measuring the duration of either light or darkness. A feature which underlines the importance of the dark period in photoperiodic floral induction in short-

day plants (SDP) is that it can be rendered ineffective by a
short light interruption. The perception of light in this
'night-break' inhibition of flowering was the first function
of phytochrome to be established in photoperiodism when it was
shown that the formation of Pfr at a particular time in the
dark period prevented flowering in SDP, and this inhibitory
effect could be prevented by subsequent FR (Borthwick et al.
1952).

It has generally been assumed that, following transfer to
darkness, Pfr disappears before the time of sensitivity to a
night-break. An early model for photoperiodic time-measurement suggested that the critical nightlength (CNL) might be
measured by the time taken for Pfr to fall to a threshold
value below which flowering was no longer inhibited (Hendricks
1960), but subsequent physiological evidence has not supported
this hypothesis. For example, the CNL in many SDP has been
shown to be essentially independent of temperature (King
1979). Moreover, an important finding has been that the CNL
is shortened only slightly when Pfr is removed photochemically
at the end of the day by exposing leaves to FR (Takimoto and
Naito 1962, Salisbury 1981, King and Cumming 1972b). Thus, it
has often been considered that dark time-measurement in SDP
involves the perception of a light-off signal through a rapid
thermal loss of Pfr and that this is then coupled to a time-
measuring system which continues to run in darkness to measure
the CNL. Under natural conditions, the light-off signal
appears to be perceived when the irradiance falls below a
threshold value (Salisbury 1981, Lumsden and Vince-Prue 1984).
There is no good evidence for the view that changes in spectral quality during twilight (e.g., an increase in FR-R ratio,
Smith 1982) plays a significant role in generating a natural
dusk signal.

Assuming that this general concept of the coupling of
light-off to time-measurement through a reduction in Pfr is
correct, many physiological studies indicate that the coupling
occurs rapidly following transfer to darkness (Vince-Prue
1983a) and, therefore, that the light-off signal is perceived
via phytochrome that is highly unstable in the Pfr form. A
problem for this model is that photoperiodic dark time-measurement takes place in mature, fully-green plants in which,
from more recent spectrophotometric studies, phytochrome is
largely present in a Pfr-stable form. An additional complica-
tion is the observation that the removal of Pfr with an end-
of-day exposure to FR may strongly inhibit flowering (Vince-
Prue 1983b). Moreover, the inhibition of flowering by FR
often persists for many hours into the dark period (Fig. 1)
indicating that this action of phytochrome in photoperiodism
involves a Pfr-stable pool. R reversibility of the FR-inhibition is readily obtained through the night, except at the time

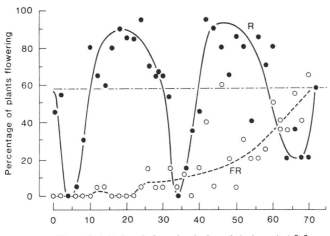

Fig. 1. Effect on flowering in <u>Chenopodium</u> <u>rubrum</u> of giving a brief exposure to either R or FR at different times during an inductive 72-h dark period (Cumming et al. 1965).

of sensitivity to a R night-break (Saji et al. 1983). Under conditions where the FR-inhibition persists for many hours in darkness, a paradox results in which at certain times exposure to either R or FR inhibits flowering. The inhibitory effect of a R night-break exhibits a circadian rhythm (Fig. 1) indicating that this action of phytochrome is coupled to the time-measuring system. Fig. 1 also shows that the inhibitory action of FR is greatest immediately following transfer to darkness and gradually declines as the night progresses; it appears, therefore, to be coupled in some way to the light period and it is not clear how closely this action of phytochrome is related to the time-measuring system. Both actions of light occur in the leaf (Knapp et al. 1986).

    The paradox that, at certain times, flowering can be inhibited by either R or FR has led to the proposal that these two actions of light involve two separate pools of phytochrome with different kinetic properties (Vince-Prue 1983b, Takimoto and Saji 1984). In one (FR-inhibition) Pfr is stable whereas in the other (night-break inhibition), Pfr is unstable. The two-pool hypothesis is achieving considerable acceptance but it is important to recognize that the evidence for it is almost entirely physiological. Although the existence of a Pfr-stable pool of phytochrome in light-grown plants is well-documented from spectrophotometric data, the continued presence of a Pfr-unstable pool (etiolated phytochrome) in mature plants is less certain. Spectrophotometric measurements using

bleached cotyledons of the SDP, Pharbitis, indicated a loss of
Pfr within a timescale consistent with physiological evidence,
i.e., within 30 min after transfer to darkness (Vince-Prue et
al. 1978), but few measurements have been made and interference by small amounts of chlorophyll is a problem in such
assays. The existence and function of Pfr reversion are also
completely uncertain. The use of sensitive immunological
probes offers good prospects for establishing if two pools of
Pfr co-exist in mature green leaves. This would be a first
step towards resolving the question of whether or not separate
pools of phytochrome are involved in the perception of photoperiodic signals. Alternative hypotheses have avoided the
concept of multiple phytochrome pools. For example, it has
been suggested that all photoperiodic signals may be perceived
via a single pool of Pfr but that these signals could involve
differentially-stable Pfr-receptor complexes (Thomas and
Vince-Prue 1987). In this model, the FR-inhibition response
depends on a highly-stable Pfr-receptor association, whereas
the night-break and light-off signals depend on one or more
Pfr-receptor complexes which rapidly dissociate after formation. Such Pfr-receptor complexes are, however, purely hypothetical at present.

B. Action of Light.

1. Rhythm Entrainment and Night-break Response. Although
several lines of evidence suggest that photoperiodic time-
measurement in higher plants is based on a circadian oscillator (Hammer and Takimoto 1964, King and Cumming 1972a, Vince-
Prue 1983b), the precise way in which this is coupled to
achieve measurement of the CNL in SDP is less well established. One approach has been to investigate the effect of
the photoperiod on the phasing (entrainment) of the photoperiodic rhythm. In dark-grown seedlings of Pharbitis, the
time of maximum sensitivity to a night-break (NBmax) was found
to occur at a fixed time (15 h) after the beginning of the
photoperiod when this was less than 6-h duration. However,
following a longer photoperiod, the time of NBmax was constant
(ca. 9 h) after the end of the photoperiod (Lumsden et al.
1982). Thus, in Pharbitis, the timing of sensitivity to a
night-break seems to be controlled by a rhythm which is
started by transfer to light (light-on signal) and continues
to run in continuous light until it reaches a specific phase
point after ca. 6 h. At this point, the rhythm appears to
become suspended for as long as the plant remains in continuous light, thereafter being released to run in darkness by the
light/dark transition. Because the rhythm is suspended in
light at a particular phase point, the time of NBmax will

always occur at a constant time after transfer to darkness: in Pharbitis, this is at about 9 h, as would be predicted from the original light-on rhythm. Very similar results have been obtained with light-grown plants of Xanthium, where the rhythm goes into suspension after about 5 h in continuous light and the NBmax then occurs at ca. 8.5 h following transfer to darkness (Papenfuss and Salisbury 1967). The Xanthium results were obtained with previously light-grown plants showing that the rhythm initiation by a light-on signal and its subsequent suspension in continuous light are relevant to normal conditions of growth as well as to dark-grown seedlings. In Pharbitis, seedlings which have received a previous exposure to light also show rhythm suspension when given a sufficiently long photoperiod (Lumsden et al. 1987).

Under natural conditions, the major determinant of flowering in SDP is the night-length. Therefore, if the night-break is to be used as an indicator of timing, it is necessary to establish that the CNL and NBmax are timed in the same way. Despite some ambiguity with intermediate durations of photoperiod, experiments with Pharbitis have established that, as with NBmax, the CNL is timed from light-on with shorter duration of light and from light-off with longer durations (Saji et al. 1984, Vince-Prue 1986). In Chenopodium, the CNL has also been shown to be a constant from light-off provided that the photoperiod is long enough (King and Cumming 1972a).

Thus, it is envisaged that the rhythm envolved in photoperiodic time-measurement is coupled to an underlying oscillator and that it is the action of light on the oscillator which initiates, suspends, or releases the observed photoperiodic rhythm of sensitivity to light (Fig. 2). In this scheme, the night-break acts at a different site - the circadian rhythm - to inhibit flowering at a specific phase-point.

2. Photoperception in Rhythm Entrainment. Relatively little information is available on the nature of the light-on signal which initiates the photoperiodic rhythm. For dark-grown Pharbitis, a saturating pulse of R was sufficient to act

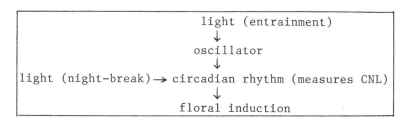

Fig. 2. Actions of light in photoperiodic floral induction in SDP.

as a light-on signal and start the flowering rhythm (King et
al. 1982). Partial reversal of a R light-on signal has been
reported for CNL timing in Pharbitis (Saji et al. unpublished
results) but, for night-break timing, the results were incon-
clusive since exposure to FR was sufficient to initiate a
rhythm in some cases (Lumsden 1984). Interpretation of the
results is also complicated by the fact that, in dark-grown
seedlings, there is a requirement for Pfr in order to achieve
flowering so that R/FR reversibility with respect to the
control of the rhythm is difficult to establish unequivocally.
However, it has been shown that only a very small amount of
light is sufficient to give a light-on signal for both CNL and
NBmax timing and that R is the most effective wavelength. It
seems highly likely, therefore, that the photoreceptor is
phytochrome.

Suspension of the flowering rhythm is achieved after
several hours in continuous light. However, the identity of
the photoreceptor involved has not been established unequivocally, although there is some circumstantial evidence for the
involvement of phytochrome. For example, R was found to be
considerably more effective than green or blue light in preventing a light-off signal (Takimoto 1967). Furthermore, R
was more effective than mixture of R+FR (Salisbury 1981),
indicating the involvement of Pfr rather than some cycling
component of the phytochrome system.

If Pfr is the active molecule, then sufficiently frequent
light pulses would be expected to substitute for continuous
light in suspending the rhythm. One approach to this question
has been to give a 24-h exposure to continuous white light in
order to suspend the rhythm and then to follow this with a 6-h
extension period during which the light conditions were varied. In Pharbitis, a fully effective extension (equivalent to
continuous light) would give NBmax at 8-9 h after the end of
the extension; a completely ineffective one (equivalent to
darkness) would give NBmax 8-9 h after the end of the 24-h
photoperiod in white light. A 1-min pulse of R given every
hour was found to be just as effective as 6-h continuous R in
delaying the time of NBmax. Less frequent pulses were less
effective but resulted in some delay compared with darkness.
The effectiveness of R pulses supports the hypothesis that
Pfr prevents the release of the rhythm, although the high
frequency necessary suggests that an unstable pool may be involved. There was no evidence of reversibility, however, indicating that the R pulses are not acting simply by maintain-
ing Pfr above some threshold value in the intervening dark
period. If Pfr is involved, coupling to the rhythm appears to
be extremely rapid. This conclusion is supported by data from
experiments with Chenopodium, where the phasing effect of a
6-h light exposure could be partially satisfied by R pulses

but these were not reversed by FR (King and Cumming 1972b).
However, it is worth noting that in none of the reversibility
experiments has it been established that the FR pulses were
sufficient to achieve photoequilibrium following a R exposure.
Since there is some evidence that continuous exposure to a
broad-band FR source (but not to FR pulses) can effectively
maintain suspension of the rhythm (Lumsden and Vince-Prue
1984), the lack of FR reversibility may be explained by failure to reduce the Pfr level sufficiently. The experiments
need to be repeated using 750 nm pulses at a fluence sufficient to saturate the photoconversion of Pfr.

There are several other unresolved problems concerning the
perception of continuous light. For example, although continuous broad-band FR was equivalent to continuous R when the FR
extension was terminated with a R pulse (Thomas and Lumsden
1984), there was no clear light-off signal in the absence of a
terminal R exposure (Takimoto 1967). Therefore, although FR
appears both to suspend the rhythm (Vince-Prue 1981) and to
maintain it in suspension (Thomas and Lumsden 1984), it is not
yet clear whether a high Pfr/P ratio is required for the generation of a light-off signal. This zeitgeber may be needed
to re-set the oscillator to a particular circadian time at the
onset of darkness rather than simply releasing it from apparent suspension. However, the interpretation of these results
is complicated by the fact that there is a Pfr-requirement for
flowering under the conditions used in all of these experiments. Consequently, it is necessary to repeat the treatments
under conditions where effects on the rhythm can be clearly
distinguished from the effect of FR to depress the flowering
response.

A further problem concerns the interpretation of the results of the R-pulse treatments. When a single pulse of R was
given at different times after transfer to darkness from continuous white light, the time of NBmax was delayed; the delay
was ca. 1 h when the light-pulse was given after 1, 2 or 3
hours in darkness, and ca. 2 - 3 h after 6 hours. From these
results, the effect of repeated pulses could be explained in
terms of the effect of each pulse to phase shift the released
rhythm, casting doubt on the interpretation that repeated
pulses maintain the rhythm in suspension (Lumsden et al.
1986). Repeated pulses may not, therefore, act in the same
way as continuous light and we cannot conclude on the basis of
the pulse extension treatments alone that Pfr is involved in
preventing release of the rhythm.

The demonstration that the photoperiodic rhythm in Pharbitis is subject to phase-shifting by a brief exposure to
light also raises questions about the action of the night-
break in the control of flowering in SDP. In addition to the
hypothesis that light inputs both to an oscillator and to a

light-sensitive phase of the flowering rhythm (Fig. 2), it has
been proposed that the only action of light is on the oscillator to control the phase of the rhythm. In this model, the
night-break response in SDP results from the fact that, at
certain times, the rhythm(s), is phase-shifted in such a way
as to inhibit floral induction. One approach to distinguish
between the two hypotheses has been to compare the sensitivity
of the night-break response with phase-shifting on the assumption that, if the night-break effect is simply a consequence
of re-phasing, the two responses will have the same dose-
relationship. In Pharbitis, the relative sensitivities of the
night-break and phase-shift have been compared at different
times after a 24-h photoperiod; the two responses were found
to have different dose-response relationships at both 6 and
8 h into the dark period, with the night-break showing an increase in sensitivity at 8 h (compared with that at 6 h) while
the sensitivity for a phase-shift decreased (Lumsden and Furuya 1987). Similar results were obtained with dark-grown seedlings following a 10-min photoperiod (Lee et al. 1987). Thus,
although the phase of the photoperiodic rhythm can be shifted
by a R pulse, a separate night-break effect can be distinguished and must result from another action of light, most
probably a direct interaction with the circadian rhythm as
proposed in Fig. 2.

The sensitivity of the phase-shift response in Pharbitis
to an extremely small input of R (Lumsden and Furuya 1987, Lee
et al. 1987) strongly indicates that phytochrome is the photoreceptor in this case and recent unpublished results of P. J.
Lumsden showing FR reversibility of the R-induced phase-shift
confirm this conclusion. This result demonstrates that the
oscillator is clearly accessible to the action of phytochrome
in the phase-shifting response and it is arguable, therefore,
that phytochrome is also the photoreceptor whch inputs to the
oscillator to suspend the rhythm in continuous light. Thus, a
tentative conclusion is that the photoperiodic response involves at least two actions of light - phase control and
interaction with the rhythm - and that both involve phytochrome.

3. The Pfr-requiring Reaction and the Effects of FR. It
is now generally agreed that a major function of the daily
photoperiod in SDP is to set the phase of the flowering rhythm
in such a way as to effect dark time measurement. However,
this is not the only function of the light period and a com-
plicating factor in many studies of photoreceptor action in
phase control is the fact that there appears to be an additional component of the flowering response for which Pfr is
required. Even after rhythm suspension has been established
in continuous light, exposure to FR before transfer to an

inductive dark period may completely inhibit the flowering response in Pharbitis. In other experiments, this effect was clearly separated from effects on timing (Lumsden and Vince-Prue 1984). The FR-inhibition of flowering is usually seen only after extremely short photoperiods and, under natural conditions with longer photoperiods and repeated cycles, the Pfr requirement appears to be satisfied in the light and does not extend into the dark period (Vince-Prue 1983b).

A different effect of FR has recently been observed in Pharbitis. When seedlings are raised in continuous white light, there is an oscillation in the response to an interruption with FR during this light period (King et al. 1986). The observed response is an increase or decrease of about 45 min in the CNL depending on when the FR exposure is given (Evans et al. 1986); night-break timing appears to be similarly affected (King 1974). The response to FR shows a 12-h rather than circadian oscillation and it is evident that this rhythm continues to run in continuous light for at least 3 cycles; it, therefore, has different properties from the circadian rhythm already discussed, which is suspended in continuous light and released on transfer to darkness. Several physiological approaches have indicated that the rhythmic response to FR is mediated through Pfr; however, in contrast to the Pfr-requiring (FR-inhibition) response already discussed, the evidence indicates that an unstable pool of Pfr is involved in the 12-h rhythm. An interruption with darkness affects timing in the same way as FR and, following transfer to darkness, the Pfr appears to fall within 40-45 min to a value sufficiently low to bring about the rhythmic effect on timing (Evans et al. 1986). The apparent kinetics of Pfr loss in this case are, therefore, very similar to those that have been suggested for an unstable Pfr pool coupled to the perception of the light-off signal for dark time-measurement.

A major question arising from the demonstration of this 12-h rhythm in the response to FR (or darkness) during the photoperiod is its relevance to the operation of the photoperiodic mechanism in natural conditions since, during a natural day, plants would be exposed mainly to random fluctuations in the FR/R ratio. However, the rhythm clearly exists, at least in Pharbitis and its input into the overall photoperiodic mechanism remains to be determined.

C. Time Measurement

A scheme that is compatible with many observations is shown in Fig. 3 (cf. B-1 of this Chapter). The key elements of this model are: 1) there is a single photoperiodic rhythm initiated by light-on; 2) this rhythm is 'suspended' in con-

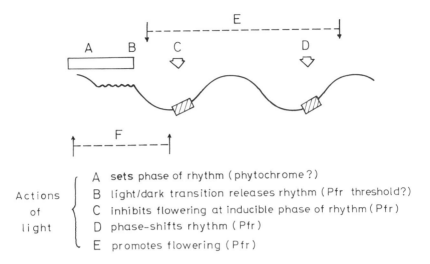

                    ⌠ A **sets** phase of rhythm (phytochrome?)
    Actions   │ B light/dark transition releases rhythm (Pfr threshold?)
       of   ⟨ C inhibits flowering at inducible phase of rhythm (Pfr)
     light   │ D phase-shifts rhythm (Pfr)
                    ⌡ E promotes flowering (Pfr)

Fig. 3. Model for the photoperiodic control of floral
induction in Pharbitis nil.

tinuous light being released at the light/dark transition to
measure the CNL; and 3) a light-on signal alone is sufficient
to induce flowering under some conditions, provided that a
second pulse of light is not given at the inducible phase
(King et al. 1982, Lumsden et al. 1982, Oota 1983). Because
of this, a skeleton photoperiod can be either inductive or
non-inductive depending on the phase of the rhythm when the
second pulse is given (Hillman 1964, Oota 1983). This has
sometimes been interpreted as the measurement of a critical
daylength (cf Oota 1983). However, in several plants (e.g.,
Chenopodium, Xanthium and Pharbitis) it has been shown that
once the duration of the daily light period exceeds a certain
value (from 5 to 12 h depending on the plant) a certain fixed
phase is achieved and timing is coupled to a light-off signal.
Thus, under normal conditions of growth with photoperiods of
several hours duration, it is evident that the characteristics
of the system lead to the measurement of a CNL, even though
the rhythm is initiated by a light-on signal.
    While this model is compatible with results obtained with
a number of intensively studied SDP, there are still some
unresolved problems. Firstly, is the model applicable to all
SDP? Certainly, light-off timing could not operate under
natural daylengths in Lemna if, as has been suggested (Oota
1985), exposure to continuous light for more than 24 h is
required to suspend the clock. (However, in earlier experiments with the same strain, Hillman (1969) concluded that only
4-6 h of light was necessary to give a new light-off signal –

a value comparable with that in other SDP and which would give
light-off timing in natural conditions). There is evidence
that in some cases, e.g. Glycine, photoperiodic timing involves the interaction of two rhythms, one phased from light-
on and one from light-off (Hamner 1969). Even in Pharbitis,
it has been proposed that dark timing involves the interaction
of two rhythms which are phased from light-on and light-off
respectively (Takimoto 1969) and, as already discussed, there
is good evidence for a rhythm, albeit not circadian, which
continues to oscillate in continuous light and which appears
to influence dark time-measurement (Evans et al. 1986).

A second problem concerns the promoting action of light on
flowering. When a long dark period is scanned with a light
interruption, a circadian rhythm of promotion of flowering is
seen as well as a circadian rhythm of inhibition (Fig 1). In
the model based on a circadian rhythm in an inducible phase,
only inhibition would be predicted (Fig. 3, reaction 3). The
promotion of flowering by R has, therefore, been attributed to
the non-rhythmic Pfr-requiring reaction. When this reaction
has not been completed, exposure to R would be expected to
promote flowering except at the inducible light-sensitive
phase of the rhythm; because of the circadian rhythm in the
inducible phase, this would lead to an apparent rhythm in
promotion by R. The broad windows of R promotion over a 60-h
period during which FR inhibits flowering in Chenopodium (Fig.
1) are not inconsistent with this interpretation and it is
clear that the timing of the inhibitory night-break effect
(inducible-phase rhythm) is much more precisely defined. Unfortunately, this particular experiment did not follow the
rhythm of sensitivity to R beyond the time that sensitivity to
FR was apparently lost after about 70 h in darkness. An
alternative explanation of the R promotion is that the photoperiodic rhythm has two light-sensitive phases. i.e., light
promotes flowering at one phase and inhibits at the other.
This model for the flowering rhythm comes close to Bunning's
original concept of photophile and skotophile phases which
alternate at 12-h intervals from a light-on signal (Bünning
1967). Experiments designed specifically to resolve this
question have yet to be undertaken.

II. PHOTOPERCEPTION IN LONG-DAY PLANTS

A. Special Problems in LDP

Long-day plants (LDP) flower when the daylength exceeds a
certain value in a 24-h light/dark cycle. Is the control
mechanism merely the mirror-image of that in SDP, the only
difference being that, at the time of night-break sensitivity,

light is necessary for induction in LDP, instead of preventing
it? This would imply the same underlying timing mechanism, a
circadian rhythm in an inducible phase, coupled to the photoperiod in the same way as in SDP and would lead to floral induction only if light (dawn) occurs before the critical nightlength is attained. However, the duration of darkness does
not seem to be the overriding factor in many LDP. For example, they often require a long daily exposure to light for
floral induction and are relatively insensitive to a brief
night-break (Vince-Prue 1983b). Light quality during the photoperiod is also important and optimal flowering is achieved
only with FR-enriched light even when the nightlength is
considerably less than the critical (Vince-Prue 1981). A
relatively high irradiance during the light period is also
often required (Friend 1969, Bodson et al. 1977). The importance of the duration, quality and irradiance of the photoperiod in controlling flowering in many LDP raises questions
about the identity and operation of the photoreceptor, as well
as the timing mechanism.

B. Time-measurement

In view of the need for long daily exposures to light, the
question arises as to whether the duration of light is measured instead of, or in addition to that of darkness. The
basis of the time-measuring system is also debatable since,
in many LDP, maximum flowering occurs in continuous light
which, in SDP, would lead to suspension of the rhythm involved
in dark timing. Circadian rhythms have been described in LDP
(Vince-Prue 1975), but the phase relationships of these
rhythms and the way that they might operate to establish
photoperiodic time-measurement are poorly understood.

A circadian periodicity in the promotion of flowering by
added FR has been demonstrated in two LDP, Lolium and Hordeum.
In both cases, the addition of FR for 4-6 h promoted flowering
compared with the background of continuous white or R light
(Vince-Prue 1975, 1983b, Deitzer et al. 1979). This rhythm
clearly persists in continuous light and contrasts with the
situation in SDP, where an essential feature of the circadian
timing mechanism is the apparent suspension of the rhythm at a
constant phase-point after a few hours in the light. In
Hordeum, the FR-response rhythm can also be phase-shifted by
FR suggesting two separate effects (Deitzer et al. 1982). FR
interacts directly with a circadian rhythm to promote flowering at a certain phase-point and also acts to phase-shift that
rhythm. This parallels the two effects of R on the circadian
timer in SDP (cf Fig. 2).

The response to FR is not the only rhythmic feature in

LDP. Although relatively insensitive to a brief night-break, many are resposive to longer exposure of a few hours. An apparent circadian oscillation in the flowering response to such a long night-break has been observed in LDP, showing that a rhythm of responsivity to light continues to run in darkness. For example, in Hyoscyamus, using a schedule of 6-h light with 66-h dark scanned by a 2-h night-break, there were 3 peaks of promotion at 24-h intervals (Clauss and Rau 1956). It might be assumed that this rhythm is entrained by light in the same way as in SDP, i.e., released at the light/dark transition to run from a phase-point established by suspension of the rhythm in continuous light. However, experiments with Lolium designed to answer this question have so far yielded ambiguous results and no clear evidence for or against such a light-off rhythm was obtained (Vince-Prue, unpublished).

A working hypothesis is that the control of flowering in LDP involves two rhythms. There is 1) a rhythm of responsivity to FR which runs in the light and may be related to the long-day requirement and 2) a rhythm of responsivity to R which is suspended in continuous light and relates, as in SDP, to the measurement of the critical nightlength. However, much further work is needed before possible direct effects of light on flowering in LDP can be separated from effects on the underlying timing mechanism.

C. Photoperception -- Is Phytochrome the Photoreceptor?

There are two definitive criteria for phytochrome as photoreceptor for particular responses but neither is completely fulfilled for the control of flowering in LDP. The first of these criteria is R/FR reversibility. Although such reversibility was demonstrated in the night-break promotion of flowering in early experiments with LDP (Downs 1956), the conditions involved the use of a relatively short dark period close to the CNL. With longer dark periods, longer exposure to light are required in order to promote flowering and FR reversibility has not been demonstrated. The second criterion for the involvement of phytochrome is an action spectrum corresponding to Pr absorption. However, when long night-breaks or day-extension treatments are used, the wavelength response curve for the promotion of flowering in LDP often shows an action maximum at around 710-720 nm (Vince-Prue 1983b). Such an action maximum is not related in any simple way to Pfr but it is similar to the action spectrum for the 'high irradiance' reaction (HIR) seen in seedling photomorphogenesis, where phytochrome is thought to be the photoreceptor. The HIR also shows a strong irradiance dependence and a shift in the action maximum from R at short exposure towards FR with longer expo-

sures. Both have been observed in LDP. For example, Brassica
campestris shows a marked irradiance dependence for flowering
(Friend 1969) and, in Hyoscyamus, there is a shift from a
maximum at 660 nm with brief night-breaks (Parker et al. 1950)
to 720 nm with 8-h night-breaks (Schneider et al. 1967).
Thus, there are a number of similarities between long-day
responses and the HIR of seedling morphogenesis.

In the HIR, the far-red peak is thought to be a function
of the complex dynamics of the phytochrome system but all
models to explain the observed characteristics of the response
depend in some way on Pfr being unstable. This assumption may
not be justified for light-grown plants and, indeed, the evidence shows that the HIR is lost following transfer to light.
In contrast, the long-day control of flowering occurs in
mature green tissues. It is not clear, therefore, what kind
or pool of phytochrome might be involved in this case and, as
with SDP, a study of the dynamics of phytochrome in mature
green leaf tissue is a necessary pre-requisite for an understanding of its role in photoperiodic induction in LDP.

What evidence is there that phytochrome is the only receptor in LDP? Because of the need for long daily exposure to
light, experiments using intermittent lighting cycles have
frequently been used to investigate the photoperception mechanism. Quite different schedules have been found to be effective in different species. In Callistephus chinensis, cycles
of one minute of light per hour had almost the same effect as
continuous light (Cockshull and Hughes 1969), which could be
interpreted in terms of a requirement to maintain the level of
a relatively stable type of Pfr above a certain critical
level. However, in Hyoscyamus, even cycles as short as 6 s
per min were not as effective as continuous light (Schneider
et al. 1967), suggesting that, if Pfr is involved, the Pfr
pool must be extremely unstable or, alternatively, that some
aspect of phytochrome cycling is required. Some sort of
photon counting mechanism is suggested from experiments with
carnation where intermittent light was as effective as continuous light only when the total fluence was the same (Harris
1972); this might be accomplished by phytochrome cycling. An
examination of the effectiveness of different pulse frequen-
cies at different temperatures might help to determine the
importance of Pfr decay during the intervening dark intervals.

The use of intermittent lighting schedules allows the
possibility of examining R/FR reversibility even under long
exposures. However, experiments involving repeated cycles of
R and/or FR over a period of several hours showed no evidence
for reversibility in Lolium, and cycles containing any sequence of R and FR, or a mixture, were more effective than
either R or FR alone (Evans et al. 1965). A problem with
these experiments is that the low irradiances used may not

have been sufficient to saturate the photoconversion, so that an intermediate level of Pfr may have been maintained in all treatments. More recent experiments, using higher irradiance (Wall and Vince-Prue unpublished results), have confirmed that any sequence of R and FR, or mixture, is more effective than R or FR alone. However, there was a limited reversibility, and sequences ending in R were more effective in promoting flowering than sequences ending in FR pointing to the involvement of Pfr. Nevertheless the results still could not be explained entirely in terms of Pfr.

An unresolved question concerns the nature of the FR promotion of flowering in LDP. One possible explanation is that the FR simply leads to a reduction in the Pfr level which is inhibitory to flowering at certain times in the photoperiodic cycle and there is some evidence for this from reversibility expertiments (Holland and Vince 1971). However, there is also evidence that continuous FR is more effective than a FR pulse followed by darkness (Vince 1965), suggesting that some factor other than a lowered Pfr level may be involved. One problem in the interpretation of the promoting effect of FR on flowering is that the sensitivity to light quality changes with time. During a 16-h day-extension, for example, FR gives maximum promotion of flowering in the early half of the extension, while R most strongly promotes flowering in the latter half (Vince-Prue 1981, 1983b). This complicates the analysis of light treatments given over the entire extension period and a better understanding may be obtained from action spectra determined for defined portions of the photoperiod.

Although, from experiments using short night-breaks or intermittent lighting schedules, there is good evidence for the action of Pfr to promote induction in LDP, it is by no means certain that only Pfr is involved. The effectiveness of a mixture of R and FR and the action maximum at 710-720 nm suggest the possibility of the involvement of phytochrome cycling. They could, however, also be explained on the assumption that an unstable form of Pfr is active in these mature green plants, even though they may have been exposed to natural daylight for many weeks before being given inductive LD cycles. Finally, the participation of other photoreceptors cannot yet be entirely ruled out.

ACKNOWLEDGMENTS

The authors wish to thank the members of the Round Table group at the Okazaki meeting for their stimulating discussion, which contributed significantly to many of the ideas expressed in this chapter.

REFERENCES

Bodson, M, King, R.W., Evans, L.T. and Bernier, G. (1977).
    Aust. J. Pl. Physiol. 4:467.
Borthwick, H.A., Hendricks, S.B. and Parker, M.B. (1952).
    Proc. Nat. Acad. Sci. USA 38:929.
Bünning, E. (1967). "The Physiological Clock." Academic Press,
    New York.
Clauss, H. and Rau, W. (1956). Z. Bot. 44:437.
Cockshull, K.E. and Hughes, A.P. (1969). Ann. Bot. 33:367.
Cumming, B.G., Hendricks, S.B. and Borthwick, H.A. (1965).
    Can. J. Bot. (1965). 43:825.
Deitzer, G.F., Hayse, R. and Jabben, M. (1979). Pl. Physiol.
    64:1015.
Deitzer, G. F., Hayse, R. G. and Jabben, M. (1982). Pl.
    Physiol. 69:597.
Downs, R. J. (1956). Pl. Physiol. 31:279.
Evans, L.T., Borthwick, H.A. and Hendricks, S.B. (1965). Aust.
    J. biol. Sci. 18:745.
Evans, L.T., Heide, O.M. and King, R.W. (1986). Pl. Physiol.
    80:1025.
Friend, D. J. C. (1969). In "The Induction of Flowering"
    (L.T.Evans, ed.), p.364. Macmillan, Melbourne.
Hamner, K. C. (1969). In "The Induction of Flowering"
    (L.T.Evans ed.), p.62. Macmillan, Melbourne.
Hamner, K.C. and Takimoto, A. (1964). Am. Nat. 98:295.
Harris, G.P. (1972). Ann. Bot. 36:345.
Hendricks, S. B. (1960). Cold Spr. Harb. Symp. quant. Biol.
    25:245.
Hillman, W.S. (1964). Am. Nat. 98:323.
Hillman, W.S. (1969). In "The Induction of Flowering" (L.T.
    Evans ed.), p. 186. Macmillan, Melbourne.
Holland, R.W.K. and Vince, D. (1971). Planta 98:232.
King, R.W. (1974). Aust. J. Pl. Physiol. 1:445.
King, R.W. (1979). Aust. J. Pl. Physiol. 6:417.
King, R.W. and Cumming, B. (1972a). Planta 103:281.
King, R.W. and Cumming, B. (1972b). Planta 108:39.
King, R.W., Schäfer, E., Thomas, B. and Vince-Prue, D. (1982).
    Pl. Cell. Env. 5:395.
King, R. W., Evans, L. T. and Heide, O. M. (1986). Pl. Cell
    Env. 9:345.
Knapp, P. H., Sawhney, S., Grimmet, M. and Vince-Prue, D.
    (1986). Pl. Cell Physiol. 27:1147.
Lee, H.S.S., Vince-Prue, D. and Kendrick, R.E. (1987). Pl.
    Cell Physiol. (in press).
Lumsden, P. (1984). "Photoperiodic Control of Floral Induction
    in Pharbitis nil." D. Phil. Thesis, Univ. Sussex.
Lumsden, P. J. and Furuya, M. (1987). Pl. Cell Physiol. (in

press).

Lumsden, P.J. and Vince-Prue, D. (1984). Physiol. Pl. 60:427.

Lumsden, P. J., Thomas, B. and Vince-Prue, D. (1982). Pl. Physiol. 70:277.

Lumsden, P.J., Vince-Prue, D. and Furuya, M. (1986). Physiol. Pl. 67:604.

Oota, Y. (1983). Pl. Cell Physiol. 24:1503.

Oota, Y. (1985). Pl. Cell Physiol. 25:923.

Parker, M.W., Hendricks, S.B. and Borthwick, H.A. (1950). Bot. Gaz. 111:242.

Papenfuss, H.D. and Salisbury, F.B. (1967). Pl. Physiol. 42:1562.

Salisbury, F.B. (1981). Pl. Physiol. 67:1230.

Saji, H., Vince-Prue, D. and Furuya, M. (1983). Pl. Cell Physiol. 67:1183.

Saji, H., Furuya, M. and Takimoto, A.(1984). Pl. Cell Physiol. 25:715.

Schneider, M.J., Borthwick, H.A. and Hendricks, S.B. (1967). Am. J. Bot. 54:1241.

Smith, H. (1982). Ann. Rev. Pl. Physiol. 33:481.

Takimoto, A. (1967). Bot. Mag. Tokyo 80:41.

Takimoto, A. (1969). In "The Induction of Flowering" (L.T. Evans ed.), p.90. Macmillan, Melbourne.

Takimoto, A. and Naito, Y. (1962). Bot. Mag. Tokyo 75:205.

Takimoto, A. and Saji, H. (1984). Physiol. Pl. 61:675.

Thomas, B. and Lumsden, P.J. (1984). In "Light and the Flowering Process" (D. Vince-Prue, B. Thomas and K.E. Cockshull, eds.), p.107. Academic Press, London.

Thomas, B. and Vince-Prue, D. (1987). In "Models in Plant Physiology/Biochemistry/Technology (D. Newman and K. Wilson, eds.), CRC Press, Boca Raton (in press).

Vince, D. (1965). Physiol. Pl. 18:474.

Vince-Prue, D. (1975). "Photoperiodism in Plants." McGraw Hill, London.

Vince-Prue, D. (1981). In "Plant and the Daylight Spectrum" (H. Smith, ed.), p.223. Academic Press, London.

Vince-Prue, D. (1983a). Phil. Trans. R. Soc. Lond. B303:523.

Vince-Prue, D. (1983b). Encyc. Pl. Physiol. NS. 16:457.

Vince-Prue, D. (1986). In "Photomorphogenesis in Plants" (R. E. Kendrick and G.H.M. Kronenberg, eds.), p. 269. Martinus Nijhoff, Dordrecht.

Vince-Prue, D. and Lumsden, P.J. (1987). In "The Manipulation of Flowering" (J. Atherton, ed.), Univ. of Nottingham 45th Easter School, April 1986 (in press).

Vince-Prue, D., King, R.W. and Quail, P.H. (1978). Planta 141: 9.

V. PROBLEMS AND PROSPECTS OF PHOTOMORPHOGENETIC STUDIES

PRIMARY ACTION OF PHYTOCHROME

Eberhard Schäfer
Biological Institute II
University of Freiburg
Freiburg i. Br., Germany

I. INTRODUCTION

The step of signal transduction between photoconversion of phytochrome and its effects on regulation of gene expression is still unknown. In the cases of Pfr mediated orientation responses, i.e. chloroplast movements in Mougeotia and polarotropism of fern chloronemata (1, 2) an action dichroism could be demonstrated. This led to the conclusion that phytochrome is located at or in the plasma membrane or is associated with the cytoskeleton. For chloroplast movements changes of Ca^{++} have been described and an involvement of calmoduline in the signal transduction chain is very probable (1). On the other hand, neither in Mougeotia nor in fern chloronemata Pfr mediated regulation of gene expression could be demonstrated whereas in the fern system effects of the blue light receptor on gene expression were clearly observed (3).

In higher plants Pfr effects on Ca^{++} have often been discussed (4). Action dichroism was reported by Marmé and Schäfer (1972) but these results could not be repeated and major influences of optical artefacts cannot be excluded (5, 6). Therefore, evidences are missing indicating that Pfr mediated changes of plasma membrane ion transport systems are integral steps of the transduction chain leading to Pfr mediated changes in gene expression.

To approach this problem two studies seem to be appropriate: 1) analysis of fast Pfr mediated reactions including localisation of phytochrome; 2) analysis of phytochrome controlled gene expression and the structure of the genes controlled by phytochrome.

II. FAST REACTIONS

The most rapid phytochrome mediated responses described

up to now are Pfr induced pelletability of phytochrome
(7-9), sequestering of phytochrome (10), and the Pfr dependent reaction leading to destruction of Pr (11).

A. Phytochrome Mediated Phosphorylation
Reactions

During the conference phytochrome mediated phosphorylation
responses were discussed: J.C. Lagarias presented data indicating that in the presence of polycations isolated phytochrome may have a kinase activity leading to autophosphory-
lation of phytochrome (12). Datta et al. (1986) observed
phytochrome mediated phosphorylation of nuclear proteins
after irradiation of isolated nuclei. The documentation of
this reaction is not so informative and clear as one should
desire it (13). Adding to this there cannot be excluded that
this reaction does not occur in vivo. There exists no solid
evidence that phytochrome is associated in vivo with nuclei.
Otto and Schäfer presented data indicating that after in vivo
irradiation a phytochrome mediated phosphorylation or de-
phosphorylation of three proteins can be observed in etiolated coleoptile tips from oat seedlings (6). These reactions
are extremely fast even at $0°C$ - the halflife was 2-59 - and
one may, therefore, speculate that they represent one of the
very first reactions, if not the first, after Pfr formation.

B. Localisation of
Phytochrome

Due to technical limitations detailed studies of Pfr
mediated sequestering of phytochrome have - until very recently - only been possible at the light microscopical level.
Using the embedding material LR white Speth et al. (1986)
and Pratt and coworkers (pers. comm.) have been able to perform studies at the electron microscopical level (14). In a
very rapid reaction after Pfr formation sequestering in small
electron microscopical dense particles can be observed. These
particles seem to aggregate to large size structures irrespective of whether the phytochrome stays in the Pfr or Pr
form. Only after this additional aggregation (15) Ubiquitin
can be localised to a larger extent in these areas (15).
   Comparison of kinetical properties and Pfr dependence
indicated that Pfr mediated pelletability, sequestering and
Pr destruction are representing the same chain of events
which means that the sequestered areas are the places of
phytochrome destruction. This view is supported by observations from R.D. Vierstra at the meeting: after Pfr formation

a ubiquitation of phytochrome can be observed.

C. Multiple Phytochrome Pools

Immunochemically different phytochrome species can be observed (16-19) whereby one of them is likely to be instable
and the other stable (20). Therefore, the question arises if
the phytochrome undergoing destruction may not be physiologically functional. Studies of phytochrome mediated threshold
responses (21) and anthocyanin accumulation led to the conclusion that the phytochrome regulating these responses is
inactivated with a halflife similar to the destruction rea-
tion. As most phytochrome mediated responses are reversible
also at least 5 min after onset of a red light pulse - a
time at which the first sequestering reaction is already
completed (10) - it seems probable that Pfr in the sequestered form is still functional.

D. Multiple Phytochrome Reactions?

In order to analyse which phytochrome species are in
fact functional the so-called null point method might be of
use (22). Several years ago this analysis has already led to
the hypothesis of different phytochrome pools (22). Unfortunately - due to the complex phytochrome kinetics and the
possibility of different actions of aged and newly formed
Pfr - such analysis cannot lead to unequivocal predictions.
Even the observations of very low and low fluence responses and their different behavior towards various stimuli
(23) cannot be taken as a clear evidence that very low and
low fluence responses reflect two different phytochrome actions. Especially the observation that phytochrome is a
dimer in vitro (24) and probably also in vivo (25) leads to
such complicated dynamical properties of various forms of
hetero dimers RF and fully transformed dimers FF that most
predictions may be doubted.
Hence we must conclude that the classical physiological
approach can only lead to limited conclusions. Still, one
should have in mind that - in spite of the fore-mentioned
physiological imponderabilities - the physiological data will
remain the basis and a test for the molecular approach, also
in the future. Especially, analysis of temporal and spatial
patterns, chloroplast nuclear interactions, and time courses
as well as fluence response relations are of importance (see
also Mohr in this volume). The analysis of the kinetics of
the loss of reversibility has also been a very useful tool in
the past. A detailed analysis of such an approach could even

lead to the conclusion that the rate of signal transduction
can be saturated although the final response is graded.

III. PHYTOCHROME CONTROLLED GENE EXPRESSION

A. Kinetic and Fluence Response Studies

Also – even though much more cumbersome – kinetic and
fluence response studies can be performed at the mRNA or
even transcription rate level (26). Analysing accumulation
of mRNA levels in parsley cell cultures after irradiation
with various fluence rates of UV light it was observed that
not the initial rate of the response but the duration is dependent on the strength of the stimulus (27). Similar observations were obtained when the rate of run off transcription
in nuclei isolated from barley seedlings were analysed after
irradiations with red or long wavelength far-red light (28).
The saturation of the initial rates is found if a certain
stimulus strength is applied, but this may be as low as
about 0.01% Pfr in the case of Pfr controlled transcription
of LHCP sequences (28).

B. Level at which Phytochrome Control Occurs

One may address the question whether light control of
gene expression is always via changes in transcription rates.
For those responses with a rapid down regulation of mRNA levels a down regulation of transcription was observed as well
(28); however, this cannot quantitatively account for the
observed changes at the mRNA level (18, 28). Also, for positively controlled gene expression an additional control be-
side transcription can be observed (Silverthorne and Tobin,
pers. comm.).
Analysis of light regulation of the rate of transcription of a specific gene should entail the test whether or
not this is due to a gene cascade mechanism. Such a test may
be performed by means of inhibitors of protein synthesis,
and one should be aware that artefacts, due to the inhibitors, cannot be excluded. Analysing the Pfr dependent induction of accumulation of LHCP mRNA Merkle and Schäfer (1987)
observed that the initial rate of accumulation is not changed
if the protein synthesis is blocked (29). This indicates that
the phytochrome action in this case uses a preexisting transduction pathway.

C. Interaction of Various Photoreceptors

It was described several times that gene expression in plants is very often controlled by several photoreceptors, i.e. UVB photoreceptor, blue light photoreceptor and phytochrome. The question should be addressed whether the photoreceptors act independently or hierarchically and whether they use common transduction chains or if they rather act independently at the chromatin level.

In recent studies with the parsley cell culture system it was possible to demonstrate at the mRNA and transcription rate level that the action of the UVB photoreceptor is essential for light regulation of chalcone synthase transcripts (27, 30). Although the blue light receptor and phytochrome do not change gene expression without the action of the UVB photoreceptor, modulations of the mRNA accumulations could be observed if blue or red light is given before, or far-red light is given after, the inductive UV light pulse (27). Up to now the parsley cell culture seems to be the only cell culture for which phytochrome action could be demonstrated. Unfortunately, a newly prepared parsley cell culture shows no longer phytochrome control of gene expression (see Ohl, unpublished).

It is a pity that - even for this well-studied system - there do, so far, not exist any data that might indicate whether or not the different signal inputs use (at least partially) the same transduction chain.

IV. IN VITRO EXPERIMENTS

Finally, yet another question remains open, namely, whether the in vitro approach would be helpful to elucidate the phytochrome action. However, a warning at this stage is necessary. Phytochrome - especially in the Pfr form - seems to be a very sticky molecule which can be copurified with almost all organelles. In vivo localisation work has never shown any localisation of phytochrome other than cytosolic; i.e., no membrane-bound phytochrome nor organelle-associated phytochrome was observed at the electron microscopical level. Unfortunately, this method is not sensitive enough to exclude other localisation of minor fractions.

There should be pointed out that localisation analysis of other than etiolated phytochrome is not possible at the moment. Pelletability assays have always led to the observation that one fraction - about 5% - is sensitive to triton X-100 treatments indicating a membrane interaction. Immunoprecipitation analysis of this phytochrome and peptide

mapping after solubilisation have not shown any differences
to the soluble phytochrome (6). Therefore, an in vitro artefact cannot be excluded. Similarly, the observation that
irradiations of isolated nuclei (13) or addition of Pfr to
isolated nuclei (28, 31) have led to potentially very interesting results and may be interpreted   by an in vitro effect which cannot occur in vivo because of different localisation of the phytochrome. Furthermore, due to possible protein kinase activity of phytochrome or copurification of a
protein kinase all experiments involving addition of ATP to
phytochrome in order to modify a biochemical response should
be interpreted with caution (12).

The results observed after addition of phytochrome to
isolated nuclei primarily show that the irradiation in vivo
has led to a different responsiveness of the in vitro system
to Pfr, at the same time indicating a gene specific modification of the system.

V.   FUTURE AIMS AND GOALS

From the preceding discussion one can conclude that the
methods for localisation analysis of phytochrome should be
improved. Above all, we need good ideas about the localisation and possible function of the "green phytochromes".

With respect to the etiolated type of phytochrome further
analysis of very fast reactions and especially the possible
role of kinase activity will be performed. This will lead
us to a complex branched transduction pathway. Whether one
or several of the pathways will directly lead to regulation
of gene expression, or whether membrane regulations are included, is still an open problem. At the moment the data
favor a directer pathway leading to gene expression with
possible membrane effects as a modulating role.

Analysis of the transduction chain and interaction of
several factors controlling gene expression will be a major
goal in the near future. Can this question be solved by continuing the classical physiological approaches? We think the
answer is no! Physiology has given and will give the backbone of precise data which should be explained in molecular
terms. In the meantime the set of data may be improved by
some molecular physiology studies.

The following points need being explained - The expression of genes is often controlled by the following factors:

(1) Temporal pattern of development (i.e. time in development
    at which expression becomes measurable);

(2) spatial pattern of development (i.e. organ specificity and often even cell specificity;

(3) control by "chloroplast development" factor;

(4) phytochrome signal (very low fluence, low fluence response and HIR);

(5) blue light receptor signal

(6) UVB light receptor signal.

Are all these points simple transcription controlling factors interacting at the chromatin level, or are there any complex cytosolic interaction patterns to be expected?

These problems can be solved, partially, by analysing patterns of DNase hypersensitive sites and - in an extremely cumbersome work - using transgenic plants. Especially, the variability in gene expression of transgenic plants will have to be taken into account. Irrespective of this, the question is to be tested if, in all cases, a transcription controlling factor plays a role.

As far as an analysis of the transduction chain is concerned, the problem seems to be even more complicated at the moment. From the photoreceptor side analysis of fast reactions may lead - with some luck - to transduction factors. From the gene side DNA-protein binding studies using gel retardation and foot printing assays will lead to the transcription factors. The gap between those loose ends is hard to bridge. Classically mutant analysis was very helpful in lower organisms to elucidate the transduction pathway. As far as we can see, it will probably be the only way-out in the search to solve this problem in higher plants: by either looking for phytochrome mutants or for transduction mutants, using light regulated promotor regions linked to a lethal gene.

ACKNOWLEDGMENTS

Research supported by Deutsche Forschungsgemeinschaft (SFB 206).

REFERENCES

1. Haupt, W., in "Phytochrome and photoregulation in plants" (M. Furuya, ed.), chapter 18. Academic Press, Japan, 1987.
2. Wada, M., and Kadota, A., in "Phytochrome and photo-regulation in plants" (M. Furuya, ed.), chapter 19. Academic Press, Japan, 1987.
3. Furuya, M., in "Encyclopedia of plant physiology" N.S. Vol. 16B (W. Shropshire and H. Mohr, eds.) p. 569. Springer Verlag, Berlin, Heidelberg, New York, Tokyo, 1983.
4. Kendrick, R.E., and Bossen, M.E., in "Phytochrome and photoregulation in plants" (M. Furuya, ed.), chapter 17. Academic Press, Japan, 1987.
5. Marmé, D., and Schäfer, E., Z. Pflanzenphysiol. 67: 192 (1972).
6. Otto, V., dissertation University of Freiburg, F.R.G. (1986).
7. Quail, P.H., and Briggs, W.R., Plant Physiol. 62: 773 (1978).
8. Pratt, L.H., and Marmé, D., Plant Physiol. 58: 686 (1972).
9. Lehmann, U., and Schäfer, E., Photochem. Photobiol. 27: 767 (1978).
10. McCurdy, D.W., and Pratt, L.H., Planta 167: 330 (1986).
11. Schäfer, E., in "Plants and the daylight spectrum" (H. Smith, ed.) p. 461. Academic Press, London, 1981.
12. Wong, Y.-S., Cheng, H.-C., Walsh, D.A., and Lagarias, J.C., J. Biol. Chem. 261: 12089 (1986).
13. Datta, N., Chen, Y.R., and Roux, S.J., Biochem. Bio-phys. Res. Comm. 128: 1403 (1985).
14. Speth, V., Otto, V., and Schäfer, E., Planta 168: 299 (1986).
15. Speth, V., Otto, V., and Schäfer, E., submitted to Planta.
16. Shimasaki, A., and Pratt, L.H., Planta 164: 333 (1985).
17. Tokuhisa, J. G., Daniels, S.M., and Quail, P.H., Planta 164: 321 (1985).
18. Quail, P.H., in "Phytochrome and photoregulation in plants" (M. Furuya, ed.), chapter 2. Academic Press, Japan, 1987.
19. Pratt, L.H., in "Phytochrome and photoregulation in plants" (M. Furuya, ed.), chapter 7. Academic Press, Japan, 1987.
20. Brockmann, J., and Schäfer, E., Photochem. Photobiol. 35: 555 (1982).

21. Mohr, H., in "Phytochrome and photoregulation in plants" (M. Furuya, ed.), X. Academic Press, Japan 1987.

22. Hillman, W.S., Ann. Rev. Plant Physiol. 18: 301 (1967).

23. Vanderwonde, W.J. in "Phytochrome and photoregulation in plants" (M. Furuya, ed.), chapter 20. Academic Press, Japan 1987.

24. Jones, A.M., and Quail, P.H., Biochemistry 25: 2987 (1986).

25. Brockmann, J., Rieble, S., Kazarinova-Fukshansky, N., Seyfried, M., and Schäfer, E., Plant Cell Environment, in press (1987).

26. Briggs, W.R., in "Phytochrome and photoregulation in plants" (M. Furuya, ed.), chapter 1. Academic Press, Japan 1987.

27. Bruns, B., Hahlbrock, K., and Schäfer, E., Planta 169: 393 (1986).

28. Mösinger, E., Batschauer, A., Apel, K., and Schäfer, E., Planta, in press (1987).

29. Merkle, Th., and Schäfer, E., submitted to Planta.

30. Ohl, S., Hahlbrock, K., and Schäfer, E., unpublished results.

31. Ernst, D., and Oesterhelt, D., EMBO J. 3: 3075 (1984).

PHYTOCHROME ACTION IN THE LIGHT-GROWN PLANT

Harry Smith and Garry Whitelam

Department of Botany

University of Leicester

Leicester. UK.

INTRODUCTION

The fact that phytochrome acts in mature light-grown plants has been known since the early investigations of Borthwick and Hendricks, whose first published action spectra for the red-light-absorbing photoreceptor in photomorphogenesis included one for the night-break inhibition of flowering in Xanthium (Parker et al, 1946). After much study of seed germination and photomorphogenesis in etiolated seedlings, the Beltsville group published evidence that phytochrome exerts direct control over extension growth in the light-grown plant, with the observation that a brief period of far-red light (FR), given to bean plants immediately after the end of the daily photoperiod, caused a substantial increase in stem and petiole extension rate (Downs, Hendricks and Borthwick, 1957). This paper demonstrating the so-called "end-of-day" effect is now often cited, and justifiably described as "classical"; it seems, however, to be rarely read! In addition to proving the involvement of phytochrome by the accepted criterion of R/FR reversibility, Downs went on to investigate the temporal nature of the sensitivity to FR, and the effects of irradiance and duration of the FR treatment. These neglected observations documented an approximate relationship between response and proportional amounts of FR and R, and implied an extensive sensitivity to FR throughout a dark period of many

hours, together with an increased growth stimulation with
increased duration, or intensity, of FR treatment. Although
probably not appearing "paradoxical" at the time, these data
were later seen not to sit comfortably with the demonstrated
instability of Pfr. The inescapable conclusion of this
physiological evidence of 30 years ago is that the
phytochrome which controls extension growth in light-
grown plants differs from the phytochrome measurable by
spectrophotometry in etiolated plants at least in the
stability of Pfr. This conclusion has been confirmed and
extended in more recent times (e.g., Vince-Prue, 1977;
Lecharny and Jacques, 1979; Lecharny, 1981), and it is now
clear that Pr and Pfr in the light-grown plant have similar,
low, in vivo turnover rates (Heim, Jabben, and Schäfer, 1981;
Brockman and Schäfer, 1982; Jabben and Holmes, 1983).

During the last 10-15 years the role of phytochrome in the
vegetative development of the light-grown plant has begun
to be extensively investigated (for reviews, see Morgan and
Smith, 1981a; Smith, 1982, 1986; Smith and Morgan, 1983;
Holmes, 1983; Jabben and Holmes, 1983; Lecharny, 1985).
Phytochrome is seen as having the fundamental function of
detecting the relative amounts of R and FR in the incident
radiation (i.e. R:FR ratio), the most important application of
this function being expressed in the strategy of shade
avoidance. This is demonstrated most effectively by plants of
shade-avoiding species, which exhibit striking developmental
changes in response to R:FR ratios that have been artificially
depressed to represent shade light quality in controlled
environments (Holmes and Smith, 1975, 1977b; Morgan and Smith,
1976, 1979; Child, Morgan and Smith, 1981; Smith, 1982). The
most spectacular of these changes is the large increase in
stem extension shown by plants of shade-avoiding species that
have been transferred from a high R:FR to a low R:FR (Morgan
and Smith, 1976, 1979). For most species examined, long-term
stem extension rate is an inverse linear function of the Pfr/P
of phytochrome established under different R:FR (e.g. Morgan
and Smith, 1976, 1978, 1979, 1981b; Whitelam and Johnson,
1982). The responses to depressions of Pfr/P at the end-of-
day are analogous to, but usually considerably smaller than,
the effect of FR during the whole day (Morgan and Smith,
1978), although Vince-Prue (1977) showed that the stimulation
of internode extension in Fuchsia by end-of-day FR was
linearly related to Pfr/P. The extensive linearity of these
published relationships between extension rate and calculated
Pfr/P is very different from the pattern seen in etiolated
seedling responses, where all types of relationships other
than linear have been reported, including exponential,
hyperbolic, biphasic, and threshold (see Smith, 1983).

With the recent convincing evidence (Abe et al 1985; Shimazaki
& Pratt, 1985; Tokuhisa, Daniels & Quail, 1985) that the bulk
of the phytochrome in the light-grown plant (Type II
phytochrome) is of a different molecular form from that which
predominates in etiolated seedlings (Type I phytochrome), a
potential explanation for the discrepancies outlined above may
now be at hand.   It is clearly, therefore, of considerable
importance to characterise the physiological nature of
phytochrome action in light-grown plants as fully as possible.
In this Chapter, we seek to review some recent observations,
and to present in preliminary form some new data, which
address the following questions:

> 1. What is the extent of the linearity between Pfr/P
>   and response?
>
> 2. How rapidly does phytochrome couple to, or
>   decouple from, the transduction chain?
>
> 3. What is the fluence rate range over which the
>   response operates?

The most effective experimental approach to these questions is
to use linear voltage displacement transducers (LVDT)
continuously to record internode extension in young, light-
grown plants.   An important point is that the seedlings used
in such experiments should be grown for at least two weeks in
WL to ensure they are fully de-etiolated.   The hypocotyl
should have completed its period of growth and all seedling
extension during the experiments should be due to the first
internode.   It is now becoming clear that, in the natural
environment, the period during which the hypocotyl extends and
the cotyledonary leaves expand and become photosynthetic, is a
crucial transition as far as phytochrome is concerned.
During this period, the Type I phytochrome of the etiolated
seedling, through the instability of the Pfr form, is rapidly
depleted as soon as there is sufficient light to establish a
photoequilibrium (Hunt and Pratt, 1979) (but see below).
Simultaneously, through a feedback mechanism mediated by Pfr,
the synthesis of Type I phytochrome is strongly inhibited
(Colbert, Hershey and Quail, 1983, 1985; Gottman and Schäfer,
1983; Otto et al, 1983; 1984).   The responses of hypocotyl
extension to variations in R:FR seem to change progressively
with age and extent of de-etiolation (Holmes et al, 1982),
perhaps because the control of extension growth is gradually
passed from Type I, to Type II, phytochrome during this
transition period.   In the experiments outlined here, mustard
(Sinapis alba L.) seedlings were grown for two weeks under
high R:FR WL, normally at 100 umolm^{-2}s^{-1}, before connexion to

the transducer; experimental FR treatments were applied directly to the growing first internode by fibre-optic probe.

Using transducer methodology with light-grown mustard seedlings, Morgan et al (1980) had already shown that extension rate was rapidly modulated by changes in R:FR acting via Pfr/P. These observations have now been extended and dealt with in a statistically meaningful manner (Child and Smith, 1987). Figure 1 shows the general nature of the changes in extension rate brought about by FR. The onset of FR is followed, after a lag of ca 10 min, by an abrupt elevation of extension rate, reaching a sharp peak at about 20 min, after which a sharp deceleration is usually observed. With the continued presence of FR a second acceleration is seen, reaching a fluctuating level intermediate between the WL rate and that of the first peak by about 60 min after the onset of FR. The first, rapid increase in extension rate we term Phase 1, whilst the later period of more-or-less steady elevated rate is Phase 2. Although no lag after FR switch-off is discernible in Figure 1 the true mean lag for deceleration of extension rate was approximately 6 min. The mean time for extension rate to return to the level before FR switch-on was ca 16 min, and this was not affected by the duration of the FR treatment. The standard response pattern as seen in Figure 1 was used as the basis for the investigation of the questions listed above.

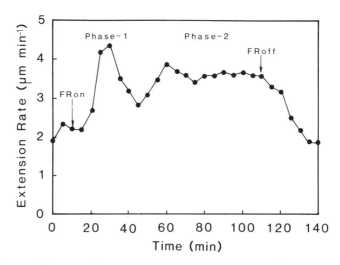

Figure 1 Effect of FR applied via fibre optic probe to the growing internode of S. alba seedlings in background white light. Each point is the mean of 10 seedlings (Child and Smith, 1987).

RELATIONSHIP BETWEEN EXTENSION RATE AND Pfr/P

Using a range of FR fluence rates against background WL of 50,
100 or 150 umolm^{-2}s^{-1}, the increment of extension rate at both
the Phase 1 peak, and over a 30 min period of Phase 2, was
obtained. In Figure 2a and 2b these values (converted to
proportional increases) are plotted against Pfr/P. As yet,
no techniques have become available for estimating Pfr/P in
green tissues, and therefore the values in Figure 2 were
measured using etiolated oat coleoptiles exposed to the light
sources in the same geometry as the experimental plants. The
range of Pfr/P, from 0.17 to 0.63, is much wider than has
previously been used in experiments on this response, and
extension rate was found to be an inverse, linear function of
Pfr/P over the whole range for each WL level and for both
phases. Moreover, analysis of variance of the slopes of the
regression lines at each background WL level showed they were
not significantly different, confirming that extension rate
was primarily determined by Pfr/P rather than fluence rate
(Child and Smith, 1987). Linear relationships between Pfr
concentration and response are not seen with etiolated
seedlings; indeed, linear relationships between effector
concentration and response are rare throughout biology, the
only common example being the dependence of reaction velocity
on enzyme concentration when all substrates are present in
non-limiting concentrations. Although there is considerable
current interest in the possibility that phytochrome is an
enzyme catalyzing a protein kinase reaction (Otto and Schäfer,
1986; Lagarias et al 1986; Wong and Lagarias, 1986) it would
be premature to attempt to derive mechanistic conclusions from
the relationships reported here.

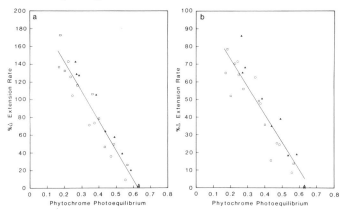

Figure 2   Relationship between extension rate response to FR
and phytochrome photoequilibrium (Pfr/P) in S. alba a, Phase
1; b, Phase 2 (Child and Smith, 1987).

In the natural environment the graded response to
increasing vegetation shade provides the plant with an
extremely fine-tuned shade avoidance mechanism. Holmes and
Smith (1977a) showed that R:FR decreases logarithmically with
increasing leaf area index (LAI). Since estimated Pfr/P
is related to R:FR hyperbolically (Smith and Holmes, 1977),
Pfr/P is, to all practical purpose, inversely related to LAI
in a linear manner. Thus, phytochrome photoequilibrium
is sensitive to increases in LAI over a wide range. The
marked sensitivity of extension rate to very small depressions
of Pfr/P even at high Pfr/P levels suggests that light-grown
plants should be capable of reacting to the reductions in
actinic R:FR caused by FR-enrichment from the reflection of
solar radiation from neighbouring vegetation. Indeed, recent
evidence from field-grown mustard plants (inter alia) has
neatly demonstrated neighbour detection via R:FR perception of
radiation reflected from nearby vegetation (Ballaré et al.,
1987).

One further point should be made in connexion with the
relationship between Pfr/P and response. Such direct
relationships as those in Figure 2 provide no evidence on
whether Pfr, or Pr, is the effector of the reponse. Thus,
the question as to whether extension is controlled via Pfr-
mediated inhibition, or Pr-mediated stimulation, or a balance
of both, must remain moot.

COUPLING OF PHYTOCHROME TO THE TRANSDUCTION CHAIN

The overall time-course as shown in Figure 1 leaves ample
opportunity for a range of potential intermediary processes to
be involved between the onset of a change in Pfr/P and the
modulation of extension rate. A lag of 6 - 10 minutes may
not, perhaps, be long enough to allow for regulation via
transcriptional control, but at least it demonstrates that
phytochrome does not interact directly with the mechanism of
cell extension. On the other hand, although the transduction
chain may be comparatively prolonged, the coupling of
phytochrome to the transduction chain seems to be very rapid.

The terminology of "phytochrome coupling to its transduction
chain" is a relatively recent fashion. In the context of
etiolated seedlings, the concept of "coupling" is identical to
that of "signal transduction", i.e., the first molecular
change induced in the responding system by Pfr, the (presumed)
active form of the photoreceptor (Haupt and Feinleib, 1979;

Shropshire, 1979; Quail, 1983).   In practical terms, the
coupling point is taken to be equivalent to the point in the
time course when loss of full FR reversibility first occurs
(Mohr, 1983).   Consequently, for induction responses in
etiolated seedlings the coupling time is readily determined by
carrying   out   escape-from-reversibility   tests,   but   for
modulations under continuous irradiation the approach is not
so clear cut.   In the experiments described here, pulses of
FR were applied against a WL background in a protocol
analogous to,   but   not   identical   with,   escape   from
photoreversibility tests (Child and Smith, 1987).

Internode extension was sensitive to pulses of FR at least as
short as one minute (Figure 3). It was also found that plants
could respond to successive brief FR treatments (Fig. 4),
although there are indications here of a gradual increase in
basal extension rate as the numbers of FR pulses accumulate
(M.Malone and H.Smith, unpublished data).   These experiments
show that brief depressions of Pfr/P cause large Phase 1
increases in extension rate that occur several minutes after
the actinic radiation has returned to the background
WL, which establishes a photoequilibrium of ca 0.63.   If

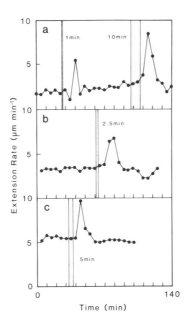

<u>Figure 3</u> Extension rate response of a single <u>S. alba</u> seedling
to brief exposure of additional FR to the growing internode
(Child and Smith, 1987).

it is assumed, in analogy with the generally-accepted views of
phytochrome action in etiolated seedlings, that it is Pfr
which "couples" with the first reaction component of the
transduction chain, the pulse experiments reported here may be
interpreted as evidence for a rapid "uncoupling" of Pfr from
its reaction partner leading to a release of extension rate
from Pfr-mediated inhibition. Symmetrical experiments using
low R:FR background WL and pulses of R have, as yet, only
been performed in a preliminary form, but they indicate that
similar, but opposite, transient depressions of extension
rate occur after a brief pulse of R (M.Malone and H.Smith,
unpublished results). It seems, therefore, that
phytochrome coupling/uncoupling occurs rapidly (i.e. at
least within 1 min) in light-grown mustard seedlings.

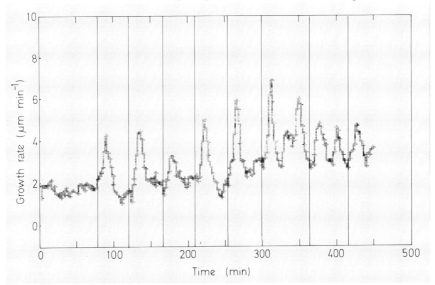

Figure 4 Repeated extension rate responses of a single S.
alba seedling to 90 s pulses of FR from fibre optic probe
given at the times indicated by the vertical lines.
Background WL was 150 umol m^{-2} s^{-1} (Data of Dr M. Malone).

Rapid escape from reversal has previously been reported for
certain inductive photomorphogenetic phenomena in light-grown
plants. The most rapid is the night-break effect in
Pharbitis flowering, in relation to which Fredericq (1964)
demonstrated what would now be regarded as coupling of Pfr
within 2 min of the onset of R. Several phenomena in
etiolated seedlings are known to escape from reversibility
within 2 to 10 s of the start of a R pulse (Quail, 1983).
The data reported here, therefore, are not particularly

remarkable in their timing, although as yet we do not know how brief a FR pulse must be before it fails to induce a detectable Phase 1-type growth effect.   The principal significance of these results lies in the difficulty which attends their interpretation on current views of phytochrome action.   The data imply that the removal of a proportion of the Pfr for a short time leads to a substantial biological effect; i.e., according to most of the models advanced to account for photomorphogenesis in etiolated seedlings, these results imply a rapid "uncoupling" of Pfr from its reaction partner leading to release of extension rate from Pfr-mediated inhibition.   In the data of Figures 3 and 4, the FR fluence rate was enough to drive Pfr/P from the WL value of ca 0.63 down to ca 0.17; i.e., a reduction of Pfr concentration of ca 73%.   At the purely physiological level, it is difficult to understand how such a transient and incomplete reduction in the concentration of an inhibitory effector would lead to a sharp peak in growth rate.

The ecological relevance of the pulse experiments is substantial.   Rapid, repeatable increases in extension rate in response to transient simulated shading mean that, in nature, shade-avoiding species such as mustard would be capable of reacting sensitively to the first signals of competition from neighbours.   As mentioned above, the first signals may not be due to direct shading, but to FR-enriched light reflected from neighbours (Ballaré et al., 1987). Holmes and Smith (1977a) showed that FR-enrichment in a sunfleck within a canopy can be substantial.   Furthermore, the experiments described here using fibre-optic probes to apply FR to growing internodes whilst the rest of the plant is in WL, simulate the perception of reflected light more effectively than they do that of light filtered through vegetation.

FLUENCE RATE COMPENSATION

In the natural environment, total fluence rate often fluctuates rapidly over a wide range.   A heavy cloud obscuring the direct solar beam could depress the above-canopy fluence rate by a factor of ten or more, causing similar proportional changes within the canopy.   If the perception of R:FR by shaded plants were substantially confounded by changes in fluence rate, then the shade avoidance strategy would be vitiated.   Consequently, on theoretical grounds alone, R:FR perception should be compensated for fluctuations in fluence rate. The photochromic property of phytochrome provides an ideal compensatory mechanism, since although cycling rate is both wavelength and fluence-rate dependent, the relative proportions of Pr and Pfr at photoequilibrium should be determined solely by the wavelength distribution of photons, and should be constant over a wide range of fluence rate.

The data used to construct Figure 2 show that, both for Phase 1 and Phase 2 responses, fluence rate compensation operates effectively over the 50-150 umolm^{-2}s^{-1} range (WL). All of the data from the three WL fluence rates, for each phase, may be plotted on single lines, with high correlation coefficients (Child and Smith, 1987). This means that, irrespective of WL fluence rate over this range, a specific depression of Pfr/P always yields the same proportional increment of extension rate. Similar conclusions were derived by Smith and Hayward (1985) for long-term extension growth rates over broadly the same fluence rate range, supporting the hypothesis that R:FR perception by phytochrome is, in principle, fluence rate compensated. Whether fluence rate compensation operates effectively in the natural environment, however, is not answered by these experiments. The 50-150 umolm^{-2}s^{-1} range cited above is reasonably typical of within-canopy fluence rates, but for a shade perception mechanism to be effective, it should be capable of operating at the much higher fluence rates found at the surface of canopies, where the first signals of vegetative shading are experienced.

The question therefore resolves itself as: does phytochrome operate at fluence rates typical of summer daylight (i.e., 2-3 mmolm^{-2}s^{-1})? Although evidence from field experiments, (e.g. Ballaré et al, 1987) might suggest that phytochrome does operate at high solar radiation levels, it could be argued that such long-term effects result principally from R:FR changes at the low fluence rates of dawn and dusk. It appears that fluence rates typical of summer sunlight have not yet been used in laboratory-based photomorphogenetic investigations, and indeed there are good theoretical reasons why a negative answer might be expected. It was shown almost 15 years ago that photoconversion intermediates accumulated at high fluence rates (Kendrick and Spruit, 1972, 1973). The intermediate meta-Rb accumulates in white light because it is the most photostable of all the intermediates, and because the succeeding relaxation step is the slowest in the cycle (Kendrick and Spruit, 1977; Rüdiger, 1980). Consequently, at daylight levels, phytochrome could become inoperative, either because the concentration of the presumed active Pfr form is too low, or because individual molecules do not remain as Pfr long enough to couple to the transduction chain. It is of considerable importance, therefore, to estimate the WL fluence rate range over which intermediate accumulation is significant, since over this range the steady-state Pfr/P will be constant, whilst the cellular concentration of Pfr will vary markedly.

THE PHOTOPROTECTION OF PHYTOCHROME

An indirect way to approach this problem is to make use of the fact that, at least for dicotyledonous plants, the loss of Type I Pfr under continuous irradiation is typically a first order reaction; thus, the exponential rate constant of Pfr breakdown at any point should be proportional to the concentration of Pfr. If intermediate accumulation increases with fluence rate, then the steady state concentration of Pfr should decrease, with a concomitant decrease in breakdown. Kendrick and Spruit (1972) had shown a deviation from first order kinetics of Pfr loss in Amaranthus caudatus at elevated fluence rates, and thus the first tests of this notion were carried out with the same species (Figure 5). It is clear from these data that first order kinetics were not seen at any fluence rate used; nevertheless, substantial inhibition of phytochrome breakdown with increasing WL fluence rate is apparent. We have termed this phenomenon "photoprotection" (Smith, Whitelam and Jackson, 1987).

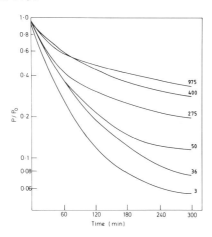

Figure 5   Photoprotection of phytochrome loss in Amaranthus caudatus seedlings by cont. WL. (Numbers on curves refer to fluence rates; data points omitted for clarity). (Smith, Whitelam and Jackson, 1987).

Photoprotection has been demonstrated in three other species to date: peas, mung beans, and oats. Although the absolute rate of phytochrome breakdown differs between the four tested species, in each case elevating the fluence rate from 2 $\mu molm^{-2}s^{-1}$ to $\underline{ca}$ 1 $mmolm^{-2}s^{-1}$ caused approximately a half log_{10} reduction in degradation rate. Immunoblots have been used to demonstrate that loss of spectral photoreversibility accompanies that of the phytochrome molecule, showing that we are not here merely dealing with some optical artefact (Smith, Jackson and Whitelam, 1987). The fluence rate at which photoprotection first becomes detectable is surprisingly low at about 50 $\mu molm^{-2}s^{-1}$, above which there is a good log-linear relationship between photoprotection and fluence rate (Figure 6). These results indicate that the availability of Type I Pfr for the cellular degradation machinery becomes increasingly restricted as the fluence rate is raised; it is not possible to distinguish between intermediate accumulation or the velocity of cycling as responsible mechanisms. However, if the availability of Pfr for the degradation machinery is restricted, it is at least plausible that its availability for its reaction partner might also be affected.

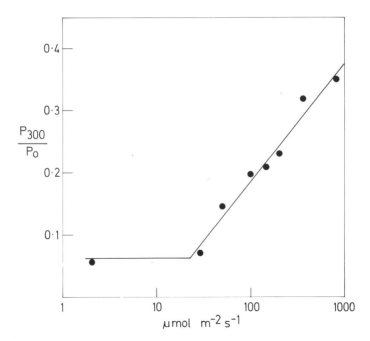

Figure 6 Relationship between fluence rate and degree of photoprotection in Amaranthus caudatus (Smith, Whitelam and Jackson, 1987).

These data have two important implications for the action of phytochrome in light-grown plants. In the first place, if the photoconversions of Type II phytochrome are similar to those of Type I, then considerable ingenuity will be needed to reconcile hypotheses for the action of phytochrome which propose a central role for the <u>concentration</u> of Pfr (or, indeed, Pr) if it is found, by direct experimentation, that FR stimulates extension at sunlight fluence rates. (Attempts to acquire evidence on this question using transducer methodology have been severely handicapped by the chronic lack of sunlight in the UK!). Secondly, these data raise the possibility that the amounts of Type I phytochrome persisting within the light-grown green plant under natural conditions may be considerably higher than has previously been thought.

CONCLUDING REMARKS

Phytochrome is a remarkably sophisticated sensor of the relative proportions of R and FR light in the natural environment. In the photosynthetically-competent light-grown plant, phytochrome provides the capability of detecting the presence, intensity and duration of vegetational shade, or of radiation reflected from neighbouring vegetation, even down to transient experiences of a minute or so, and even though the total amount of light may be fluctuating massively due to weather conditions. Coupling and uncoupling of phytochrome to and from its transduction chain occur quickly, probably within less than a minute, but although the lag of 6-10 min between a change in R:FR and the onset of response seems short, undoubtedly ample opportunity is thereby provided for a range of potential intermediary steps. The extensive linearity of response with Pfr/P is ecologically desirable, but begs the question of mechanism. Few of these character-istics fit with the classical picture of phytochrome derived from studies on etiolated plants, emphasising the need for further intensive analysis of the physiological properties of phytochrome in light-grown plants. At the same time, the possibility that high-fluence-rate photoprotection may allow significant proportions of Type I phytochrome to be present in light-grown plants should not be disregarded.

ACKNOWLEDGMENTS

The authors wish to acknowledge the excellent technical help of Michael Jackson and Jill Atkin. Drs M. Malone and R. Child are thanked for the inclusion of unpublished data.

REFERENCES

Abe, H., Yamamoto,K.T., Nagatani, A., Furuya, M. (1985) Plant
    Cell Physiol. 26:1387-1399
Ballaré, C.L., Sanchez, R.A., Scopel, A.L., Casal, J.J.,
    Ghersa, C.M. (1987) Plant, Cell Environ. (in press)
Brockman, J., Schäfer, E. (1982) Photochem. Photobiol. 35:555-
    558
Child, R., Morgan, D.C., Smith, H. (1981) New Phytol. 89:545-
    555
Child, R., Smith, H. (1987) Planta (in press)
Colbert, J.T., Hershey, H.P., Quail, P.H. (1983). Proc. Natl.
    Acad. Sci. USA 80:2248-2252
Colbert, J.T., Hershey, H.P., Quail, P.H. (1985) Plant Molec.
    Biol. 5:91-102
Downs, R.J., Hendricks S.B., Borthwick, H.A. (1957) Bot. Gaz.
    118:199-208
Fredericq, H. (1964) Plant Physiol. 39:812-816
Gottman, K., Schäfer, E. (1983) Planta 157:492-400
Haupt, W., Feinleib, M.E. (1979) In: Encycl. Plant Physiol. N.
    S. Vol 7: Physiology of Movements, pp 1-9 Haupt, W.,
    Feinleib, M.E., eds. Springer, Berlin
Heim, B., Jabben, M,. Schäfer, E. (1981) Photochem. Photobiol.
    34:89-93
Holmes, M.G. (1983) Phil. Trans. Roy. Soc. B303:503-521
Holmes, M.G., Beggs, C.J., Jabben, M., Schäfer, E. (1982)
    Plant, Cell Environ. 5:45-51
Holmes, M.G., Smith, H. (1975) Nature 254:512-514
Holmes, M.G., Smith, H. (1977a) Photochem. Photobiol. 25:539-
    545
Holmes, M.G., Smith, H. (1977b) Photochem. Photobiol. 25:551-
    557
Hunt, R.E., Pratt, L.H. (1979) Plant Physiol. 64:332-336
Jabben, M., Holmes, M.G. (1983) In: Encycl. Plant Physiol.
    N. S. Vol 16B: Photomorphogenesis, pp 704-722
    Shropshire, W., Jr., Mohr, H., eds. Springer, Berlin
Kendrick, R.E., Spruit, C.J.P. (1972) Nature 237:281-282
Kendrick, R.E., Spruit, C.J.P. (1973) Photochem. Photobiol.
    18:134-144
Kendrick, R.E., Spruit, C.J.P. (1977) Photochem. Photobiol.
    26:201-204
Lagarias, J.C., Wong, Y-S., Berkelman, T.R., Elich, T.D.,
    Kidd, D.G., McMichael, R.W. Jr. (1986) Proc. XVI
    Yamada Conf. Phytochrome and Photomorphogenesis. pp
    36-37
Lecharny, A. (1981) These Doct. Etat, University Pierre-et-
    Marie-Curie, Paris. (cited in Lecharny, 1985)

Lecharny, A. (1985) Physiol. Veg. 23:963-982

Lecharny, A. Jacques, R. (1979) Planta 146:575-577

Mohr, H. (1983) In: Encycl. Plant Physiol. N. S. Vol 16A: Photomorphogenesis, pp 336-357 Shropshire, W., Jr., Mohr, H., eds. Springer, Berlin

Morgan, D.C., O´Brien, T., Smith, H. (1980) Planta 150:95-101

Morgan, D.C., Smith, H. (1976) Nature 262:210-212

Morgan, D.C., Smith, H. (1978) Planta 142:187-193

Morgan, D.C., Smith, H. (1979) Planta 145:253-258

Morgan, D.C., Smith, H. (1981a) In: Encycl. Plant Physiol. N. S. Vol 12A Physiological Plant Ecology pp 109-134 Lange, O.L., Nobel, P.S., Osmond, C.B., Ziegler, H. eds. Springer, Berlin

Morgan, D.C., Smith, H. (1981b) New Phytol. 88:239-248

Otto, V., Mösinger, E., Sauter, M., Schäfer, E. (1983) Photochem. Photobiol. 38:693-700

Otto, V., Schäfer, E. (1986) Proc. XVI Yamada Conf. on Phytochrome and Photomorphogenesis p106

Otto, V., Schäfer, E., Nagatani, A., Yamamoto, K.T., Furuya, M. (1984) Plant Cell Physiol. 25:1579-1584

Parker, M.W., Hendricks, S.B., Borthwick, H.A., Scully, N.J. (1946) Bot. Gaz. 108:1-26

Rüdiger, W. (1980) In: Structure and Bonding 40, pp 101-140 Hemmerich, P., ed. Springer, Berlin.

Quail, P.H. (1983). In: Encycl. Plant Physiol. N. S. Vol 16A: Photomorphogenesis, pp 178-257 Shropshire, W., Jr., Mohr, H., eds. Springer, Berlin

Shimazaki, Y., Pratt, L.H. (1985) Planta 164:333-344

Shropshire, W., Jr., (1979) In: Encycl. Plant Physiol. N. S. Vol 7: Physiology of Movements, pp 10-41 Haupt, W., Feinleib, M.E., eds. Springer, Berlin

Smith, H. (1982) Ann. Rev. Plant Physiol. 33:481-518

Smith, H. (1983) Phil. Trans. Roy. Soc. London B303:443-452

Smith, H., Hayward, P. (1985) Photochem. Photobiol. 42:685-688

Smith, H., Holmes, M.G. (1977) Photochem. Photobiol. 25:547-550

Smith,H., Morgan, D.C. (1983) In: Encycl. Plant Physiol. N. S. Vol 16B: Photomorphogenesis, pp 491-517 Shropshire, W., Jr., Mohr, H., eds. Springer, Berlin

Smith, H., Jackson, M.G., Whitelam, G. (1987) Planta (in preparation).

Tokuhisa, J.G., Daniels, S.M., Quail, P.H. (1985) Planta 164:321-332

Vince-Prue, D. (1977) Planta 133:149-166

Whitelam, G.C., Johnson, C.B. (1982) New Phytol. 90:611-618

Wong, Y-S., Cheng, H-C., Walsh, J.C., Lagarias, J.C. (1986) J. Biol. Chem. 261: (in press)

BLUE LIGHT RESPONSES OF PHYCOMYCES

Edward D. Lipson[1]

Department of Physics
Syracuse University
Syracuse, New York

I. INTRODUCTION

The fungus Phycomyces blakesleeanus exhibits various
responses to blue light, including phototropism, the transient light-growth response, light-induced carotene
synthesis (photocarotenogenesis), and light-induced sporangiophore initiation (photophorogenesis) (Galland and Lipson,
1984; Cerdá-Olmedo and Lipson, 1987). Phototropism and the
light-growth response occur in the sporangiophore. Photocarotenogenesis occurs in both the sporangiophore and the
filamentous mycelium but is generally measured in mycelial
preparations. Photophorogenesis is strictly a mycelial response. Besides these light responses, the sporangiophore
also responds to gravity, stretch, wind, chemicals, and
nearby objects (avoidance response).

The giant unicellular sporangiophore of Phycomyces is
well known for its phototropism, which depends on a lens
effect within the cylindrical growing zone (upper few milli-
meters just below the spherical sporangium). In addition,
for symmetrically applied stimuli, the growth rate can be
modulated by changes in the stimulus intensity. The light-
growth response (Foster and Lipson, 1973) ensues when the
intensity is stepped or pulsed from one level to another;
the elongation rate of the sporangiophore, normally 3 mm/h,

---

[1]Supported by NSF grant DMB-8316458 and NIH grant GM29707.

becomes modulated transiently. This light-growth response is
related to phototropism. Both responses occur over a range
of ten decades of blue light intensity from 10^{-9} to 10
W m^{-2}. To manage this enormous range, these light responses
incorporate photosensory adaptation, i.e range adjustment of
sensitivity (Lipson and Block, 1983; Galland and Russo,
1984). The kinetics and overall range of light and dark
adaptation in Phycomyces are similar to those in visual
adaptation.

As a fungus, Phycomyces lacks many other prominent pigments that are abundant in plants (chlorophylls, phytochrome, etc.). Its sensitivity range, especially for phototropism and the light-growth response (but also for photo-
differentiation) is unparalleled: it is 10^5 times more sensitive than the well known phototropism of the Avena coleoptile, one of the most sensitive phototropic systems in
higher plants.

II. BEHAVIORAL GENETICS

Mutants with abnormal phototropism (genotype mad) have
been classified phenotypically and genetically in a simple
branched scheme (Bergman et al., 1973; Cerdá-Olmedo, 1977)
with three classes and seven genes (madA to madG). Night-
blind mutants affected in genes madA, madB, and madC are
associated with the photoreceptor input. The phototropic
threshold for madA mutants is elevated by several thousand-
fold above the wild type value of 10^{-9} W m^{-2}. For madB and
madC mutants, it is elevated by almost a millionfold
(Bergman et al., 1973; Ootaki et al., 1974; Lipson and
Terasaka, 1981). Mutants affected in the four genes madD
through madG (stiff mutants) are defective in the growth
control output common to all tropisms. They show normal
phototropic thresholds, but have very slow bending responses. Further, madA and madB mutants are abnormal for both
photocarotenogenesis and photophorogenesis, whereas madC
mutants are normal for these mycelial photoresponses. A
family of double mutants has been constructed for all 21
pairwise combinations of the 7 genes madA to madG (Lipson et
al., 1980; Lipson and Terasaka, 1981). Hypertropic (madH)
mutants show enhanced tropisms (Lipson et al., 1983) and
unusual genetic properties (López-Díaz and Lipson, 1983a,b).
These hypertropic mutants behave, in most ways, opposite to
the stiff mutants; on this basis, they appear to be affected
near the output of the sensory pathway common to all

tropisms. However, their photogravitropic action spectra
(Galland and Lipson, 1985b) and the wavelength dependence of
their adaptation kinetics (Galland et al., 1984) are abnormal too, suggesting that the hypertropic mutants are also
defective near the photoreceptor input.

III. BLUE LIGHT PHOTORECEPTORS

The photoreceptors for blue light responses remain generally unidentified in Phycomyces and other blue-light sensitive organisms (Senger, 1984; 1987). Flavoproteins are the
favored candidates in most cases. Phototropic action spectra
(Galland and Lipson, 1985b) indicate that madB and madC
mutants are probably affected in the photoreceptor system.
These mutants can therefore serve as a genetic assay for
components of the photoreceptor complex. The kinetics of
phototropic adaptation, which enable Phycomyces to respond
to light over the 10 decade range from 10^{-9} to 10 W m^{-2},
depend on wavelength (Galland et al., 1984) and are altered
in madA, madB, and madC mutants (Galland and Russo, 1984);
these and other results (Lipson, 1975) suggest that adaptation is mediated by the photoreceptor complex.

Not only do action spectra for phototropism differ between wild type and mutant strains, but also they differ even
for wild type alone, depending upon which method and
conditions are used. Specifically, the action spectra are
markedly different when obtained by the phototropic balance
method (Galland and Lipson, 1985a) or the photogravitropic
equilibrium method (Galland and Lipson, 1985b). Further, the
shapes of phototropic balance action spectra themselves
depend critically upon the wavelength and intensity (fluence
rate) of the reference beam. In addition, it was found
previously that, in the low intensity region, the effects of
blue light on photogravitropic equilibrium could be counteracted by green or red light that by itself would be photo-
tropically inactive (Löser and Schäfer, 1980; 1986).

Recent results on phototropism kinetics (Galland and
Lipson, 1987) have revealed two photosystems specialized for
low and high intensity ranges (below and above 10^{-6} W m^{-2},
respectively). In these experiments, when dark-grown, dark-
adapted sporangiophores were exposed to continuous light,
they showed a two step bending response. The two bending
components showed different kinetics (including those for

phototropic adaptation), different action spectra, and different optimum intensity ranges.

These results together indicate clearly that multiple receptor pigments must operate in _Phycomyces_ phototropism. Some of the complexity might be attributable to photochromic or photocycling pigments.

IV. SYSTEM ANALYSIS OF LIGHT-GROWTH RESPONSE

The transient light-growth response (Delbrück and Reichardt, 1956; Foster and Lipson, 1973), which is related to phototropism (Bergman et al., 1973; Galland et al., 1985), has been studied extensively by system identification and analysis methods in the framework of the Wiener theory (Marmarelis and Marmarelis, 1978). The experiments employ special test stimuli called "Gaussian white noise" (Lipson, 1975; Poe and Lipson, 1986; Poe et al., 1986a,b) and "sum of sinusoids" (Pratap et al., 1986a,b; Palit et al., 1986); the former is a stochastic test signal while the latter is deterministic. Both allow the determination of temporal functions called Wiener kernels, which together with the Wiener series represent the nonlinear input-output relation of the system (Marmarelis and Marmarelis, 1978; Victor and Shapley, 1980).

The results of this work indicate that the latency and other kinetic parameters of the light-growth response depend significantly on wavelength and temperature. Further, certain mutants, especially those affected in genes _madB_ and _madC_ (see above), have highly altered response kinetics. Analysis of a family of single and double mutants affected in the seven genes _madA_ through _madG_ show substantial interactions among various gene products, including interactions between those in the input or night-blind class (_madA_, _madB_, and _madC_) and those in the output or "stiff" class (_madD_, _madE_, _madF_, and _madG_). This evidence of extensive interactions calls for reconsideration of the original classification scheme (above), which had suggested a simple linear transduction pathway. In particular, the results suggest an integrated sensory transduction complex, presumably located in the plasma membrane of the sporangiophore that handles both the photoreception and growth modulation processes underlying the light-growth response and phototropism.

V. BIOCHEMISTRY

To help identify flavoproteins and other proteins involved in these blue-light responses, we have employed the
techniques of two-dimensional polyacrylamide gel electrophoresis and column chromatography. In plasma-membrane enriched fractions from short synchronous sporangiophores, two
flavoproteins with covalently linked flavin have been detected on gels (Pollock et al., 1985a,b). One of these proteins, denoted FP_1, shows a characteristically short lifetime (Pollock and Lipson, 1985). The other flavoprotein,
FP_2, was absent in gels from some madC strains but present
in others and in gels for wild type. This apparent discrepancy was resolved subsequently in segregation studies with
the conclusion that the gene coding for FP_2 is closely
linked to the madC gene (Garcés et al., 1986); this work
showed that FP_2 was not the madC gene product after all.
Another protein (without covalent flavin) seems to be the
product of the madB gene, because it is consistently absent
in gels for all madB strains studied (Pollock et al.,
1985b).

We have recently extended this two-dimensional gel electrophoresis work to proteins from stiff mutants (Trad, C.
H., Garcés, R., and Lipson, E. D., in prep.). In comparisons
of wild type, single mutants, and double mutants, reproducible differences have been found for two of the four
"stiff" genes, namely madE and madF. In madE strains (all of
which carry the same allele), the relative densities of two
nearby spots are clearly and consistently reversed in comparison to $madE^+$ strains. Spot E1 (52 kdal, pI 6.65) is much
darker in wild type and the other $madE^+$ strains, while spot
E2 (50 kdal, pI 6.65) is darker in madE strains. In madF
strains, a spot F1 (53 kdal, pI 6.1), which is present in
wild type and other $madF^+$ strains, is missing. This may
represent the madF gene product.

Nitrate reductase has been proposed as a blue light
photoreceptor in Neurospora crassa for a particular light-
induced conidiation response under starvation conditions
(Klemm and Ninnemann, 1978). We found in Phycomyces that
this enzyme complex cannot serve such a role, because
Phycomyces evidently lacks nitrate reductase. The evidence
for this claim is that Phycomyces is unable to grow on
medium containing nitrate as the sole nitrogen source and
that no nitrate reductase activity could be detected in

<u>Phycomyces</u> under a variety of assay conditions (Garcés et
al., 1985).

VI. SPECTROPHOTOMETRY

Light-induced absorbance changes (LIAC) and other differential spectrophotometric techniques (Horwitz et al.,
1986a,b) offer spectroscopic approaches to the isolation and
identification of blue-light photoreceptors. With our
special interest in <u>madB</u> and <u>madC</u> as likely photoreceptor
mutants (see above), we reinvestigated an old result in the
literature (Berns and Vaughn, 1970) that indicated a LIAC
modification in sporangiophore extracts from a <u>madC</u> strain
(C. Trad, B. Horwitz, and E. Lipson, submitted). Because of
differences in our instrumentation, we altered the protocols
somewhat. All strains studied carried a <u>carA</u> marker to
eliminate the bulk carotene; further residual carotene was
eliminated in the fractionation steps leading to the 10,000
x g supernatant. Absorbance changes were sampled between 430
and 500 nm, because these wavelengths corresponded to
extrema in LIAC difference (light-minus-dark) spectra. The
basic protocol included a sequence of three actinic
exposures: blue, near ultraviolet, and again blue.
Absorbance changes were measured before, during, and after
each actinic exposure.

The actual difference found between <u>madC</u> and control
(<u>madC</u>$^+$) strains was as follows: the absorbance change to the
two blue pulses was negative (i.e. the absorbance decreased
at 430 nm relative to 500 nm) for both strains. The LIAC for
the near-ultraviolet actinic pulse (middle pulse) was
positive. The key differences were that the LIAC were larger
in the <u>madC</u> mutants and showed only partial regeneration
rather than the essentially complete regeneration observed
in the controls over the time span of several minutes.

Although we lack a specific explanation for these
LIAC's, we consider this a very promising result for the
following reasons: (a) <u>madC</u> mutants are probable photoreceptor mutants, (b) the result repeats for all three
control and eight <u>madC</u> strains analyzed, (c) the effect is
obtained <u>in vitro</u>, so that further purification should be
possible with monitoring by this spectrophotometric assay.
If the responsible molecules can be suitably purified, this
will offer a good opportunity for isolation and identif-

ication of one or more key components of the photoreceptor complex.

In a separate study that did not involve actinic light (i.e. LIAC) effects, we sought subtle spectral differences between wild type and mad mutant strains. Only with essentially complete elimination of the screening pigment ß-carotene by a carB mutation (in both mad and behaviorally normal control strains) could we begin to resolve subtle differences that might be attributable to flavin photoreceptors. Experiments were conducted in vivo both on sporangiophores and mycelium. To eliminate a green screening pigment (probably phenolic acid polymers) from sporangiophores, potassium iodide was added to the growth medium.

Surprisingly, the main difference we found in absorbance spectra was not in the night-blind mutants but rather in one of the "stiff" (output) mutants, specifically a madE mutant. Recent results from system analysis of the light-growth response (Poe et al., 1986b) suggested that the mad gene products are organized in a molecular complex, allowing for interaction between input (madA, madB, and madC) and output (madD to madG) gene products (see above). In this context, it is perhaps not so surprising that a spectrophotometric difference would show up in a stiff mutant.

Although the most notable difference in these absorbance spectra was found for madE, an interesting light effect was found as well that distinguished madB and madC mutants (photoreceptor mutants, according to phototropic action spectra) from control strain(s). For the control strain L91 (carB, but with normal behavior), light-grown sporangiophores show substantial suppression of a 480 nm peak in the second derivative spectra in comparison to dark grown sporangiophores. This light suppression fails to occur in L131 (carB madB) and L130 (carB madC), but does occur normally in strain L26 (carB madE) just as in the control.

VII. CONCLUSION

Recent physiological work on phototropism and the light-growth response have indicated the action of multiple receptor pigments and have suggested that the sensory transducers coded by the mad genes may be arranged in an integrated molecular complex. Biochemical and spectrophotometric analyses of wild type and mutant strains have provided some

tantalizing clues that may allow the isolation and identification of receptor pigments and other molecules involved in
Phycomyces responses to blue light.

ACKNOWLEDGMENTS

I am indebted to Paul Galland for comments on this
manuscript. The following past and present members of my
research group were active participants in the recent
research results reported here: Paul Galland, Benjamin
Horwitz, Rafael Garcés, John Pollock, Promod Pratap,
Anuradha Palit, and Chafia Trad.

REFERENCES

Bergman, K., Eslava, A., and Cerdá-Olmedo, E. (1973). Mol.
    Gen. Genet. 123:1.
Berns, D. S. and Vaughn, J. R. (1970). Biochem. Biophys.
    Res. Commun. 39:1094.
Cerdá-Olmedo, E. (1977). Ann. Rev. Microbiol. 31:535.
Cerdá-Olmedo, E. and Lipson, E. (1987). "Phycomyces" Cold
    Spring Harbor Laboratory, Cold Spring Harbor, New York,
    in press.
Delbrück, M. and Reichardt, W. (1956). In "Cellular
    Mechanisms in Differentiation and Growth" (D. Rudnick,
    ed.), p. 3. Princeton University Press, Princeton, N.J.
Foster, K. W. and Lipson, E. D. (1973). J. Gen. Physiol.
    62:590.
Galland, P. and Lipson, E. (1984). Photochem. Photobiol.
    40:795.
Galland, P. and Lipson, E. (1985a). Photochem. Photobiol.
    41:323.
Galland, P. and Lipson, E. (1985b). Photochem. Photobiol.
    41:331.
Galland, P. and Lipson, E. (1987). Proc. Natl. Acad. Sci.
    (U.S.A.) 84:104.
Galland, P. and Russo, V. E. A. (1984). J. Gen. Physiol.
    84:101.
Galland, P., Pandya, A., Lipson, E. (1984). J. Gen. Physiol.
    84:739.
Galland, P., Palit, P., and Lipson, E. (1985). Planta
    165:538.
Garcés, R., Pollock, J. A., and Lipson, E. D. (1985). Plant
    Science: 40:173.

Garcés, R., Chmielewicz, C., and Lipson, E. D. (1986). Mol. Gen. Genet. 203:341.

Horwitz, B., Trad, C., and Lipson, E. (1986a). Plant Physiol. 81:726.

Horwitz, B., Trad, C., and Lipson, E. (1986b). Photochem. Photobiol. 44:207.

Klemm, E. and Ninnemann, H. (1978). Photochem. Photobiol. 28:227.

Lipson, E. (1975). Biophys. J. 15:989.

Lipson, E. D. and Block, S. M. (1983). J. Gen. Physiol. 81:845.

Lipson, E. D. and Terasaka, D. T. (1981). Exp. Mycol. 5:101.

Lipson, E. D., Terasaka, D. T., and Silverstein, P. S. (1980). Mol. Gen. Genet. 179:155.

Lipson, E. D., López-Díaz, I., and Pollock, J. A. (1983). Exp. Mycol. 7:241.

López-Díaz, I. and Lipson, E. D. (1983). Mol. Gen. Genet. 190:318.

López-Díaz, I. and Lipson, E. D. (1983). Curr. Genet. 7:313.

Löser, G. and Schäfer, E. (1980). In "The Blue Light Syndrome" (H. Senger, ed.), p. 244. Springer-Verlag, Berlin, Heidelberg, New York.

Löser, G. and Schäfer, E. (1986). Photochem. Photobiol. 43:195.

Marmarelis, P. Z., and Marmarelis, V. Z. (1978). "Analysis of Physiological Systems - The White-Noise Approach," Plenum Press, New York, London.

Ootaki, T., Fischer, E. P., and Lockhart, P. (1974). Mol. Gen. Genet. 131:233.

Palit, A., Pratap, P., and Lipson, E. (1986). Biophys. J. 50:661.

Poe, R. and Lipson, E. (1986). Biol. Cybern. 55:91.

Poe, R., Pratap, P., and Lipson, E. (1986a). Biol. Cybern. 55:99.

Poe, R., Pratap, P., and Lipson, E. (1986b). Biol. Cybern. 55:105.

Pollock, J. and Lipson, E. (1985). Photochem. Photobiol. 41:351.

Pollock, J., Lipson, E., and Sullivan, D. (1985). Planta 163:506.

Pollock, J., Lipson, E., and Sullivan, D. (1985). Biochem. Genet. 23:379.

Pratap, P., Palit, A., and Lipson, E. (1986a). Biophys. J. 50:645.

Pratap, P., Palit, A., and Lipson, E. (1986b). Biophys. J. 50:653.

Senger, H. (ed.) (1984). "Blue Light Effects in Biological Systems," Springer-Verlag, Berlin, Heidelberg, New York.

Senger, H. (ed.) (1987). "Blue Light Responses: Phenomena
    and Occurrence in Plants," CRC Press, Boca Raton,
    Florida.
Victor, J., and Shapley, R. (1980). Biophys. J. 29:459.

PROBLEMS AND PROSPECTS OF BLUE AND ULTRAVIOLET LIGHT EFFECTS

Horst Senger

Fachbereich Biologie/Botanik
Philipps-Universität
Marburg, West Germany

Edward D. Lipson

Department of Physics
Syracuse University
Syracuse, New York, USA

I. INTRODUCTION

The best known pigment controlling photomorphogenesis in plants is phytochrome. Besides this pigment there are a number of photoreceptors absorbing in the blue and UV region of the spectrum that trigger and regulate a variety of reactions. These so-called blue-UV light effects are prevalent in cyanobacteria, fungi and algae, but extend over mosses and ferns to higher plants. The variety of photoreceptors and their significance in photoregulation decreases from lower to higher organisms.

It seems that during evolution of plants, which started in the water, several light receptors were tested, that absorbed the blue-UV light which readily penetrate the water. When plants conquered the land, a red light absorbing photomorphogenic pigment, i.e. phytochrome, was developed and the importance of blue-UV receptors was diminished.

Blue light effects have been reported since 1864 (Julius Sachs, 1864). In recent years several reviews and books have dealt with this interesting topic (Briggs and Iino, 1983; Dörnemann and Senger, 1984; Gressel and Rau, 1983; Horwitz and Gressel, 1986; Kowallik, 1982; Senger, 1980,1984,1987; Senger and Schmidt, 1986; Schmidt, 1984).

In this chapter we report on the contributions and discussions on blue-UV phenomena and related action of red light made during two round table sessions at the Yamada Conference on "Phytochrome and Plant Photomorphogenesis" in Okazaki.

315

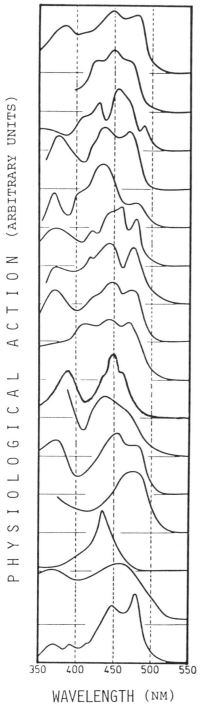

Phototropism of *Phycomyces*
(Lipson *et al.*, 1984)

Light-induced absorbance change
(LIAC) (Widell *et al.*, 1983)

Hair whorl formation of *Acetabularia* (Schmidt, 1984)

Photoreactivation of nitrate reductase (Roldan and Butler, 1980)

Germination of spores of *Pteris
vittata* (Sugai *et al.*, 1984)

Perithecial formation in *Gelasinospora reticulispora*
(Inoue and Watanabe, 1975)

Formation of 5-aminolevulinic
acid (Oh-hama and Senger, 1978)

Phototropism in *Avena*. 10 degrees
(Shropshire Jr. and Wihrow,
1958)

Phototropism in *Avena*. 0 degrees
(Shropshire Jr. and Withrow,
1958)

Respiration enhancement in *Scenedesmus* (Brinkmann and Senger,
1978a)

Inhibition of indole acetic
acid (Galston and Baker, 1949)

Chloroplast rearrangment in
Funaria (Zurzycki, 1967)

Cortical fibre reticulation in
Vaucheria (Blatt and Briggs,
1980)
DNA-photoreactivation (Saito
and Werbin, 1970)

Loss of carbohydrate in *Clorel-
la* (Kowallik and Schätzle,
1980)

Carotenogenesis in *Neurospora*
(De Fabo *et al.*, 1976)

II. THE PHENOMENA:

Blue-UV light causes a remarkable variety of effects.
These can be roughly classified as follows:
(i)   directional (e.g. phototropism, photomovement),
(ii)   morphogenic (e.g. spore germinations),
(iii)   metabolic (e.g. pigment biosynthesis, enzyme
           activation) and
(iiii)   ecological responses (e.g. stomatal control,
           algal zonation). A selection of such blue light
           controlled responses is summarized in Fig. 1
           (from Senger and Schmidt, 1986).

III. PHOTORECEPTORS:

The central question in blue light research is that about
the nature of the photoreceptors. There is no doubt that a
number of different photoreceptors trigger blue light effects. One group of photoreceptors absorbing with peaks
around (370),420,450 and 480 nm has been designated as "cryptochrome". But a great number of photoreceptors identified by
their absorption spectra differ from these typical cryptochrome spectra (cf. Fig. 1). Generally blue light receptors
absorb in the region from 350-500 nm. Bordering are UV-A
receptors (320-400 nm), often combined with blue light receptors. The photomorphogenic response to green light is observed (Schneider and Bogorad, 1987; Senger et al., 1980). A
very specific blue-UV receptor system is the "mycochrome"
(Kumagai, 1978, 1984) so far only found in certain lower
fungi.
The minimum fluence of blue light necessary to trigger
blue light reactions ranges from 10^{-9} J m^{-2} for phototropism
of *Phycomyces to* 10^{-5} Jm^{-2} for the induction of stimulation
of graviresponsiveness in corn roots. The latter value is of
the same order of magnitude as the high irradiance response
of phytochrome.
The chemical nature of blue-UV photoreceptors remains
unknown. From action spectroscopy and other indirect evidence, the flavoproteins seem to be the most likely candidates.

Fig. 1: Generalized action spectra of blue and ultraviolet light triggered responses. Modified after Senger and
Schmidt (1987).

But carotenoproteins are not ruled out either. Hemoproteins and protoporphyrins might extend the variety of effects to the green region and quinones and other compounds to the UV region of the spectrum.- The problem remains that no operational system except the final reaction at the end of the transduction line exists to identify the blue light receptor and this indication is lost when cells are fractionated. The light induced absorbance change at about 430 nm (LIAC) might only apply in reactions in which cytochromes undergo flavin-mediated reductions (Widell et al., 1983).

Even action spectra themselves are not decisive since they might be distorted by tissue internal absorption and scattering effects, by screening through other pigments, or even by energy transfer from "light harvesting pigments".

The locations and concentrations of blue light photore-ceptors are generally unknown. Their concentration must be very low, judging from photoreceptors of other reactions. Relative values indicate a concentration difference of 4000 times within 2 mm of the tip of the *Avena* coleoptile (Briggs, 1963). Some of the photoreceptors, at least those of direc-tional responses, must be bound to plasma membranes (Widell, 1987). Some photoreceptors show dicroism under polarized blue light (Zurzycki, 1980). For the phototropic reaction of *Phy-comyces* a photochromic photoreceptor system has been sugges-ted (Löser and Schäfer, 1980).

Some of the blue light responses are of very complex nature and it was discussed whether one or more receptors participate. This is particularly the case for phototropic reactions of *Phycomyces* (Galland and Lipson, 1985a,b) and maize coleoptiles (Iino, 1987) as discussed below.

The close cooperation between blue-UV receptors and the phytochrome system has been demonstrated in several higher plants, it is of multiple and complex nature (Mohr et al., 1984; Mohr, 1987). The findings on higher plants have recent-ly been extended to green algae (Kraml et al., 1987). Chloroplast movement in both the filamentous *Mougeotia* and the unicellular *Mesotaenium* demonstrates a particular mode of reaction depending on the sequence of excitation of the phytochrome and the blue light receptor. A synergistic inter-action between UV-B (280-320 nm) and red light stimulates anthocyanin synthesis in apple skin disks (Arakawa et al., 1985,1986).The red light component is due to photosynthetic activity and could be eliminated by DCMU. Conidial develop-ment in certain fungi is controlled by the mycochrome system (Kumagai, 1978;1984). This system consists of a conidiation inducing near UV component and a conidiation inhibiting blue light effect, which has a cryptochrome action spectrum (Kuma-gai, 1986a,b). The two receptors can reverse each others reaction by multiple alternate excitations.

IV. SENSORY TRANSDUCTION

The knowledge about reactions in the sensory transduction chain between blue light absorbance and the final event is scarce. One of the more extensively investigated examples is the light-growth response and phototropism of the xantho-phycean alga *Vaucheria* (Kataoka, 1975, 1981). Unilateral or symmetrical irradiation of the apex tip with a blue light pulse causes within two minutes a phototropic bending or a positive light-growth response, respectively (Kataoka, 1987). In both cases a very rapid and large promotion of a growth-relating electric current influx into the apex was detected by a vibrating electrode technique (Kataoka and Weisenseel, 1987).

In the blue light response of stomata, a proton pump located on the guard-cell plasma membrane (most likely an electrogenic H^+-ATPase) has been proven to be a component of the photosensory transduction system (Shimazaki et al., 1986).

The fungus *Neurospora* has a circadian rhythm of conidia-tion, accompanied by a circadian oscillation of cAMP (Hasunuma, 1984). The application of *Neurospora* strains with various contents of cyclic nucleotides, changing culture conditions and exposure to light demonstrates a connection between cAMP and cGMP and conidiation (Hasunuma et al., 1987a,b). The same correlation was established for different species and strains of *Lemna*. From these results the authors conclude that cyclic nucleotides may be involved in the signal transduction chain of photocontrolled circadian rhythms.

Many experiments suggest that the photoreceptors are bound to plasma membranes and that at least part of the signal transduction takes place in these membranes. In bean hypocotyls (Hartmann et al., 1981; Hartmann and Hock, 1985) and liquid tissue cultures of mosses (Vanderkhove et al., 1984), a stress- and light-triggered turnover of phospholi-pids could be detected. The physiological modification of a phosphatidylinositol specific phospholipase and its degrada-tion products are believed to play an important role in signal transduction (Pfaffmann et al., 1987; Pfaffmann and Hartmann, 1987). In this context, it was demonstrated that liquid tissue cultures of mosses are an excellent tool to study photomorphogenesis (Hartmann and Jenkins, 1984; Hartmann et al., 1986).

   In a kinetic model of the sensory transduction process, a
molecular component is considered to exist in two intercon-
versible forms, A and B. If A is the physiologically active
form, then B is the inactive form. The A -- B conversion is a
light reaction, whereas the B -- A conversion is a dark
(thermal) reaction. This model has so far been applied successfully to the blue-light response of stomata (Iino et al.,
1985), the blue-light-induced proton extrusion by *Vicia*
guard-cell protoplasts (Shimazaki et al., 1987), the phototropic response of maize coleoptiles (Iino, 1986) and the
blue light stimulation of fern protonemal cell division (Iino
et al., in preparation). Although in the case of the stomatal
response, the light-dependent A -- B reaction could be approximated by first-order kinetics (Iino et al., 1985), the
coleoptile phototropism could be best explained with second-
order kinetics (Iino, 1987). The dark reaction (B -- A) could
be explained in either case by first-order kinetics, with the
estimated first-order rate constant of about 0.001 s^{-1} (Iino
et al., 1985; Iino, 1987).This hypothetical substance (possibly an enzyme) is considered to be a common component in
various blue-light responses; it is not clear whether or not
this component is the blue-light receptor itself.

V. GENETIC ASPECTS

   A very promising approach to shed more light on the
nature of the blue light receptors and their signal transduction chains derives from the application of photoreceptor
mutants and genetic analyses. The study of various mutants of
Phycomyces and their phototropic response to blue light in
combination with biochemical analysis allows an insight into
the mechanism of blue light reactions. This aspect is covered
in the preceding chapter.

VI. SPECIAL PHENOMENA

   a. Phototropism. Phototropism and the light-growth response of *Phycomyces* sporangiophore had been believed to be
controlled by a single blue-light photoreceptor (cryptochrome). However, the following observations support the

hypothesis that multiple interacting photorceptor pigments
are acting: a) action spectra of photogravitropic equilibrium
differ from action spectra of phototropic balance (Galland
and Lipson, 1985a), b) the shape of phototropic balance
action spectra depends on the intensity range (Galland and
Lipson, 1985b), c) the behavior of certain double mutants
with defects in the photogravitropic action spectrum is unexplained by the single receptor hypothesis (Galland and Lipson, 1985a), d) phototropic dark-adaptation kinetics depend
on wavelength (Galland et al., 1984) contrary to what is
expected in a single-photoreceptor system, e) reciprocity
fails differently in the low and intermediate intensity ranges (Galland et al., 1985), f) blue light in the low intensity region can be antagonized by other wavelengths (green or
red) that are themselves phototropically inactive (Löser and
Schäfer, 1986).- The problem of multiple photoreceptors was
reexamined by measuring phototropism of dark-grown sporangiophores. Two photosystems, which control the different bending
components, can be distinguished by their intrinsic kinetic
properties, by their action spectra, and by the intensity
ranges for which they are optimized. Below 10^{-6} W m^{-2} (450
nm) phototropism is controlled by a low-intensity photosystem
(LP) and above 10^{-6} W m^{-2} by a high-intensity photosystem
(HP). LP is more effective at 334, 347 and 550 nm, while HP
is more effective at 383 nm. The dark adaptation kinetics
associated with LP are approximately half as fast as those
associated with HP. However, the light adaptation kinetics
for LP are approximately three times faster than the kinetics
associated with HP. Both photosystems are complex and each
probably involves more than a single receptor pigment (or
else a single pigment with photchromic or photocyclic properties).

Studies of phototropism in maize coleoptiles result in a
single photoreceptor system theory. The fluence-response
curve of the first positive curvature obtained with a pulse
stimulation is bell shaped and can be explained by a model
based on a single photosystem (Iino, 1986). First positive
curvature is found not only in coleoptiles of *Avena* and maize
but also in a number of dicotyledons; in addition the effective fluence range is very similar among those plants when
they are gown under red light, i.e. phytochrome responses are
saturated and maintained at steady states (Baskin and Iino,
1986). The action spectra for first positive curvature of
Avena coleoptiles and alfalfa (*Medicago sativa*) hypocotyls
(Baskin and Iino, 1986) are very similar. The fluence-response curves obtained between 300 and 500 nm for alfalfa first-
positive curvature were congruent. These results suggest that
first positive curvature is mediated by a single and common

photosystem. Even if multiple photoreceptors are involved in this system, they must be acting on a *common primary photoreaction* (Baskin and Iino, 1986). This single-photosystem argument is restricted to first positive curvature of coleoptiles. It is possible that under continuous stimulation as in the case of *Phycomyces* more than one photosystem operates for the induction of phototropic responses.

Light and temperature affect sporangiophore initiation in the fungus *Pilobolus* (Kubo and Mihara, 1986). In young sporangiophores, positive or negative phototropism is observed after irradiation with blue light or UV respectively. This is explained by the lens effect under blue light and its absence under UV. Blue and UV light cause different lag periods. Two different photoprocesses are probably involved.

Several piloboloid mutants have been obtained from the fungus *Phycomyces* (Koga and Otaki, 1983a,b). A mutant (genotype *pil*) ceases elongation, expands radially and shows negative phototropism upon lateral irradiation (Koga et al., 1984). Since all other responses are like in the wild type, it is concluded that the reverse phototropism results from a loss of the convergent lens effect in the sporangiophore with a larger diameter.

b. <u>Photomovement</u>. Some of the extensively investigated blue light-induced phenomena are the movements of microorganisms (Nultsch and Häder 1987) and of chloroplasts (Haupt 1982, Schönbohm, 1980).

Phototaxis in the unicellular wall-less alga *Dunaliella* (Halldal, 1958) consists of two photoresponses, one permissive and one inhibiting or competing. Application of various compounds that influence flavin reactions indicates that flavins are the photoreceptor for the inhibitory or competing response, where rhodopsin is discussed as photoreceptor of the permissive reaction (Wayne et al.,1986).

Photoorientation of chloroplasts in gametophytes of the fern *Adiantum* is controlled by phytochrome and a blue-light receptor (Yatsuhashi et al., 1985). The mechanism of the chloroplast movement could be detected by a video-tracking system and application of filament staining in the presence and absence of cytochalasin B. It was concluded that active movement in the light occurs through the actinomyosin system (Kadota et al 1986).

Likewise in *Adiantum* phytochrome and a blue light receptor collaborate in the green algae *Mesotaenium* and *Mougeotia* (Haupt 1982, Schönbohm 1980). In *Mougeotia* red as well as blue light induce chloroplast movement, but in *Maesotaenium* both wavelengths are ineffective over a range of several

minutes. Only subsequent or simultaneous irradiation with red
and blue light trigger a movement in *Mesotaenium*, indicating
a mutual dependence and cooperation betwen the two photoreceptor systems (Kraml et al., 1987).

Chloroplast movement can be induced in *Mougeotia* by polarized nanosecond dye-laser pulses (Scheuerlein and Braslavsky, 1987). By variation of polarization, number and sequence
of flashes the photochromic nature of the phytochrome system
can be proven. This method might be suitable to test whether
or not the blue light system might be photochromic as has
been suggested (Löser and Schäfer, 1984).

c. Morphogenesis. Light dependent spore germination in
ferns is a long studied morphogenic effect (Sugai and Furuya,
1967). It was shown that red light induction could be reversed by UV and blue light. The inhibitory effect of blue light
can be counteracted by respiratory inhibitors (Sugai et al.,
1982; Sugai and Furuja, 1983). When the wavelength dependence
of the light, which was counteracted by respiratory inhibitors was measured, it became obvious that blue and near UV
light were counteracted whereas UV (260 nm) was not. These
results suggest that at least two photoreceptors in the UV
and blue region of the spectrum participate in the inhibition
of fern spore germination.

Blue light causes a rapid inhibition of stem elongation
of cucumber hypocotyls grown under dim red light. For the
action of the blue light photoreceptor system, Ca^{++} ions are
required without being the direct cause of inhibition. Ascor-
bate abolished the blue light effect, perhaps by acting as a
redox buffer. Blue light causes the cell wall localized
peroxidase to become more accessible to extraction, but with
a lag time incompatible with the hypothesis that the increase
in peroxidase activity is the cause of the reduced elongation
rate (Shinkle and Jones, 1985).

Development from etioplasts to chloroplasts in a pigment
mutant (C-2A`) of *Scenedesmus* is light dependent (Senger and
Bishop, 1972). Respiration, biosynthesis of protein (Brinkmann and Senger 1978) and 5-aminolevulinic acid (Oh-hama and
Senger, 1978) are enhanced by blue light. Chlorophyll b and
the light-harvesting complex protein are preferentially formed in blue light. Wild-type cells of *Scenedesmus* grown under
high and low intensities of white light are very similar to
those grown under red and blue light respectively (Humbeck et
al., 1987). This coincides with the fact that the light-
harvesting complex is preferentially formed in blue light.
Therefore, blue light mimics the shade conditions for algae,
whereas red light is similar to shade conditions for higher

plants.

VII. CONCLUDING NOTE

The presentation of a variety of new results and the discussion of the "blue light syndrome" in relation to the
phytochrome system facilated by the XVI Yamada Conference
will be most stimulating for the future research.

ACKNOWLEDGEMENTS

We thank Ms Ilse Krieger and Dr. Donat-Peter Häder for
the preparation of the manuscript and the Deutsche Forschungsgemeinschaft for financial support.

REFERENCES

Arakawa, O., Hori, Y. and Ogata, R. (1985). Relative effectiveness and interaction of ultraviolet-B, red and blue
light in anthocyanin synthesis of apple fruit. Physiol.
Plant. 64:323.

Arakawa, O., Hori, Y. and Ogata, R. (1986). Japan Soc. Hort.
Sci. 54:424.

Baskin, T. and Iino, M. (1987). An action spectrum in the
blue and ulraviolet for phototropism in alfalfa. Photochem.
Photobiol. (in press).

Blatt, M.R. and Briggs, W.R. (1980). Blue-light induced
cortical fiber reticulation concomitant with chloroplast
aggregation in the alga *Vaucheria sessilis*. Planta 147:355.

Binder, B.J. and Anderson, D.M. (1986). Green light-
mediated photomorphogenesis in a dinoflagellate resting cyst.
Nature 322:654.

Briggs, W.R. (1963). The phototropic responses of higher
plants. Annu. Rev. Plant Physiol. 14:311.

Briggs,   W.R.,   Iino,   M.   (1983).   Blue-light-
absorbing photoreceptors in plants. Phil.Trans. R. Soc.
B303:345.

Brinkmann, G. and Senger, H. (1978a). The development
of structure and function in chloroplasts of greening
mutants of *Scenedesmus* IV. Plant & Cell Physiol. 19:1427.

Brinkmann, G. and Senger, H. (1978b). Light-dependent formation of thylakoid membranes during the development
of the photosynthesic apparatus in pigment mutant C-2A' of
Scenedesmus obliquus. In "Chloroplast Development in Plant
Biology" (G. Akoyunoglou and J.H. Argyroudi-Akoyunoglou,
eds.), p. 201. Elsevier/North Holland Biomedical Press, Amsterdam, New York, Oxford.

De Fabo, E.C., Harding, R.W. and Shropshire, Jr.W. (1976).
Action spectrum between 260 and 800 nanometers in *Neurospora
crassa.* Plant Physiol. 57:440.

Dörnemann, D. and Senger, H. (1984). Blue-light photoreceptor. In "Techniques in Photomorphogenesis" (H. Smith and M.G.
Holmes, eds.),p.279, Academic Press, New York.

Galland,   P. and Lipson, E.D. (1985a). Action spectra
for phototropic balance in *Phycomyces blakesleeanus*: Dependence   on reference wavelength and intensity range. Photochem. Photobiol. 41:323.

Galland,   P. and Lipson, E.D. (1985b). Modified action spectra of photogeotropic equilibrium in *Phycomyces blakesleeanus* mutants with defects in genes madA, madB, madC, and
madH. Photochem. Photobiol. 41:331.

Galland, P., Palit, A. and Lipson, E.D. (1985). Phycomyces: Phototropism and light-growth response to pulse stimuli. Planta 165:538.

Galland, P., Pandya, A.S. and Lipson, E.D. (1984). Wavelength dependence of dark adaptation in *Phycomyces* phototropism. J. Gen. Physiol. 84:739.

Galston,   A.W.   and Baker, R.S. (1949). Studies on the physiology of light action. II. The photodynamic action of
riboflavin. Am. J. Bot. 36:773.

Gressel,   J., Rau, W. (1983). Photocontrol of fungal development, Photomorphogenesis. In "Encyclopedia of Plant Physiology, New Series, Vol. 16 (W. Shropshire jr. and H. Mohr, eds.), p. 603. Springer Verlag, Berlin, Heidelberg, New York, Tokyo.

Halldal, P. (1958). Action spectra of phototaxis and related problems in volvocales, ulva-gametes and dinophyceae. Physiol. Plantarum 11:118.

Hartmann, E., Beutelmann, P., Vanderkhove, R., Euler, and Kohn, G. (1986). Moss cell cultures as sources of arachidonic and eicosapentaenoic acids. FEBS 3471, Vol. 198, No. 1.

Hartmann, E. and Hock, K. (1985). Fatty acids in protoplasts. In "The Physiological Properties of Plant Protoplasts" (P.E. Pilet, ed.), p. 190. Springer-Verlag, Berlin, Heidelberg.

Hartmann, E., Jeck, U. and Grasmück, I. (1981). Effect of light on the composition of mitochondrial lipids from hypocotyl hooks of bean seedlings (*Phaseolus vulgaris* L.). Z. Pflanzenphysiol. 103:427.

Hartmann, E. and Jenkins, G.I. (1984). 9. Photomorphogenesis of mosses and liverworts. In "Experimental Biology of Bryophytes", p. 203, Academic Press, London.

Hasunuma, K. (1984). Circadian rhythm in *Neurospora* includes oscillation of cyclic 3', 5'-AMP level. Proc. Japan Acad. 60:Ser. B, 377.

Hasunuma, K., Funadera, K., Shinohara, Y., Furakawa, K. and Watanabe, M. (1987). Circadian oscillation and light-induced changes in the concentration of cyclic nucleotides in *Neurospora*. Current Genetics (in print).

Hasunuma, K., Shinohara Y., Funadera, K. and Furukawa, K. (1987). Biochemical characterization of mutants in cyclic nucleotide metabolism showing rhytmic conidiation in *Neurospora*. Adv. in Chronobiology, in press.

Haupt, W. (1982). Light mediated movement of chloroplast. Ann. Rev. Plant Physiol. 33:205.

Horwitz, B.A. and Gressel, J. (1986). Properties and working mechanism of the photoreceptors. In "Photomorphogenesis in Plants" (R .E. Kendrick and G.H.M. Kronenberg, eds.), p. 159. Martinus Nijhoff/Dr. W.Junk Publishers, Dordrecht, The Netherlands.

Humbeck, K., Bauer, B. and Senger, H. (1987). Influence of intensity and quality of light on the molecular organization of the photosynthetic apparatus in *Scenedesmus* Planta, in press.

Iino, M. (1987). Kinetic modelling of phototropism in maize coleoptiles. Planta, in press.

Iino, M., Ogawa, T. and Zeiger, E. (1985). Kinetic properties of blue-light response of stomata. Proc. Natl. Acad. Sci. USA 82:8019.

Inoue, Y. and Furuya, M. (1975). Perithecial formation in *Gelasinospora reticulispora*. Action spectra for the photoinduction. Plant. Physiol. 55:1098.

Kadota, A., Murata, T. and Wada, M. (1986). Photoorientation of chloroplasts in *Adiantum* gametophytes as analyzed by video-tracking system. Abstract, XVI Yamada Conference, Okazaki.

Kataoka, H. (1975). Phototropism of *Vaucheria germinata* I. The action spectrum. Plant & Cell Physiol. 16:427.

Kataoka, H. (1981). Expansion of *Vaucheria* cell apex caused by blue or red light. Plant & Cell Physiol. 22:583.

Kataoka, H. (1987). Light-growth response of *Vaucheria*. Plant & Cell Physiol. 28 (1).

Kataoka, H. and Weisenseel, M.H. (1986). Blue light-promoted current influx in *Vaucheria*. Planta, submitted.

Koga, K. and Ootaki, T. (1983). Complementation analysis among piloboloid mutants of *Phycomyces blakesleeanus*. Experimental Mycology 7:161.

Koga, K. and Ootaki, T. (1983). Growth of the morphological, piloboloid mutants of *Phycomyces blakesleeanus*. Experimental Mycology 7:148.

Koga, K., Sato, T. and Ootaki, T. (1984). Negative phototropism in the piloboloid mutants of *Phycomyces blakesleeanus* Bgff. Planta 162:97.

Kowallik, W. (1982). Blue light effects on respiration. Annu. Rev. Plant Physiol. 33:51.

Kowallik, W. and Schätzle, S. (1980). Enhancement of carbohydrate degradation by blue light. In "The Blue Light Syndrome" (H. Senger, ed.), p. 344. Springer-Verlag, Berlin.

Kraml, M, Büttner, G. and Haupt, W. (1987).Coaction f red and blue light in chloroplast movement of *Mesotaenium*. Planta, submitted.

Kubo, H. and Mihara, H. (1986). Effects of light and temperature on sporangiophore initiation in *Pilobolus crystallinus* (Wiggers) Tode. Planta 168:337.

Kumagai, T. (1978). Mycochrome system and conidial development in certain fungi imperfecti. Photobiol. 27:371.

Kumagai, T. (1984). Mycochrome system in the induction of fungal conidiation. In "Blue Light Effects in Biological Systems" (H. Senger, ed.), p. 29. Springer-Verlag, Berlin.

Kumagai, T. (1986). Light changes fungal life. Bioscience and Biotechnology 24:20.

Kumagai, T. (1986). Suppressive effect of pre-irradiation with blue light on near ultraviolet light induced conidiation in *Alternaria tomato*. Canad. J. Bot. 64:896.

Lipson, E.D., Galland, P. and Pollock, J.A. (1984). Blue light receptors in *Phycomyces* investigated by action spectroscopy, fluorescence lifetime spectroscopy, and two-dimensional gel electrophoresis. In "Blue Light Effects in Biological Systems" (H. Senger, ed.), p. 228. Springer-Verlag, Berlin.

Löser, G. and Schäfer, E. (1980). Phototropism in *Phycomyces*: a photochromic sensor pigment? In" The Blue Light Syndrome (H. Senger, ed.), p. 244. Springer-Verlag, Berlin.

Löser, G., Schäfer, E. (1986). Are there several photoreceptors involved in phototropism of *Phycomyces blakesleeanus*? Photochem. Photobiol. 43:195.

Mohr, H. (1987). Mode of coaction between blue-UV light and light absorbed by phytochrome in higher plants". In "Blue Light Responses: Phenomena and Occurence in Plants (H. Senger, ed.). CRC Press, Inc., Boca Raton, Florida.

Mohr, H., Drumm-Herrel, H. and R. Oelmüller (1984). Coaction of phytochrome and blue-UV light photoreceptors. In

"Blue Light Effects in Biological Systems (H. Senger, ed.), p. 6. Springer-Verlag, Berlin.

Nultsch, W. and Häder, D.P (1987). Photomovement in motile microorganism. Photochem. Photobiol., in press.

Oh-hama, T. and Senger, H. (1978). Spectral effectiveness in chlorophyll and 5-aminolevulinic acid formation during regreening of glucose-bleached cells of *Chlorella prototothecoides*. Plant & Cell Physiol. 19:1295.

Pfaffmann, H., Drobes, B., Brightman, A., Hartmann, E. and Morre, D.J. (1987). Phosphphatidylinositol specific phospholipase C of plant stems: membrane associated activity concentrated in plasma membranes. Plant Physiol., in press.

Pfaffmann, H. and Hartmann, E. Phosphatidylinositol specific phospholipase C of etiolated beans *Phaseolus vulgaris* L. FEBS Letters, in press.

Roldan, J.M. and Butler, W.L. (1980). Photoactivation of nitrate reductase from *Neursopora crassa*. Photochem. Photobiol. 32:375.

Sachs, J. (1964). Wirkungen des farbigen Lichts auf Pflanzen. Botan. Z. 22:353.

Saito, N. and Werbin, H.(1970). Purification of a blue-green algal deoxyribonucleic acid photoreactivation enzyme. An enzyme requiring light as physical cofactor to perform its catalytic function. Biochemistry 9:2610.

Scheuerlein, R. and Braslavsky, S.E. (1987). Induction of chloroplast movement in the alga *Mougeotia* by polarized nanosecond dye-laser pulses. Photochem. Photobiol., in press.

Schmid, R. (1984). Blue light effects on morphogenesis and metabolism in *Acetabularia*. In "Blue Light Effects in Biological Systems (H. Senger, ed.),p. 419. Springer-Verlag, Berlin.

Schmidt, W. (1984). Blue light physiology. Bioscience, 34:698.

Schneider, H.A.W. and Bogorad, L. (1978). Light-induced, dark-reversible absorbance changes in roots, other organs, and cell-free preparations. Plant Physiol. 62:577.

Schönbohm, E. (1980). Phytochrome and non-phyto-chrome dependent blue light effects on intracellular movements in fresh-water algae. In "The Blue Light Syndrome" (H. Senger, ed.), p. 69, Springer-Verlag, Berlin.

Senger, H., ed. (1980). The Blue Light Syndrome. Springer-Verlag, Berlin.

Senger, H., ed. (1984). Blue Light Effects in Biological Systems. Springer-Verlag, Berlin.

Senger, H., ed. (1987). Blue Light Responses: Phenomena and Occurrence in Plants. Vol. I and II. CRC Press, Inc., Boca Raton, Florida.

Senger, H. and Bishop, N.I. (1972). The development of structure and functin in chloroplasts of greening mutants of *Scenedesmus*. I. Formation of chlorophyll. Plant & Cell Physiol. 13:633.

Senger, H., Klein, O., Dörnemann, D. and Porra, R.J. (1980). The action of blue light on 5-aminolaevulinic acid formation. In "The Blue Light Syndrome" (H. Senger, Ed.), p. 541.Springer-Verlag, Berlin.

Senger, H. and Schmidt, W. (1986). Diversity of photoreceptors. In " Photomorphogenesis in Plants" (K. Kendrick and G.H.M. Kronenberg, eds.),p. 137. Martinus Nijhoff/Dr. W. Junk Publishers, Dordrecht, The Netherlands.

Shimazaki, K., Iino, M. and Zeiger, E. (1986). Blue light-dependent proton extrusion by guard-cell protoplasts of *Vicia faba*. Nature 319:326.

Shinkle, J.R., Russel, L.J. (1986). Rapid inhibition of cucumber stem elongation by blue light: Involvement of calcium and cell wall redox state. Abstract, XVI Yamada Conference, Okazai.

Shropshire, Jr. W. and Withrow, R.B. (1958). Action spectrum of phototropic tip-curvature of *Avena*. Plant Physiol. 33:360.

Sugai, M. and Furuya, M. (1967). Photomorphogenesis in *Pteris vittata*. I. Phytochrome-mediated spore germination and blue light interaction. Plant Cell Physiol. 8:737.

Sugai, M. and Furuya, M. (1983). Action spectrum in ultraviolet and blue light region for the inhibition of red-light-induced spore germination in *Adiantum capillus-veneris* L. Plant & Cell Physiol. 26:953.

Sugai, M., Tomizawa, K., Watanabe, M. and Furuya, M. (1984). Action spectrum between 250 and 800 nanometers for the photoinduced inhibition of spore germination in *Pteris vittata*. Plant & Cell Physiol. 25:205.

Vandekerkhove, O., Euler, R., Kohn, G. and Hartmann, E. (1984). Influence of stress conditions on the fatty acid patterns of the moss *Leptobryum pyriforme*. Journ. Hattori Bot. Lab. No. 56:187.

Wayne, R., Roux, St. and Thompson, G. (1986). Photomovement in *Dunaliella*. Abstract, XVI Yamada Conference, Okazaki.

Widell, S., Caubergs, R.J and Larsson, C. (1983). Spectral characterizaton of light reducible cytochrome in a plasma membrane-enriched fraction and in other membranes from cauliflower in fluorescences. Photochem. Photobiol. 38:95.

Widell, S. and Larsson, Ch. (1984). Blue light effects and the role of membranes. In "Blue Light Efects in Biological Systems" (H. Senger, ed.), p. 177. Springer-Verlag, Berlin.

Widell, S. and Larsson, Ch. (1987). Plasma membrane purification. In "Blue Light Responses: Phenomena and Occurrence in plants" (H. Senger, ed. CRC Press, Inc., Boca Raton, Florida, in press.

Yatsuhashi, H., Kadota, A. and Wada, M. (1985). Blue- and red-light action in photoorientation of chloroplasts in *Adiantum* protonemata. Planta 165:43.

Zurzycki, J. (1967). Properties and localization of the photoreceptor active in displacement of chloroplasts in *Funaria hygrometrica* I. Action spectrum. Acta Soc. Bot. Pol. 36:133.

Zurzycki, J. (1980) Blue light-induced intracellular movements. In "The Blue Light Syndrome" (H. Senger, ed.), p. 50. Springer-Verlag, Berlin.

FUTURE STRATEGY IN PHOTOMORPHOGENESIS RESEARCH

Hans Mohr

Biological Institute II
University of Freiburg
Freiburg i. Br., Germany

I. INTRODUCTION

Plant growth and development is controlled by light via the sensor pigment phytochrome (1). Phytochrome-mediated responses in any system require the presence of active phytochrome (Pfr, the far-red absorbing form of phytochrome) and the competence of the system to respond to Pfr. This competence varies endogenously in course of development (2).

Within the limits of competence, responsiveness of a plant toward Pfr depends on the quality and quantity of the ambient light. A higher plant measures light throughout the spectrum, and this information - obtained via phytochrome, cryptochrome, and UV-B photoreceptor - determines the responsiveness toward Pfr or, in other words, the efficiency of Pfr action (3). As an example of general significance, anthocyanin formation in milo seedlings occurs only in white and blue/UV light (BL/UV) while red light (RL) and far-RL (FR) are totally ineffective. However, after a BL/UV pretreatment, the participation of phytochrome can be demonstrated. Thorough analysis (4) shows that BL/UV cannot mediate induction of anthocyanin synthesis in the absence of Pfr. Rather, the action of BL/UV must be considered to establish responsiveness of the anthocyanin-producing mechanism to Pfr. Pfr operates via two different channels. As the effector of the terminal response it brings about appearance of anthocyanin. The second function of Pfr is to contribute to the responsiveness to the effector Pfr in mediating anthocyanin synthesis (Fig. 1).

The functional relation between input and output in phytochrome-mediated responses can be written as (5):

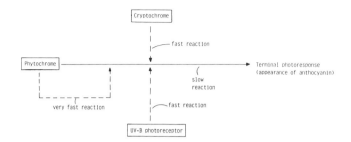

Fig. 1. Mode of coaction between BL/UV and light absorbed by phytochrome in light-mediated anthocyanin formation in
the milo seedling. ➔ , Temporal sequence of events set in
motion by the effector Pfr, leading to the terminal response;
— ➔ light-dependent reactions that determine the effectiveness of the effector Pfr (in other words, the responsiveness of the anthocyanin-producing mechanism towards Pfr). The
point of action of UV-B, relative to the action of BL/UV-A,
remains undecided at present (4).

$$R = f ([Pfr], s, c) \quad \text{whereby}$$

R = extent of the response, [Pfr] = amount (activity) of
phytochrome, s = responsiveness (a variable depending on the
light conditions [4]), c = extent of competence (a light-
independent variable, determined endogenously [2]).

In developmental biology "competence" means that a cell
or a tissue is able to respond to a specific (inductive) stimulus with a specific response. As a classical example, an
epidermal cell of a mustard cotyledon acquires competence
towards Pfr with respect to anthocyanin synthesis approximately 27 h after sowing (25° C). Competence disappears approximately 36 h later (6).

II. GENE EXPRESSION

At present - and probably in foreseeable future - emphasis
in research is on Pfr controlled gene expression. Even though
gene expression is often measured on the level of transcription or abundance of mRNA, full gene expression means the
appearance of a final direct gene product - a protein - active at its physiological site of action (7), e.g. the LHCP
apoprotein in the thylakoid membrane, SSU as integral part
of RuBPCase in the plastidic matrix. In principle, there are
many steps between the initiation of transcription and the
accumulation of the gene product at its functional location
where gene expression could be regulated. However, it seems

that the primary site of photocontrol is at the level of transcription (8). Irrespective of details of mechanism, in molecular work on phytochrome action (genic) competence means that a particular gene expression is affected by Pfr.

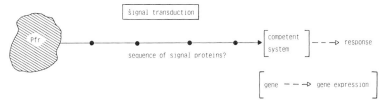

Fig. 2.   Scheme to illustrate the discussion about competence and signal transduction.

A.   A basic scheme

To introduce some research topics of interest for the immediate future I refer to Fig. 2. In recent attempts to understand the molecular mechanism by which phytochrome regulates gene expression, research is mainly directed to defining structural properties of the phytochrome molecule potentially related to its regulatory function (9). Efforts to define those factors which determine the spatial and temporal patterns of competence (2) have not yet reached the molecular level. However, recent work with transgenic plants is directed to defining regulatory functions of cis-acting DNA sequences involved in photocontrol of genes (10-12).

B. Phytochrome

It would be trivial to predict that research directed to defining structural properties of the phytochrome molecule potentially related to its regulatory function will continue at a high rate. Emphasis will probably be directed towards the molecular properties of the newly discovered 118 000 – kDa phytochrome species (13) as an essential step in defining its functional role in green tissue.

C. Competence

Efforts to define those factors which determine the spatial and temporal patterns of competence (2) have not yet reached the molecular level – what is the chromatin structure of competent genes? – even though competence as a problem has been widely recognized, including work with transgenic plants (14).

However, three aspects have recently been recognized as

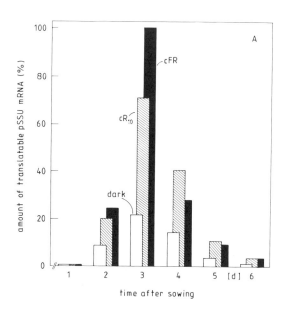

Fig. 3A. Time courses of the amount of translatable mRNA
for the precursor (p) of SSU in darkness (□), cR_{10} (▨)
and cFR (■). Mustard seedlings were harvested at the time
points indicated on the abscissa, and the RNA of the cotyledons was isolated. After _in_ _vitro_ translation, the pSSU was
immunoprecipitated. In all assays the same amount of translation products was used as a system of reference. The amount
of translatable pSSU mRNA, operationally the amount of immunoprecipitated pSSU, of 3 d old FR-grown mustard cotyledons
was considered 100% (15) c, continuous.

essential for future work: 1. Time course of competence. -
The temporal pattern of competence is endogenously controlled.
Under steady state conditions with regard to phytochrome
(continuous red or far-red light, cR or cFR) the time course
of competence - as gauged by the level of translatable mRNA -
is specific for particular proteins, e.g. SSU _vs_. LHCP (Fig.
3A,B). In both cases we observe an up and down in levels of
translatable mRNAs. However, the peak positions differ by approximately 12 h, with $mRNA_{LHCP}$ peaking later. Since $mRNA_{LHCP}$
level in mustard cotyledons was previously found to have a
much shorter half-life than $mRNA_{SSU}$ (16) the later peaking of
$mRNA_{LHCP}$ level compared to $mRNA_{SSU}$ must be attributed to differences in the time course of competence. 2. Modulation _vs_.
induction. - In considering the mechanism of phytochrome action, it was recognized that in case of SSU and LHCP phyto-

chrome has a modulatory rather than an inductive function. In fact, the present data (Fig. 3) confirm the previous sugges-

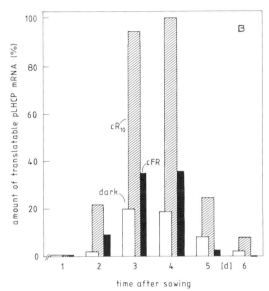

3B. Time courses of the amount of translatable mRNA for the precursor (p) of LHCP in darkness (□), cR_{10} (◩) and cFR (■). Mustard seedlings were harvested at the time points indicated on the abscissa, and the RNA of the cotyledons was isolated. After *in vitro* translation, the pLHCP was immunoprecipitated (15).

tion (17) that control of RuBPCase and NADP–GPD appearance by phytochrome is a modulation of a process which is turned on by a light-independent endogenous factor. In the case of RuBPCase synthesis in mustard cotyledons, competence towards phytochrome begins only 12 h after the first appearance of translatable $mRNA_{SSU}$ (Fig. 4). 3. Competence-determining factors. – 'Factors' which determine competence could not be identified so far, except that a 'plastidic factor' was postulated as an indispensable prerequisite for expression of the SSU and LHCP genes in light and in darkness (16, 19). The experimental evidence allowed the conclusion that it·is the 'plastidic factor' which determines competence of phytochrome-controlled gene expression in those nuclear genes which are involved in plastidogenesis (and nitrate assimilation) (20). Since the 'factor' was indeed found to be short-lived, but insensitive toward inhibition of intraplastidic translation, its molecular nature (probably not a protein) has remained elusive so far (16).

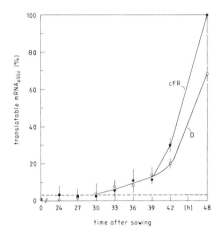

Fig. 4. Time course of the level of translatable mRNA of
pSSU in cotyledons of far-red light (●) and dark-grown
(○) mustard seedlings. Equal amount of total RNA from cotyledons of different ages (abscissa) were used for in vitro
translation. For immunoprecipitation the same amount of radioactively labelled protein was used. - 100% = radioactively
labelled pSSU immunoprecipitate from seedlings grown for 48 h
in continuous far-red light. Values are means of 4 independent experiments. Bars indicate estimates of SEM (18).

D. Signal transduction

As stated above, the (molecular) nature of the signal
transduction chain is not known. As a working hypothesis it is
assumed (see Fig. 2) that signal transduction occurs via a
sequence of signal proteins, possibly through rapidly occurring changes in phosphorylation status (21). While the goal
of research is clear - the precise association of DNA-binding
proteins with localized regions of DNA is crucial for regulated expression of the genome - progress towards identification of sequences and regulatory proteins responsible for
regulation has been slow so far.

Experimental progress has recently been made only insofar
as storage of a phytochrome-derived signal could be demonstrated in control of appearance of nitrate-induced nitrate
(NR) and nitrite (NIR) reductases in the cotyledons of the
mustard seedling. This offers a chance to investigate the
stored signal in biochemical (molecular) terms.

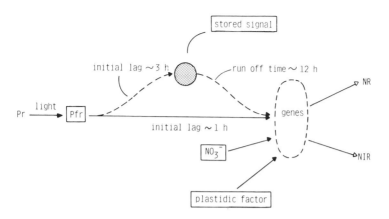

Fig. 5. A scheme to illustrate the essential results obtained in experiments on storage of phytochrome-derived signals in modulation of nitrate-induced formation of nitrate
(NR) and nitrite (NIR) reductases by phytochrome (22).

Figure 5 summarizes the major conclusions (22): Three
factors are involved in control of appearance of NR and NIR
in the cotyledons of mustard seedlings: a plastidic signal
(see above), nitrate and phytochrome. The three factors operate in a hierarchy: Without the plastidic signal nitrate cannot perform enzyme induction and without nitrate the light
effect - operating through phytochrome - cannot express itself. Application of nitrate leads to an induction of NR and
NIR in dark-grown seedlings, and this induction can strongly
be promoted by a light pretreatment - operating through phytochrome - prior to nitrate application. This light treatment is totally ineffective - as far as gene expression is
concerned - if no nitrate is given. Storage of the light signal requires at least a 3 h light pretreatment and is saturated after 12 h of light, while the break-down of the signal
store in darkness occurs with a 12 h period. A re-illumina-
tion leads to a renewed signal storage, however a significant
lag-phase (as observed during the first illumination) is no
longer detectable.

In the presence of nitrate the small rate of enzyme appearance in darkness can be promoted by light with an initial
lag-phase of 1 h ('Initial lag phase' is the duration of time
between the onset of light and the first detectable change of
the enzyme level). Since at least 3 h of light is necessary
to bring about a storage of the light signal this observation
means that signal transduction - the sequence of events between Pfr and the gene to be expressed - takes place much
faster than signal storage. Since a storage of the phyto-

chrome-derived signal can also be observed in seedlings
grown on nitrate, i.e. under conditions where nitrate permits
the formation of NR and NIR, it is concluded that treatment
with light - operating through phytochrome - leads (a) to
control of enzyme formation (modulation of gene expression
in the presence of nitrate, the initial lag is 1 h) and (b)
to a storage of the light signal (the initial lag is 3 h).
Storage of the phytochrome-derived signal occurs in the absence as well as presence of nitrate. At present we consider
signal storage as taking place in a reaction sequence which
parallels - as indicated in Fig. 5 - the direct signal transduction sequence (22). Thus, in the presence of nitrate,
phytochrome exerts a twofold action: (i) It modulates - via
the direct pathway of signal transduction - nitrate-induced
gene expression. (ii) It causes the storage of the light
signal.

     Storage of the light signal was measured for NR and NIR.
The process shows enzyme-specific differences: in the case
of NR a high capacity for signal storage is observed at 60
and 84 h after sowing while in the case of NIR the capacity
for signal storage is totally gone at 84 h after sowing. It
was concluded that the action of the stored signal on appearance of NR is different from the action of the stored
signal on appearance of NIR (22). The conclusion that action
of the stored signal is specific for particular genes is compatible with the finding that signal transduction chains
have been highly conserved in evolution. This is indicated
by work with transgenic plants and mice. As an example from
plants (14): a LHCP gene from pea can be expressed in a light-
'inducible' manner in tobacco plants whereby 0.4 kb of the
upstream flanking sequences of this gene are sufficient for
both organ-specific and light-regulated expression of the
chimaeric constructs. An example from mice (23): A murine
ß-thalassemia was corrected by the transfer of cloned mouse
or human ß-globin genes into the mouse germ line. Both introduced genes produced functional ß-globin genes curing the
anemia and associated abnormalities of the red blood cells.

                         III. FUTURE STRATEGY

     Emphasis on molecular physiology - including work with
transgenic plants - will be necessary to achieve further progress with regard to the 'nature' of the competent state of
genes (or chromatin), 'nature' of the signal transduction
chain(s), and action of the final signal (-protein) on control elements of competent genes.

Future research on these topics requires molecular physiology, i.e. the simultaneous application of subtle physiological approaches and sophisticated molecular methods. Limits to premature reductionism must seriously be considered,
as exemplified by the ambiguities in interpreting the effect
of applied phytochrome on 'run off' transcripts of isolated
nuclei (24) or the difficulties which arise from the requirement for a 'plastidic factor' in expression of those nuclear
genes which are involved in plastidogenesis (16).

It would be misleading to conceive phytochrome-controlled
expression of particular genes as isolated events, under-
standable eventually in terms of competence of a gene, signal
transduction to that gene and control of expression. Rather,
in many (most?) cases control by Pfr is part of a control
system (25). In exploring these aspects more deeply phytochrome research could contribute essentially to cell and developmental biology in general. Some case studies were chosen
to indicate the potential of this system's approach.

A. Hierarchical control

As an example - described more in detail above (22) -
appearance of nitrate (NR) and nitrite (NIR) reductases is
controlled by three factors: a plastidic signal, nitrate and
Pfr. These three factors operate in a hierarchy: without the
plastidic signal nitrate cannot induce NR; without nitrate
Pfr cannot affect gene expression (even though the Pfr-de-
rived signal can be stored to a large extent in the absence
of nitrate).

B. Coarse control by Pfr of synthesis
of a gene product vs. fine tuning during
assembly of the product into structures
of a higher order

A recent example is LHCP (Fig. 6) (25). Translatable
LHCP mRNA is formed in darkness and in far-red light (strong
phytochrome action but no chlorophyll accumulation). However,
LHCP protein is not detectable under these circumstances
because of the lack of Chl. It was shown that competence of
gene expression towards Pfr is determined endogenously also
in the case of LHCP since appearance and disappearance of
LHCP mRNA shows the same time course in R and FR-grown seedlings independently of whether or not the protein accumulates
in the thylakoid membrane (15). Appearance of protein, on the
other hand, is not only limited by mRNA but depends on the
availability of Chl (15).

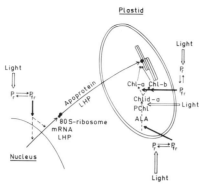

Fig. 6. An illustrating model for the effect of light
on the formation of chlorophyll and on the synthesis of the
'light-harvesting chlorophyll a/b binding protein' (LHCP)
(modified from 26).

C. Push-and-pull control in
carotenoid accumulation (Fig. 7)

Accumulation of carotenoids is promoted to some extent
by phytochrome action in the absence of chlorophyll. However,
accumulation of larger amounts of carotenoids requires chlorophyll. In the milo shoot the composition of coloured carotenoids (violaxanthin and lutein as major constituents,
traces of ß-carotene) is the same in darkness and in continuous far-red light (cFR) (recall that cFR strongly activates
phytochrome while significant chlorophyll synthesis does not
take place) (27). On the other hand, the pattern changes dramatically in white or red light with increasing amounts of
chlorophyll (lutein and ß-carotene dominate, ß-carotene

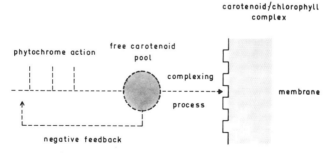

Fig. 7. A scheme to illustrate the proposed regulatory
mechanism involved in light-mediated carotenoid accumulation
(28).

showing the strongest relative increase). These recent data
support the original scheme (Fig. 7) (28) which suggests
interplay of 'push and pull' regulations in controlling the
accumulation of carotenoids. The coarse push control by
phytochrome is non-specific with regard to carotenoid pattern.
It is fine tuning in connection with pull regulation (chlorophyll formation and assembly of carotenoids into holocomplexes) which determines the final pattern of carotenoids. The
genetic information for the mature carotenoid pattern resides in the apoproteins of the holocomplexes. This process
which involves coarse control through phytochrome as well as
fine tuning during assembly of constituents - including protein folding - offers an access to the 'code of morphogenesis',
the grammar of assembly (29).

D. Rapid interorgan signal transmission

In attached mustard (Sinapis alba L.) cotyledons (Fig. 8)
graded control of chlorophyll (Chl) synthesis by phytochrome
(Pfr) and threshold control by Pfr of the 'potential capacity'
to photophosphorylate are totally different phytochrome actions, even though both controls are essential for the build
up of the same functional complex, the machinery for photophosphorylation in the mesophyll chloroplasts (Fig. 9) (30).

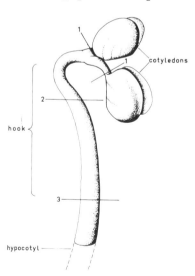

Fig. 8. Upper part of a mustard seedling as it appears
during the period of experimentation on 'potential capacity to
phosphorylate'. 'Hook' and 'cotyledons' are defined by the
numbered bars.

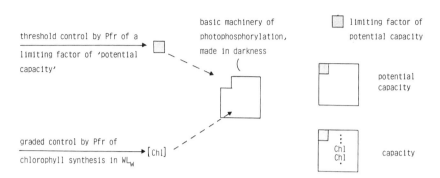

Fig. 9. A sketch to illustrate the twofold action of phytochrome (Pfr) on development of the capacity for photophosphorylation in mesophyll cells of mustard seedling cotyledons (30).

The action of Pfr on Chl formation is slow, the control by Pfr is graded and not affected by the separation of hook and cotyledons. In other words, the Pfr which controls Chl accumulation is located inside the cotyledons.

On the other hand, the action of Pfr on the capacity of photophosphorylation operates only, if the hypocotylar hook is connected to the cotyledons for at least 2.5 minutes after the end of the 1 min inductive light pulse (30). Partial irradiations of hook and cotyledons demonstrate that in the case of induction of 'potential capacity' transmission of the Pfr signal occurs from the hypocotyl hook to the cotyledons. Pfr within the cotyledons is totally ineffective in inducing 'potential capacity' even though cotyledon Pfr strongly affects Chl synthesis (Fig. 10). A 3 min red light (RL) pulse suffices for <u>full</u> transmission of the signal. This agrees with the observation that full escape from reversibility occurs within 2.5 min after the onset of RL. The threshold value of Pfr in the hook to induce appearance of 'potential capacity' in the cotyledons is 1.25% Pfr (based on Ptot at 36 h after sowing = 100%), and is thus the same as previously found in threshold control of lipoxygenase (LOG) synthesis in the same system (31). Thus control by Pfr of LOG synthesis (suppression) and 'potential capacity' formation (induction) in the mustard cotyledons seem to share the same threshold reaction in the hook and the same signal transmission, although the terminal responses are totally different (Fig. 11).

A basipetal transmission of the Pfr signal from the hook to the hypocotyl can be shown in Pfr controlled inhibition of hypocotyl lengthening, which is also an all-or-none threshold process (32). The threshold control determines the

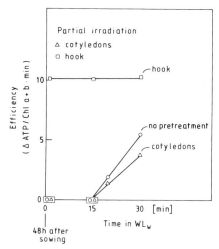

Fig. 10. Effect of partial irradiation of hook (□) or cotyledons (△) with a saturating red light pulse (20 s, 660 nm, 5 Wm^{-2}, diameter of light beam 0.5 mm) on the time course of efficiency of photophosphorylation in weak white light (WL$_w$, 0.24 Wm^{-2}). The red light pulses were given 47 h after sowing. Onset of WL$_w$ was at 48 h after sowing (data obtained by H. Oelze-Karow). Irradiation of the hook totally eliminates the lag in appearance of photophosphorylation while irradiation of the cotyledons is without effect. The steeper rise of the controls (no pretreatment) must be attributed to the higher chlorophyll content of the pre-irradiated cotyledons which decreases efficiency as long as the 'factor' (see Fig. 9) is limiting efficiency.

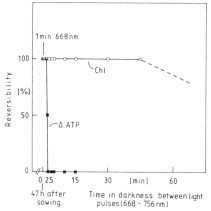

Fig. 11. Time courses of escape from full reversibility for Pfr action on the rate of Chl accumulation and for development of the capacity for photophosphorylation (30).

growth inhibition following a light pulse treatment. The
threshold value of about 0.05% Pfr based on Ptot at 36 h
after sowing = 100% shows that the growth response is extremely sensitive towards Pfr. The threshold control by Pfr
of hypocotyl growth operates only if the hook is connected
to the hypocotyl. While excision of the cotyledons reduces
the rate of hypocotyl lengthening, this operation has no effect on the threshold control mechanism per se, i.e. on the
duration of the growth inhibition. The threshold value of Pfr
which must be undercut to allow resumption of growth remains
the same (32).

The physico-chemical basis of the described rapid inter-
organ signal transmission is unknown. To explore this phenomenon is a tremendous challenge to plant physiology. This
is one reason why I plead for a (partial) swing back in phytochrome research to whole plant physiology.

E. (Photo-) Morphogenesis

Morphogenesis, i.e. the appearance in space and time of
an organized whole, is encoded in a distributed form throughout the genome. In connection with plastidogenesis we have
already noticed that it requires at least two different types
of regulation - one operating at the level of gene expression,
the other operating in connection with molecular assembly -
to make a holocomplex (see Fig. 6). Much the same could be
said when we ask what does it take to make a flower or a
seed, make a hand or a foot, make a liver or a brain. The
specifications for these structures are scattered throughout
the genome. To make these structures is not a neat, sequential process, like the linking together of amino acids in
a protein. Rather, a large number of linear production lines
are going on simultaneously, and it is the 'grammar of assembly' which determines the specific putting together of
the products (29). Moreover, the study of morphogenesis is
not only the molecular biology of the cell and its organelles
but also - and predominantly - a (molecular) study of the
interactions between cells, a wide field of research which
we have barely touched so far.

ACKNOWLEDGMENTS

Research supported by Deutsche Forschungsgemeinschaft
(SFB 206).

REFERENCES

1. Shropshire, W., and Mohr, H. (eds.), Encyclopedia of Plant Physiology, Vol. 16A,B, "Photomorphogenesis". Springer, Heidelberg – New York, 1983.

2. Mohr, H., in 1., p. 336.

3. Mohr, H., in "Photomorphogenesis in Plants" (R.E. Kendrick and G.H.M. Kronenberg, eds.), p. 547. Martinus Nijhoff publishers, Dordrecht, 1986.

4. Oelmüller, R., and Mohr, H., Proc. Natl. Acad. Sci. U.S.A. 82: 6124 (1985).

5. Mohr, H., and Schäfer, E., Phil. Trans. R. Soc. Lond. B 303: 489 (1983).

6. Oelmüller, R., and Mohr, H., Planta 161: 165 (1984).

7. Lamb, C.J., and Lawton, M.A., in 1., p. 213.

8. Ellis, J.R., Bioscience Reports 6: 127 (1986).

9. Quail, P.H., Hershey, H.P., Christensen, A.H., Daniels, S.M., Jones, A.M., Lissemore, J.L., Colbert, J.T., Sharrok, R.S., Idler, K., Barker, R.F., and Murray, M.G., Photochem. Photobiol. 43, Supplement: 119 S (1986).

10. Simpson, J., Montagu, van H., and Herrera-Estrella, L., Science 233: 34 (1986).

11. Nagy, F., Kay, S.A., Bontry, M., Hsu, M.Y., and Chua, N.B., EMBO Journal 5: 119 (1986).

12. Fluhr, R., Kuhlemeier, C., Nagy, F., and Chua, N.H., Science 232: 1106 (1986).

13. Vierstra, R.D., and Quail, P.H., in "Photomorphogenesis in Plants" (R.E. Kendrick and G.H.M. Kronenberg, eds.), p. 35. Martinus Nijhoff Publishers, Dordrecht, 1986.

14. Simpson, J., Timko, M.P., Cashmore, A.R., Schell, J., Montagu, van M., and Herrera-Estrella, L., EMBO-Journal 4: 2723 (1985).

15. Schmidt, S., Drumm-Herrel, H., Oelmüller, R., and Mohr, H., Planta (in press).

16. Oelmüller, R., Levitan, I., Bergfeld, R., Rajasekhar, V.K., and Mohr, H., Planta 168: 482 (1986).

17. Oelmüller, R., and Mohr, H., Planta 161: 165 (1984).

18. Oelmüller, R., Dietrich, G., Link, G., and Mohr, H. Planta 169: 260 (1986).

19. Oelmüller, R., and Mohr, H., Planta 167: 106 (1986).

20. Rajasekhar, V.K., and Mohr, H., Planta 168: 369 (1986).

21. Otto, V., and Schäfer, E., in "Proceedings of the XVI Yamada Conference/Phytochrome and Plant Photomorpho- genesis" (M. Furuya, ed.), p. 106. Yamada Science Foundation, Okazaki, 1986.

22. Schuster, C., Oelmüller, R., and Mohr, H., Planta (submitted).

23. Marx, J.L., Science 228: 1516 (1985).

24. Ernst, D., and Osterhelt, D., EMBO Journal 3: 3075 (1984).
25. Mohr, H., in "Regulation of Chloroplast Differentiation" (G. Akoyunoglou and H. Senger, eds.), p. 623. Liss, New York, 1986.
26. Apel, K., and Kloppstech, K., in "Chloroplast Development" (G. Akoyunoglou, ed.), p. 653. Elsevier, Amsterdam, 1978.
27. Oelmüller, R., and Mohr, H., Planta 164: 390 (1985).
28. Frosch, S., and Mohr, H., Planta 148: 279 (1980).
29. Mohr, H., in "Molecular Biology and Living Organisms – Problems of Reductionism and System Theory in the Life Sciences" (P. Hoyningen-Huene and F.M. Wuketits, eds.), p.... . Reidel, Dordrecht, 1987.
30. Oelze-Karow, H., and Mohr, H., Photochem. Photobiol. 44: 221 (1986).
31. Oelze-Karow, H., and Mohr, H., Photochem. Photobiol. 18: 319 (1973).
32. Oelze-Karow, H. (personal communication).

Index